International Conference on Managing Business Through Web Analytics

Soraya Sedkaoui •
Mounia Khelfaoui • Rafika Benaichouba •
Khalida Mohammed Belkebir
Editors

International Conference on Managing Business Through Web Analytics

Springer

Editors
Soraya Sedkaoui
Université Djilali Bounaama Khemis Mili
Khemis Miliana, Algeria

Rafika Benaichouba
Université Djilali Bounaama Khemis Mili
Khemis Miliana, Algeria

Mounia Khelfaoui
Université Djilali Bounaama Khemis Mili
Khemis Miliana, Algeria

Khalida Mohammed Belkebir
Université Djilali Bounaama Khemis Mili
Khemis Miliana, Algeria

ISBN 978-3-031-06970-3 ISBN 978-3-031-06971-0 (eBook)
https://doi.org/10.1007/978-3-031-06971-0

This Springer imprint is published by the registered company Springer Nature Switzerland AG
The registered company address is: Gewerbestrasse 11, 6330 Cham, Switzerland

Preface

The price of light is less than the cost of darkness.

(Arthur C. Nielsen)

Data is the raw material of the future – and this especially applies to business. Companies have increasingly incorporated data analysis into their overall strategy to understand their customers better and achieve greater success. Large companies such as Amazon, Google, Netflix, Facebook, Twitter, Walmart, and eBay, among others, are showing how future business decision-making can be automated and personalized on the basis of real-time and continuous data analysis.

Until a few years ago, data analytics machine learning and algorithms were exclusively a topic of scientific discussions, but today they are increasingly finding their way into everyday-life products. At the same time, the amount of data produced and available is growing due to increasing digitization, the integration of digital measurement and control systems, and the automatic exchange between devices (the Internet of Things).

Algorithms and data analysis, and thus the systematic collection, are increasingly forming the basis of the business playground. The potential of data-driven business is promising, but the professional handling of data raises many questions in practice. Technical requirements and data protection aspects, but, above all, the choice of the proper methods, present companies with major challenges.

Data is now being integrated into day-to-day operations rather than being used only for specific projects. This integration is driven by the business user and takes place in two ways: first, through big data capabilities on top of existing business applications; secondly, through the creation of new analytical tools designed explicitly for business units, with a graphical interface and intuitive handling.

The search for improved performance and positioning of organizations grows as the "digital presence" assumes strategic importance in regularly obtaining better visibility of specific audiences. Online business dominance continues to grow, and no business can afford to be absent today from the online universe. Having a website is much more than just putting the pages online. If companies want to generate results for their business, they must constantly optimize. Therefore, the process of webpage analysis is essential in the digital world.

Choosing the right channels and content to build relevance in the recipient's eyes is a major challenge for companies. The growth of corporate use of the *Web* consolidates interactive sites with the growth of users of Facebook, Twitter, LinkedIn, Flicker, and others and sharing content with collaborative production via wikis and recommendation websites.

In this context, it is important to understand the attitudes of consumers who, with mobile access to the Web and connectivity in social media, share their shopping experiences and increase m-commerce and collective purchasing groups. In order to better understand, control, and improve online actions, metrics are essential analyses. Metrics are performance measures that must be aligned with the organization's strategy.

The development of metrics for multiple media increases the amount of useful information about the consumption profile, enabling the optimization of digital strategies for targeted audiences. The measurement and continuous monitoring of business actions lead to knowledge of consumer behavior, not only in variables such as frequency, recency, and purchase value but also in terms of engagement and interaction with the product and brand.

In order to measure the efficiency of the business performance, appropriate tools are required: the so-called "Web analytics" tools, which deal with the evaluation of business and marketing success. Metrics models and Web analytics provide a basis for developing digital strategies that keep an eye on consumption in real-time.

Web analytics is an emerging concept that reflects the increasing importance of data generated by the Web. It has been growing ever since the World Wide Web development. Over time, it has evolved from a simple HTTP (Hypertext Transfer Protocol) traffic logging function to a more comprehensive set of a usage data tracker, analyzer, and report functions.

Web analytics is the monitoring of websites so that companies can better understand the complex interactions between the actions of visitors and the offers that their websites have and provide information to increase customer loyalty and sales.

Web analytics is used for different business contexts, including traffic monitoring, e-commerce optimization, marketing/advertising, Web development, analyzing and improving online customer experience and website profitability, information architecture, website performance improvement, and Web-based campaigns/programs.

There are currently many unanswered questions in theory and practice that affect the intersection of professional data use and successful business presence. Best practice examples, understandable information on the legal basis, or Web analytics implementation in individual application areas are few and far between.

Cognitive applications and services, along with the evolution toward data-driven intelligence enabled by digital technologies, are positioned as the most relevant categories in future digital development. Against this background, the idea arose in early 2020 to bring together the perspectives of experienced representatives from science and practice in an international event: The *International Conference on Managing Business through Web Analytics (ICMBWA 2021).*

On the 13th of October 2021, in the Faculty of Economics at the University of Khemis Miliana, Algeria, we celebrated and shared knowledge on this exciting field. However, like many other conferences, we have had to adapt our practice in response to the impact of COVID-19.

This *International Conference* provided an important international forum to share knowledge, ideas, and results in theory, methodology, and applications manner, and discuss the role of Web analytics in formulating and orienting business strategies.

ICMBWA2021 was the meeting point for academics engaging in Web analytics and related fields to analyze all the opportunities offered by the new normal to implement measures based on data and algorithms. A space where a community was created, some solutions were discovered and discussed, and knowledge was shared. Throughout this day, many national and international academics participated in synthesizing, connecting the discussed ideas, and translating them into valuable knowledge.

Perspectives, technologies, and fields of application were presented in an implementation-oriented and theory-led manner – that was the ambitious idea. We are glad that we have been able to implement this project with many colleagues who (like us) stand equally for Web data and analytics.

All their contributions, presented in this book, provide an overview of the possibilities and limitations of Web analytics. Perspectives, technological aspects, and fields of application are explained and practically illuminated by economics, mathematics, and technology authors. In this way, decision makers receive a well-founded presentation of the current and future development of Web analytics as well as important basic reading.

This book provides the theoretical foundations and practical implications for the digital transformation of our economy, production, and customer relationships and thus offers a solid foundation for practitioners, academics, and scientists. The presentations are based on both business and technical aspects to provide a reference process model for structuring Web analytics in the company.

The basic Web analytics activities and dynamics, social network analysis and algorithms, management processes, marketing and implementation, and other special topics and application fields for Web analytics are explained. Numerous practical examples also inspire many activities in the digital world.

The book introduces the essential aspects of analyzing Web data using methods and algorithms for business in a condensed form, presents machine learning and the most important algorithms in a comprehensible form using the business analytics technology framework, and shows application scenarios from various industries.

The content of this book shows that there is no doubt that data has value and that some companies extract a better return than others on the data they have. However, value has many dimensions and is highly dependent on the context in which the data is or may be used. For this reason, it is necessary to think of data as an asset in the same way that it is usual to consider others such as capital, human resources, or some intangibles such as patents, trademarks, and more.

The content of this book was wholly built based on the important contributions of all participants in this conference. For this, we would like to thank all the authors who actively contributed to this international conference, and the colleagues from university and practice who provided valuable impulses for this field. Special thanks also go to the Springer team, who has accompanied the book's creation in a competent and motivating manner.

Khemis Miliana, Algeria

Soraya Sedkaoui
Mounia Khelfaoui
Rafika Benaichouba
Khalida Mohammed Belkebir

Contents

About the Editors

Soraya Sedkaoui is a senior lecturer in econometrics, statistics, forecasting techniques, and probability at Khemis Miliana University. Her main areas of activity are big data analytics, computer science, machine learning, and the development of algorithms and models for business applications. She has many years of professional experience as a management consultant for strategic issues at SRY Consulting (Montpellier, France). She was a researcher at the University of Montpellier's TRIS lab in France (2011–2017), where she participated in a teaching and training program on big data and data mining. She contributed to the European project "Internet Economics: Methods, Models, and Management (2017)" in collaboration with Pr. Gottinger. She also contributed to creating many algorithms for business applications, such as the algorithm of Snail 2016. Dr. Sedkaoui's previous books and research have been published in several refereed editions and journals.

Mounia Khelfaoui is a professor at the University Djilali Bounaama Khemis Miliana in Algeria. She graduated from the University of Algiers 3 with a PhD in economics and HDR in environmental economics. With experience in research, she is a member of the research laboratory "Industry, Organizational Development of Enterprises and Innovation." Her research focuses on sustainable development, particularly CSR, the sharing economy, and the circular economy. She has written on CSR and sustainable development for several publications and conferences. She has published in various journals and conferences dealing with CSR and sustainable development.

Rafika Benaichouba is a senior lecturer at the University of Khemis Miliana and a member of the "entrepreneurship and local development laboratory" at the same university. She is specialized in banking and finance with many years of teaching. She has also published many research papers in the related fields.

Khalida Mohammed Belkebir is a senior lecturer at the University of Khemis Miliana since 2007, and a member of the "Industry, Organizational Development of Enterprises and Innovation" laboratory. Her field of interest includes management, entrepreneurship, and SMEs.

Part I
Web Analytics Activities and Dynamics

Web Data Conceptual Framework: Integration, Cleaning, Analysis, Visualization, and Security

Check for updates

Fethia Benhadj Djilali Magraoua and Saliha Hafifi

Abstract The rapid growth of the web in the last decade makes it the largest publicly accessible data source in the world. The amount of data/information on the web is huge and still growing. The web has many unique characteristics, making mining helpful information and knowledge fascinating and challenging. The coverage of the information is also extensive and diverse.

Keywords Web data · Integration · Cleaning · Analysis · Visualization · Security

1 Introduction

With millions of customers now online, the importance of websites in influencing their purchasing decisions is significant. With the company's website having the potential to ideally become a single all-encompassing access point to all the stakeholders – customers, investors, employees, and external partners, the management of their perceptions and the website has become important for business success. The unique characteristics of the Internet, such as intense competition, immediate access to product and service information, instant price comparisons, and the ease with which customers can leave an e-commerce website, force companies to concentrate on the management and measurement of this critical customer interface. Knowledge discovery is a term used in databases to describe the process of analyzing data (KDD). Discovery of useful patterns or knowledge from data sources is a common definition. Data mining is a multidisciplinary field involving machine learning, statistics, databases, artificial intelligence, information retrieval, and visualization. Numerous data-mining projects can be found in the market today. For example, supervised learning (also known as classification), unsupervised learning (also

F. Benhadj Djilali Magraoua (✉) · S. Hafifi
University of Djilali Bounaama Khamis Miliana, Khemis Miliana, Algeria
e-mail: f.ben-magraoua@univ-dbkm.dz; s.hafifi@univ-dbkm.dz

S. Sedkaoui et al. (eds.), *International Conference on Managing Business Through Web Analytics*, https://doi.org/10.1007/978-3-031-06971-0_1

known as clustering), association rule mining, and sequential pattern mining are among the most common. The web data principles are not a specific set of technologies, but rather simple principles and patterns.

2 Ethical Issues in the Analysis of Web Data

As a result, data mining raises significant ethical concerns because individuals who aren't made aware of the collection and use of their personal information aren't given the option to give their consent or withhold it. On the Internet, it's common practice to gather data invisibly. When personal data is misused or used for a purpose other than the one for which it was provided, mining the web can put people at risk (secondary use). This knowledge, on the other hand, has numerous benefits. Planning and control applications benefit greatly from the knowledge gleaned through data mining. Web data mining has a number of specific advantages, such as enhancing the intelligence of search engines. Analyzing a web user's online behavior and turning it into marketing knowledge are other ways to use web data mining in the context of marketing intelligence.

There may be ethical concerns when mining web data that does not include personal information, such as data on automobiles or animals. This chapter, on the other hand, focuses on web data mining that includes some personal data. Only harm to individuals will be examined; any harm to organizations, animals, or other subjects is outside the scope of this investigation. For now, we'll focus on the private sector's web data mining applications. Because of this, personal data mining on the web will be analyzed from an ethical standpoint. This technique has a lot going for it, and we think it has a lot of good qualities and potential. Web data mining is appealing to businesses for a variety of reasons. Consumer data and government records, for example, could be used to determine who might be a new customer and other useful information. In the most general sense, it can increase profits by actually selling more products or services or minimizing costs. Marketing intelligence is needed to accomplish this. It's possible to use this intelligence to better your marketing strategies, competitor analyses, or interactions with clients and customers. The different kinds of web data related to customers will then be categorized and clustered to build detailed customer profiles. This helps companies retain current customers by providing more personalized services and contributes to the search for potential customers. That web data mining can be very beneficial to businesses is beyond dispute. To ensure that this technique will be further developed in a properly thought-out way, however, we shall focus on its possible objections. For a well-informed development and well-considered application, being aware of all the potential dangers is critical. The different ways in which privacy is threatened are the dangers of web data mining. To structurally analyze the many different ways to mine the web, it's important to distinguish between the various types of web data mining. Web structure data, such as hyperlink structure, can be distinguished from actual data on web pages and across web documents and weblog data regarding the users who browsed web pages. We shall divide web data mining into three categories:

1. Analyzing content data found in web documents falls under the heading of "content mining." This can be anything from a picture to an audio file. Content mining, on the other hand, will only refer to text mining in this study.
2. Structure mining is a subcategory of link mining. Its goal is to examine the relationships between various web documents.
3. In the field of data mining, this is known as "usage mining." Data logged by users when they interact with the web is analyzed by usage mining.

"Log mining" is a term used to describe the process of extracting data from web server logs. When used in conjunction with some form of content mining to decipher the contents of hyperlinks, structure mining can be even more valuable.

3 Security, Privacy, Access Control, and Sharing

Unlike other data mining techniques, web usage mining has a unique application. We'll talk about it separately because it has a different set of advantages and challenges values in a different way.

3.1 Privacy Threatened by Web Data Mining

The use of personal data in web data mining can disrupt some important normative values. This is what we are going to discuss in this section. People's (informational) privacy may be violated, which is a clear ethical concern. Protecting the privacy of users of the Internet is an important issue. Privacy, on the other hand, is conceptually weak. In today's society, the term "privacy" covers a wide range of social practices and concerns. The philosophical and legal debates surrounding privacy will be omitted from this chapter. We'll use a looser (and more common) definition of informational privacy for the purposes of this discussion. In order to maintain one's privacy online, one must be able to manage the information that is made public about oneself. Information about an individual's privacy is safeguarded using this term. When information about an individual is obtained, used, or disseminated without their knowledge or consent, their privacy may be violated. Web mining privacy concerns frequently fall into this category. As a result, we'll be focusing our attention on this section. It is in this context that the term "privacy" will be used throughout the rest of this chapter. However, the value of "individualism" may be violated if people are judged and treated based on patterns found through web data mining. To begin, let's examine the connection between privacy and individualism more closely. The privacy of an individual may be violated when data is gleaned from the web through web data mining. People's privacy may be violated if their data is categorized and grouped into profiles before being used for decision-making. In this case, however, the discovered information is no longer linked to specific individuals, and no direct sense of privacy violation because the profiles do not contain "real" personal data is violated when the data is anonymized before being produced. Group profiles, on the other hand, can be used as if they were personal data, resulting in the unfair evaluation of individuals – known as individualization (see the following section). Privacy can be thought of as a stepping-stone to other fundamental values. As Vedder (2000: 452) puts it, "... privacy is a servant of many master values." Categorical privacy, which would allow group characteristics that are applied as if they were individual characteristics to be considered personal data, could be a solution. Such a solution, according to Tavani, is not appropriate because it may necessitate the creation of new privacy categories as new technologies are introduced.

3.2 Individuality

One way to describe the quality of individuality is to say that it is the quality of being an individual or of having a distinct personality from others. Individualism is a strong Western value. The core values of being an individual and expressing one's individuality are widely held in Western countries. A tendency to judge and treat people on the basis of group characteristics rather than their own individual characteristics and merits can result from profiling through web data mining. 10 A person's sense of self is jeopardized if group profiles are used as a basis for policymaking or if they are made public in some other way. As a result, individuals will be treated less like individuals and more like members of a group. The risk is heightened when profiles contain personal information that should be kept private and are, for example, used in allocation procedures to make decisions. People may be stigmatized or discriminated against simply because they are members of a group or because they have certain characteristics. The use of factors like race and religion in making decisions can be both inappropriate and discriminatory. Non-distributive group profiles pose an even greater threat because not every member of the group shares every characteristic of the group. Using probabilities, averages, and other statistical concepts, non-distributive group profiles often obscure personal information. It is no longer possible to identify an individual from the anonymized information because it no longer contains any data that can be used to identify them.

4 Web Analytics Features, Benefits, and Limitations

All of these advantages demonstrate that web data mining is an extremely valuable technique that is being developed and applied on a large and growing scale. However, there are some serious threats to some of the most important values in the web data mining field, and this is likely to cause a lot of tension. Unfortunately, many business professionals who use web data mining do not see any ethical issues with it. Twenty web data miners were interviewed in order to get a better understanding of current practices and the attitudes of web data miners toward ethical issues. Using interviews and a literature review, we can conclude that people prefer to discuss the benefits of web data mining rather than the possible risks. According to them, web data mining does not pose a real threat to privacy and other values because of a variety of reasons. Arguments in favor of data mining's near-limitless use can be broken down into six categories, each of which contains valuable information. The purpose of this brief discussion is to demonstrate that the arguments presented here do not support the use of data mining indefinitely.

4.1 Limitation

Web data mining itself does not raise any new ethical issues for discussion or investigation.
Laws and online privacy statements guarantee the confidentiality of personal information.
Because so many people have opted to give up their privacy, why not make use of it?
Most of the information gathered is of a non-personal nature or is used to create anonymous profiles.
There are fewer unsolicited marketing approaches as a result of web data mining.
Personalization leads to individualization instead of de-individualization.

5 Information Diffusion on the Web

With the advent of web, social networks have become an important medium for the dissemination of information in the Internet. The process of information dissemination in social networks has been studied using a number of information diffusion models. Due to the collaborative nature of the networks and limited accountability of the users, the media is often misused for spread of rumors and misinformation. In this chapter, we have a proposed a novel information diffusion model for the spread of misinformation using evolutionary game theory and evolutionary graph theory. The proposed model could be used to analyze as well as predict the spread of misinformation. It also provides a framework to study the effects of multiple campaigns in the network which would enable us to estimate the efficacy of launching countercampaigns against the spread of misinformation. We have used extensive simulation to support our claim.

Sharing information, such as news and rumors, is a key feature of social networks. Using natural connections, information can be disseminated in written, oral, or electronic form. With the widespread adoption of the Internet and the World Wide Web, the physics of information diffusion has changed. It used to be difficult for people to spread information in a large community because of the high costs associated with deploying the necessary technology to reach a large number of people. This stumbling block has been largely dismantled, thanks to the widespread availability of high-speed Internet. Due to its importance in social interactions and day-to-day life, information diffusion has been one of the primary research topics in the field of social network studies. It has only been in the last 20 to 30 years or so that there has been a shift toward actively participating in and shaping the flow of information and innovation. We can reason about the spread of information by modeling the diffusion of information in networks.

6 Important Approaches to Web Measurement

Businesses now view websites as more than just another channel or storefront or a simple informational portal for their customers. If you don't have an effective website, you're missing out on a lot of business opportunities. It is much easier for businesses to make adjustments and enhance their operations if they can get early and frequent feedback on how their website is performing from the point of view of its users. Several instruments and methodologies were developed to measure the website performance, usability, and quality in information systems, marketing, and operations management literature. This study reviews the literature in web quality measurement and employs a 25-item instrument developed by Aladwani and Palvia to measure the user-perceived web quality. It attempts to test the factorial validity of the instrument in Australian context using structural equation modeling technique. Analysis revealed that the data set do not fit the Aladwani and Palvia's model well enough.

6.1 Background and Literature Review

Many businesses are known only through their websites on the Internet. Whatever the size of a company, whether it's a sales brochure, or a customer contact point, or the sole distribution channel for the product, creating and maintaining an effective website is essential in today's business world. Though in the early days of Internet commerce, websites were expected to provide some entertainment to the customers, it is now considered irrelevant in today's business environment, except in some entertainment service websites. E-commerce websites' ultimate goal is to draw in potential customers and encourage them to make a purchase. In order to have a successful website, it must reflect the company's value proposition and meet the needs of its customers. If you think about it, the business strategy and operational policies of any given company are reflected in their website. Attracting and converting visitors into customers begins with the quality of the website and the way it interacts with its users. Despite the fact that companies have spent a lot of money advertising their websites, only 3.5% of the unique visitors buy something. Customer loyalty and recurring revenue can be increased by providing a superior online experience. Order fulfillment is a top priority for customers who use the Internet to gather information or make a purchase. For companies that conduct business online, additional complexity is required in terms of security, backup, and redundancy. Quality support for various functions, such as information search, transactions for purchasing goods and services, and post-sale support, is critical to the effectiveness and overall quality of a website. Operations management literature's concept of quality as "fitness for use" and the role of users or consumers in determining it are adopted in information systems research as well.

6.2 *Validity and Correlations*

Aladwani and Palvia's four-dimensional construct of perceived web quality and its validity are examined by looking at the relationships between scale ratings and Amazon users' overall quality ratings (Table 1).

It can be seen from the table above that the correlations between all four factors range from 0.526 to 0.657 and are all significant. However, these four factors/constructs have a significant correlation with the Amazon website's overall quality rating (global quality), which ranges from 0.298 to 0.574. Users' perceived web quality (the sum of all 25 items' scores) has a strong correlation with each of the 4 factors, with coefficients ranging from 0.777 to 0.869. Additionally, there is a statistically significant correlation between the overall quality rating of the website and the perceived web quality index. As a result, the accuracy and reliability of the test are further supported by this research.

Table 1 Correlations among factors and statistics

Factors	Technical adequacy	Content quality	Specific content	Appearance	User-perceived web quality	Overall quality of Amazon
Standard deviation	0.76	0.916	0.98	0.96	0.76	1.02
Content quality	657					
Specific content	.601	621				
Appearance	619	.526	.526			
User-perceived web quality	.869	.823	.799	.777		
Overall quality of Amazon	.574	493	.412	.298	.517	
Mean	5.02	4.85	5.03	4.62	4.83	4.94

7 Conclusion

The web was a hot topic in some research fields, but it had yet to be popularized and become a common technology. Some areas of the Semantic Web are already taking advantage of its potential, such as search engines and the metadata embedded in the pages of websites that want to be better understood by robots. The Semantic Web is currently in an intermediate stage. The Semantic Web is still in its infancy, and the development of applications that can take advantage of this new model is just beginning. Massive data publishing is still going on, but only a small number of applications are taking advantage of it. The importance of applications, or apps, as they are known today, is undeniable in the world of mobile devices. It's no surprise that many people in big cities use published data mined by an app when they're looking for a cab.

References

Amazon: Amazon.com launches web services [Press Release] (2002, July 16). Retrieved from http://phx.corporate-ir.net/phoenix.zhtml?c=176060&p=irol-newsArticle&ID=503034. Google Scholar

Rice, M.: What makes users revisit a web site. Market. News. **31**(6), 23 (1997)

Web Analytics: Definition and Reality in Algeria

Sarra Bouguesri and Fatma Mana

Abstract By the rising of information technology tools, all business companies have their own website, but most of them only use it as advertisement tools or news tools. However, web analytics are designed to change that and help entrepreneurs in gathering useful data from their websites. The aim of this research is to investigate the concept of web analytics in Algeria. This chapter shows that data analytics is extremely important to the Algerian companies because it helps them to optimize their marketing campaigns. When it comes to measuring the success of a marketing campaign and determining which campaigns are most effective, Google provides an easy-to-use tool that generates a unique tracking code (URL) for any link to a website.

Keywords Web analytics · Algerian companies · Web analytics tools

1 Introduction

Companies today struggle to manage increasing levels of information, searching for ways to leverage this critical business data to empower their employees. Web analytics integrate data from operational systems and the web, turning data into critical knowledge about how to run their business more effectively and be more responsive to dynamic market conditions. Despite the fact that online marketing is no longer a new phenomenon, many entrepreneurs are still hesitant to use it to its full potential. Businesses have their own websites, but most use them only for marketing

S. Bouguesri (✉) · F. Mana
Faculty of Economic and Commercial Sciences and Management Sciences, Hassiba Benbouali University, Chlef, Algeria
e-mail: s.bouguesri@univ-chlef.dz; f.mana@univ-chlef.dz

S. Sedkaoui et al. (eds.), *International Conference on Managing Business Through Web Analytics*, https://doi.org/10.1007/978-3-031-06971-0_2

or news purposes. In order to remedy this, web analytics was developed in order to assist business owners in collecting useful information from their websites.[1]

Since the advent of the World Wide Web, web analytics has been on the rise. The web analytics market and industry are also booming, with a slew of tools, platforms, employment opportunities, and businesses. With an annual growth rate of more than 15%, the market was expected to reach 1 billion in 2014.[2]

Web analytics technologies are usually categorized into on-site and off-site web analytics: On-site web analytics refers to data collection on the current site, and off-site analytics is usually offered by third-party companies such as Twitalyzer (http://twitalyzer.com) or Sweetspot (http://www.sweetspotintelligence.com). Additionally, it incorporates information from a variety of other resources such as surveys, market research reports, competitor comparisons, and the like.

According to **Guangzhi Zheng and Svetlana Peltsverger** (2015), web analytics has different usages:

It improves website/application design and user experience by helping to identify user interest/attention areas and improving web application features.
It optimizes e-commerce and improves e-CRM on customer orientation, acquisition, and retention.
It tracks and measures success of actions and programs such as commercial campaigns; web analytics must differentiate between a wide variety of traffic sources, marketing channels, and visitor types.
It identifies problems and improves performance of web applications.

However, the problem of this research is to investigate the concept of web analytics in Algeria. So, the aims of this research are:

To define the concept of web analytics
To determine the different processes of web analytics
To investigate the most used tools in the web analytics
To examine the use of web analytics in Algeria

The hypothesis of this study is as follows: There is a weak use of web analytics in the Algerian companies.

---

[1]Cao Truong, Hoang Phuong Nguyen Thi and Huyen Trang, ***Web analytics tools and benefits for enterprise***. Lahti University of Applied Sciences Degree programme in Business Information Technology (2017), Bachelor's Thesis in Business Information Technology, 79 pages.

[2]Guangzhi Zheng and Svetlana Peltsverger (2015), ***Web Analytics Overview***, Encyclopedia of Information Science and Technology, Third Edition, IGI, p. 2

2 Definition of Web Analytics

The researchers define web analytics in different ways; the most important are as follows:

Web analytics is the methodological study of **online/offline** patterns and trends. It is a technique that you can employ to collect, measure, report, and analyze your website data. It is normally carried out to analyze the performance of a website and optimize its web usage.[3]

As defined by Ivan Bekavac and Daniela Garbin Praničević, web analytics encompasses all four components listed above in an effort to better understand and optimize user experience on a website.[4]

An understanding of how to generate revenue from a website, how to create an appropriate user experience, and how to improve a company's competitive advantage is provided by web analytics.[5]

In order to improve the online experience of visitors, web analytics is a tool that analyzes qualitative and quantitative data on the website. This helps the company achieve its goals more efficiently and effectively. Web analytics is an information technology tool that collects, stores, analyzes, and graphically presents data collected from websites. Web analytics techniques are being applied to find and organize information from the web that is useful for data visualization and dashboarding.

Nabil Alghalith (2015) said that web analytics is a technology that helps organization in managerial planning and decision-making and helps it to gain a competitive advantage.[6]

Web analytics is the study of user behavior on web pages. In other words, web analytics are techniques that assess quantitative data such as web traffic, surveys, sales transactions, and others to improve the performance of marketing activities. Web analytics is the measurement, collection, analysis, and reporting of web data for the purpose of understanding and optimizing web use.

Web analytics provides data about the website as well as the visitors. The web analytics will provide businesses with information about the audience like number of visitors to the site, audience behavior ("What did they see when they visited the site?" and "How did they get to the site?"), and campaign-related data ("Which marketing campaign is more effective?" and "Which campaign brings more visitors to the site?").[7]

---

[3]Tutorials Point (I) p. 2, 2015, www.tutorialsPoint.com

[4]Ivan Bekavac and Daniela Garbin Praničević. (2015), ***Web analytics tools and web metrics tools: An overview and comparative analysis***, Croatian Operational Research Review, CRORR 6374.

[5]Ivan Bekavac and Daniela Garbin Praničević, Opcit, p. 375.

[6]Nabil Alghalith (2015), ***Web Analytics: Enhancing Customer Relationship Management***, Journal of Strategic Innovation and Sustainability Vol. 10(2), p. 12.

[7]Cao Truong, Hoang Phuong Nguyen Thi. Opcit, p. 79.

In general, we can say that web analytics is the process of collecting, processing, and analyzing website data.

3 Web Analytics Process

Waisberg and Kaushik identified the following steps in a web analytics process:[8]

2-1 **Objective (goal) determination** which differs according to the company, for example, for the commercial objectives, the goal of a website is to help customers buy products by providing them with all the information they need to make an informed decision.
2-2 **KPIs (key performance indicators) definition**: that show the particular progress or detect lagging in achieving goals.
2-3 **Data collection** in a database for subsequent data analyses.
2-4 **Data analysis**: it includes observing and transforming previously collected data in order to discover useful information that supports future decisions.
2-5 **Change implementation**: Refers to the save of information and makes change if it is.

These steps are summarized in the Fig. 1.

4 Benefits of Web Analytics

There are different benefits of web analytics, most of which are illustrated as follows:[9]

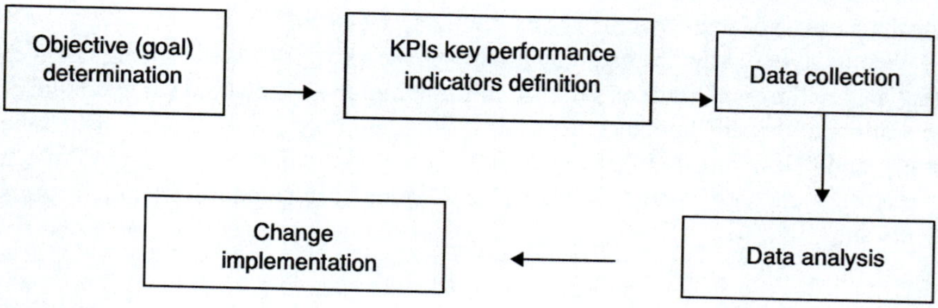

Fig. 1 Web analytics process

[8]Ivan Bekavac and Daniela Garbin Praničević, Opcit, p. 376.
[9]https://www.linkedin.com/pulse/5-key-benefits-using-website-analytics-business-shekhar-pawar/visited 7/06/2020.

Knowing the Visitors: Measure online traffic by measuring the number of users and visitors the company has on its website at any given time, the cause to use the website, and the time spending on the website. An easy-to-understand breakdown of all sources of traffic and website conversions will be provided by the analytics.

Tracking Bounce Rate: In web analytics, a "bounce rate" refers to the percentage of visitors who leave a website without taking any action. When a website has a high bounce rate, it means that users aren't finding what they're looking for and that the company needs to work on improving the user experience.

Optimizing and Tracking of Marketing Campaigns: In order to track the success of different marketing campaigns, either online or offline, unique and specific links are created for each campaign. Tracking offline-to-online campaigns is now possible with the help of Google Campaign URL Builder. If a company wanted to track the results of an event or mailing campaign, for example, it could share its unique link.

Identifying the Right Audience and Capitalizing on That Audience: Companies can use web analytics to identify and target the most appropriate audiences for their products and services.

With this information, companies will be able to craft effective marketing campaigns that make their customers feel good.

Optimizes and Enhances Websites and Web Services: The company will find potential problems on its website and its services.

Conversion Rate Optimization (CRO): The goal of CRO is to get people to do what they've been asked to do. "CRO" is calculated by dividing the number of users by the number of goals received. Examples about these conversions are as follows:

- Every stage of the sales process (add to carts, purchases, product views, etc.)
- Leads
- Newsletter signups
- Registrations
- Video views
- Brochure downloads
- Clicks on text links
- Bids and offers
- Event registrations
- Spent time on a website
- Shares on social media
- Contacts from contact forms

The improvement of conversion rates with web analytics allows the company the improvement of its website's profitability and return on investment.

Tracking Business Goals Online: Companies can use web analytics to set and monitor specific goals. By actively tracking goals, you can respond more quickly to certain events by using data.

Improve the Results from Google Ads and Facebook Ads: Google ads and Facebook ads are tools that enable the company to improve its results and increase the efficiency of its ads. Using analytic data, a company can see how many clicks, conversions, and how the ads are being received by the target audience are generated by online advertising. Use of remarketing in advertising is made possible by web analytics.

Starting is Easy: For many businesses and websites, the use of Google Analytics is easy because it is simple to install on any platform. It gives the company quickly an overview of how its online business is performing.

New Creative Ideas: By the analyzing data, it is possible to find new perspectives within the business model. Tracking the data will provide more insights about trends and customer experiences within the business of the company.

Find Out How to Optimize Your Website: Observe what type of device and browser are being used to access the company's website. When a company's target audience is increasingly using mobile devices to access its website, it may be necessary for the company to develop a mobile app.

Discover the Needs of the Customer: Web analytics data helps the companies to see the levels of traffic and which are most popular web pages of the company in which they can use this information to then tailor sales pitch or update web pages to convert more traffic.

Padma Jyothi et al. (2017) illustrate some examples of the benefits of web analytics as follows:[10]

It's a good way to keep tabs on how people are using a website.

Key metrics such as unique visitors, unique sessions, top-performing website content, the performance of different traffic sources, and much more can be viewed in real time.

Get a clear picture of where your visitors come from, as well as the most popular sources of traffic.

Utilizing a website, count the total number of visitors.

Visitor segmentation into new/returning and referral sources.

The ability to determine which pages were most frequently visited by visitors.

Check the average time spent on each page by visitors.

Visitors can be identified by the links they click on.

5 Web Analytics Tools

There are different tools for tracking general traffic and even more specific goals. Some of that are free and others are paid. The most common tools like Google Analytics can be used by everyone that runs a website. If other tools are truly

[10]U. Padma Jyothi, Sridevi Bonthu and B. V. Prasanthi. (2017), ***A Study on Raise of Web Analytics and its Benefits***, International Journal of Computer Sciences and Engineering Vol. 5(10), p. 61

required for a business, they should only be used after careful consideration. In this part, we will summarize some of these tools:

5.1 *Google Analytics*

As the most widely used web analytics service in the world, Google Analytics provides a wide range of features to help businesses better understand how users interact with their websites and, as a result, better plan their online strategies.[11]

Google Analytics is the foundation of web analytics. It offers varied features for businesses to get a hold over user behavior on website. Suraj Chande illustrates some features of Google Analytics:

Advertising Reports: All digital channels' conversion rates and returns can be seen in Google Analytics Ad reports.
Campaign Measurement: A user can use this tool to evaluate each campaign's performance and then decide whether or not to strengthen a weaker campaign.
Cost Data Import: Users compare the cost details on the digital marketing channels to allow them to make better decisions on marketing programs.
Advanced Segments: Using Advanced Segments, businesses can isolate and analyze subsets of the web traffic, like paid traffic.

In real time, users can see how many people are visiting their website and what they're looking at right now.

An audience's demographics and frequency of visits and the sources from which they come to a website can all be determined using Google Analytics, which is a free service.

Google Analytics Alerts and Intelligence Events are used to are used to generate an alert when it notices sudden spike in traffic from certain geographical area, it helps in making investment decisions about marketing and sales.

5.2 *Google Tag Manager*

It is a simple tool to use which allows installing various web analytics and marketing tools and their management without coding. It helps to measure the website's events quickly using analytics, and it also helps to manage multiple scripts or tags on a website. Google Tag Manager improves site speed.

[11] Suraj Chande, Google Analytics – Case study, January 2015, https://www.researchgate.net/publication/271447580.

Table 1 Web analytics tools: advantages and disadvantages

Web analytics tools	Advantages	Disadvantages
Webtrends Analytics	Provides detailed information, excellent heat map feature, and access to real-time data	(Relatively) high price
FireStats	Easy to use, downloadable data raw logs, real-time data	Not recommend for beginners due to install requirements
StatCounter	Access to real-time data, provides two levels of analysis	Outdated user interface
eTracker	Tracks visitor mouse movement, survey options	(Relatively) high price
IBM Unica NetInsight	Very flexible and customizable reports, customizable dashboards	(Relatively) high price
AWStats	Reveals how much time visitors spend on site, processes raw log files, open-source	Not possible to provide an in-depth analysis, neither to measure user activity
GoSquared	Pinging feature that reveals how long visitor stayed on site	Monthly page view limit

Source: Ivan Bekavac and Daniela Garbin Praničević, Web analytics tools and web metrics tools: An overview and comparative analysis, (2015) Croatian Operational Research Review 373 CRORR 6, p. 378

5.3 *Facebook Pixel*

Web analytics can be improved by using this tool, which provides a new perspective on data. In addition to keeping track of everyday activities like purchases and sales leads, Pixel also records financial data. It creates better Facebook advertising campaigns (https://engaiodigital.com/what-is-web-analytics/).

Other web analytics tools analyzed by Ivan Bekavac and Daniela Garbin Praničević are illustrated in the following Table 1.

6 The Benefits of Web Analytics Tools for Business Growth[12]

The web analytics tools help the company to analyze the performance of its website using the traffic of websites. This means that increased website traffic indicates the site's development, whereas decreased website traffic indicates the site's inactivity or decline. This helps in planning the marketing campaigns consequently.

Web analytics tools can be used to identify visitors who return frequently, as well as those who are unique. According to the number of repeat visitors, your website is

[12]https://www.businesswire.com/news/home/20180810005198/en/Major-Benefits-Web-Analytics-Tools-Business-Growth visited 24/08/2020.

doing well at retaining current visitors, but not so well at appealing to new visitors. As a result of this, it may be possible to develop new ways to attract new visitors to the website.

Web analytics tools help in dividing traffic into groups like referral, organic, and social which help to progress the website's performance:

Organic traffic is found using a search engine, and knowing this gives you a better idea of where your website stands in the results.

Referral traffic link is coming from a different website, either one with a connection to yours or one for which you've done a guest post.

Social traffic is from a variety of social media platforms via your shared posts.

Web analytics tools help also the company to understand how many visitors are exiting in website instantly after arriving. Lower bounce rate specifies that its website is able to engage and occupy the visitors for a longer time.

7 The Use of Web Analytics in Algeria

According to this graph, the top ten African countries from 2009 to 2016 are shown in the following table. There has been a noticeable decrease in the number of visitors from specific countries since 2013. Another thing to note is that in some years, several countries fail to make it to the top ten. Tanzania in 2009, Egypt in 2016, Zimbabwe in 2009 and 2012, Algeria in 2009 and 2011, and Namibia in 2013 are the most recent examples. According to this, there is an ever-changing ranking (Fig. 2).[13]

According to the statistics of Google Analytics updated on **23 September 2020**, the websites using Google Analytics as a tool of web analytics in Algeria are shown in Table 2. As we see in this table, not all the companies in Algeria use Google Analytics; this means that in Algeria there is a lack usage of the web analytics tools and the most used tool is Google Analytics.

In applying web analytics to business objectives, four main categories of metrics are used: website usability, traffic sources, visitor profiles, and conversion statistics. Website usability evaluates items such as page views, time on sight, and click paths to determine how user-friendly or user-relevant a website is. Traffic source metrics identify traffic origination points, such as referral websites or even offline advertising campaigns. Visitor profiles data from visitor profiles can provide information such as geographical origination of traffic, the time of day users most frequently visit, or what keywords are used in reaching the sight. Conversion statistics measure which visitors are new, returning, or abandoning the site, as well

[13]Shadrack Katuu. (2018) ***Using Web Analytics to Assess Traffic to the Mandela Portal: The Case of African Countries***, NEW REVIEW OF INFORMATION NETWORKING, VOL. 23, NO. 1–2, p. 8.

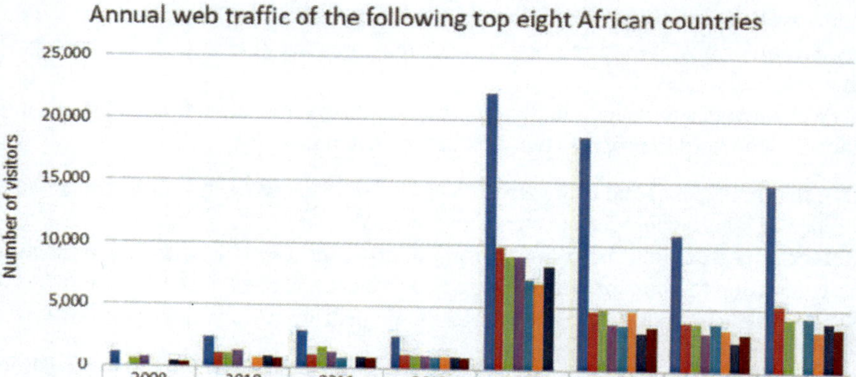

	2009	2010	2011	2012	2013	2014	2015	2016
Nigeria	1,174	2,429	2,968	2,604	22,089	18,683	10,826	14,978
Tanzania		1,189	1,128	1,182	9,836	4,713	3,958	5,301
Ghana	614	1,189	1,660	1,085	9,093	4,900	3,862	4,293
Egypt	775	1,418	1,340	1,063	9,088	3,770	3.002	
Zimbabwe			902	1,049	7,211	3,698	3,814	4,361
Algeria		823		1,032	6,926	4,776	3,429	3,361
Uganda	407	916	981	988	8,280	3,035	2,329	4,000
Namibia	345	782	866	974		3,575	2,976	3,535

Fig. 2 Annual web Traffic top eight African countries. (Source: Shadrack Katuu, Using Web Analytics to Assess Traffic to the Mandela Portal: The Case of African Countries, NEW REVIEW OF INFORMATION NETWORKING, 2018, VOL. 23, NO. 1–2, p 8)

as which are actually completing sales. Web analytics programs can arrange relevant data into a convenient dashboard for regular metric monitoring.[14]

8 Conclusion

Every business must always measure and optimize their website and product in the current digital era. Web analytics data helps businesses achieve those goals with real-time customer-focused data by analyzing customer data. Companies today struggle to manage increasing levels of information, searching for ways to leverage this critical business data to empower their employees. In order to better run your business and respond to changing market conditions, it is essential to use web analytics and business intelligence solutions to combine data from operational systems and the web.

[14]Nabil Alghalith, ***Web Analytics: Enhancing Customer Relationship Management***, http://www.na-businesspress.com/JSIS/AlghalithN_Web10_2_.pdf , acceded 09/24/2020. p. 3.

Table 2 Websites using Google Analytics in Algeria

	Website
	poste.dz
	liberte-algerie.com
	airalgerie.dz
	algerietelecom.dz
	interieur.gov.dz
	elkhadra.com
	ooredoo.dz
	mobilis.dz
	condor.dz
	radioalgerie.dz
	djezzy.dz
	pme-dz.com
	univ-setif.dz
	cerist.dz
	faf.dz

Source: https://trends.builtwith.com/analytics/Google-Analytics/Algeria acceded in 24/09/2020

This chapter shows that data analytics is extremely important to the Algerian companies for a variety of reasons:[15]

Managers can understand the behavior of visitors and can use that data to optimize their website more effectively.
Data analytics can track traffic sources, links, and more.
Data analytics provides data based on the changes made to the site, helping to evaluate the effect of change.
Data analytics improves online marketing strategies and thus creates more opportunities for customers to lead to the site.
Data analytics helps to find the most appropriate marketing techniques for business.
It also helps to plan the marketing strategy, optimize the SEO campaign, and design the website.

[15] Cao Truong, Hoang Phuong Nguyen Thi and Huyen Trang, opcit p. 79.

Data analytics helps businesses optimize their marketing campaigns. For example, Google offers a tool for generating custom tracking code (URL) for any link to a website, which helps marketers measure the performance of their campaign and which campaigns drive the best counter.

Web Analytics plays an important role in the success of the performance of the Algerian companies especially the online business, so the Algerian managers must take their tools in consideration to achieve a competitive advantage despite the fact that the application of the web analytics is not easy and requires high technological capabilities and human skills.

References

Alghalith, N.: Web analytics: enhancing customer relationship management. J. Strat. Innov. Sustain. **10**(2), 12 (2015)

Alghalith, N.: Web analytics: enhancing customer relationship management, p. 3. http://www.na-businesspress.com/JSIS/AlghalithN_Web10_2_.pdf. Accessed 24 Sept 2020

Bekavac, I., Praničević, D.G.: Web analytics tools and web metrics tools: an overview and comparative analysis. Croatian Oper. Res. Rev. **6** (2015)

Chande, S.: Google Analytics – Case Study (2015, January). https://www.researchgate.net/publication/271447580

https://www.businesswire.com/news/home/20180810005198/en/Major-Benefits-Web-Analytics-Tools-Business-Growth. Visited 24 Aug 2020

https://www.linkedin.com/pulse/5-key-benefits-using-website-analytics-business-shekhar-pawar/. Visited 7 June 2020

Katuu, S.: Using web analytics to assess traffic to the Mandela portal: the case of African countries. New Rev. Inf. Netw. **23**, 1–2 (2018)

Padma Jyothi, U., Bonthu, S., Prasanthi, B.V.: A study on raise of web analytics and its benefits. Int. J. Comput. Sci. Eng. **5**(10) (2017)

Truong, C., Thi, H.P.N., Trang, H.: Web analytics tools and benefits for enterprise. Lahti University of Applied Sciences Degree programme in Business Information Technology, Bachelor's Thesis in Business Information Technology (2017)

Tutorials Point: www.tutorialsPoint.com (2015)

Zheng, G., Peltsverger, S.: Web analytics overview. In: Encyclopedia of Information Science and Technology, 3rd edn, IGI, Hershey (2015)

Ontology-Based Data Access to Web Analytics

Abdelmoutia Telli, Abdelouahab Belazoui, and Nourelhouda Dekhili

Abstract Web analytics requires integration and aggregation of heterogeneous, distributed, and static data. Ontology-based data access (OBDA) is a popular approach to query databases that use ontology to expose data by abstracting from the technical schema-level information of the underlying data in a conceptually transparent way. OBDA approach has a great potential to facilitate such tasks. For this aim, OBDA should be extended to become analytics of web. In this chapter, we propose an ontology with mapping and query language for OBDA which can be aggregated to analytical functions of web data. In addition, we developed query optimization techniques that allow to efficiently process analytical tasks of web data.

Keywords OBDA · Web analytics · Description logic · Conjunctive query

1 Introduction

Ontology-based data access (*OBDA*) (Calvanese et al. 2009) is an access to knowledge access approach stored by an abstraction layer in multiple data sources that mediates between data sources and consumers of data. This layer uses an ontology to provide a uniform conceptual scheme that defines the problem domain of the underlying data regardless of how and where the data is stored and declarative mappings to define how the ontology corresponds to the data by applying elements of the ontology to data source queries. The ontology and mappings are used to translate ontological queries into data queries over data sources, i.e., ontological queries. The ontology and mappings offer a declarative, modular, and query-independent specification of both the conceptual model and its relationship to data

A. Telli (✉) · N. Dekhili
University of Biskra, Biskra, Algeria
e-mail: a.telli@univ-biskra.dz

A. Belazoui
LAMIE Laboratory, University of Batna 2, Batna, Algeria
e-mail: a.belazoui@univ-batna2.dz

S. Sedkaoui et al. (eds.), *International Conference on Managing Business Through Web Analytics*, https://doi.org/10.1007/978-3-031-06971-0_3

sources, as well as abstracting from data storage and access details; this simplifies development and maintenance and enables simple integration with current data management infrastructure. *OBDA* is assumed to be a central component in the modern information systems age. In the *OBDA* model, an ontology describes a high-level global schema of data sources and offers a vocabulary for user questions. In the language of the data sources, an *OBDA* framework rewrites those questions and ontologies and then assigns the actual query assessment to an effective query answering system such as a relational database management system or a data log engine (Kontchakov et al. 2013).

However, the incorporation of web data from various heterogeneous sources in this framework includes working with diverse data structures, schema languages, and query languages. There is a strong need for integrative procedures to have reliable access to numerous heterogeneous web data sources for advanced data mining algorithms.

This chapter introduces the data analytics field and illustrates the importance of following an approach where comprehensive network metrics compilation and analysis are conducted. In this new area, it explores current literature and typically advances the notion of advanced analytics. This study's primary purpose is outlined as follows:

- Developed a semantic approach for the web data integration and consolidation of multiple web analytics data sources
- Implemented an *OWL* Ontology (Dean and Schreiber 2004) for web analytics

The rest of the chapter is organized as follows: Sect. 2 provides the needed background on ontology-based data access (*OBDA*). Section 3 presents some elementary concepts on web analytics. Section 4 introduces our idea about *OBDA* aware analytics web, and Sect. 5 concludes the chapter.

2 Ontology-Based Data Access

The key objective of the modeling process, which began in the 2000s, is the creation of intelligent structures for data processing from database sources (Calvanese et al. 2009). In order to show data on *RDF* graphs from a relational database, the central concept is to provide declarative mapping requirements for domain ontology axioms. These *RDF* maps are triple-materialized. The premise is that triples do not materialize and stay imaginary, and then in a second stage where they are executed, query processes are created. Therefore, by preventing recursion and property chains, the typical technique used is query rewriting (Calvanese et al. 2017).

We will find the full collection of detailed advantages of *OBDA* systems in (Kogalovsky 2012). In a nutshell, they are structured to involve domain-specific patterns. This means that the logical schema used in the model in combination is also the specific framework added to the Structured Query Language (*SQL*) database

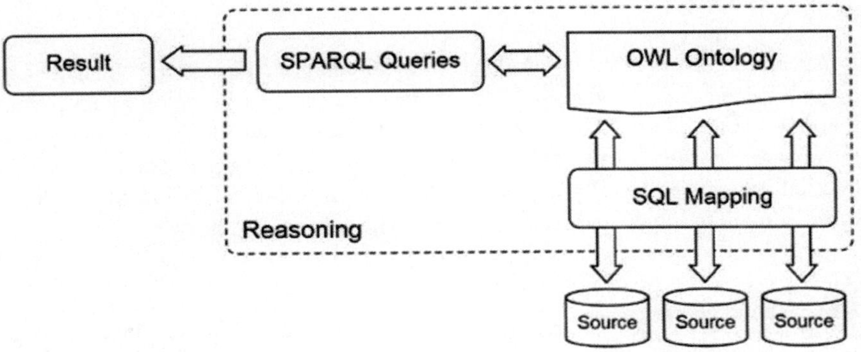

Fig. 1 Extended OBDA architectural model

method. Impedance mismatch is the key problem when mixing relational logic with schemas.

2.1 A Three-Level Architecture of OBDA

OBDA's main concept is to provide users with access to information from their data sources through a three-tier architecture, consisting of ontology, sources, and mapping between the two, where ontology is a systematic definition of the area of interest and is the center of the framework. *OBDA* offers a semantic end-to-end connection between users and data sources via this framework, enabling users to explicitly query data scattered over multiple distributed sources, via the ontology's familiar vocabulary: the user formulates *SPARQL* queries over the ontology that are translated into *SQL* queries over the underlying relational databases via the mapping layer (Fig. 1).

2.1.1 Ontology Layer

In architecture, the ontology layer is the way of following a declarative approach to data integration and, more broadly, data governance. A systematic and high-level overview of both its static and dynamic dimensions, described by the ontology, defines the organization's domain knowledge base. We obtain reusability of the gained information by making the representation of the domain clear, which is not done when the global schema is just a unified definition of the data sources underlying it.

2.1.2 Mapping Layer

By specifying the relationships between the domain principles on the one hand and the data sources on the other hand, the mapping layer links the ontology layer with the data source layer. These mappings are not only used for the running of the information system but can also be a valuable metadata asset in situations where data information is broadly dispersed into different pieces of metadata which are often difficult to obtain and seldom comply with universal standards.

2.1.3 Query Layer

The mapping layer links the ontology layer with the data source layer by defining the relationships between domain concepts, on the one hand, and data sources on the other.

2.2 *Advantages of the OBDA Approach*

2.2.1 End User Data Access

In *OBDA*, by means of an abstract representation of the environment, the information system client may communicate with the system, avoiding the man-in-the-middle, represented by the *IT* expert. Based on the principles of the environment, users should ask questions rather than the constructs of the data sources. The *OBDA* method is responsible for converting the initial query into a query to be tested at the root by considering the ontology and the mappings to the data sources.

2.2.2 Integration

OBDA can be seen as a type of integration of knowledge, where the normal global schema is replaced by the application domain conceptual paradigm, formulated as an ontology. In this approach, the integrated view that the system presents knowledge consumers in is not just a data structure that accommodates multiple sources of data, but a semantically rich description, as well as the relationships between such concepts, of the related concepts in the field of interest.

2.2.3 Extensibility

The *OBDA* solution would not demand that the data sources be completely implemented at once. Instead, additional data sources or new components may be introduced incrementally after constructing even a rough skeleton of the domain model, as they become usable, or as needed, thus amortizing the cost of integration. The overall architecture can therefore be assumed to be the gradual phase of recognizing and describing the domain, the data sources available, and the relationships between them. The aim is to enable both ontology and mapping to evolve in such a way that the structure continues to work as it progresses.

2.2.4 Documentation

Ontology and the corresponding mapping of data sources provide a common basis for the documentation of all the data within the organization, with clear advantages for the management and management of the information system.

2.3 *Logical Description for OBDA with Databases*

The key notion of *OBDA* with databases is the rewriting of queries. The user formulates a query q in the vocabulary of a given ontology $T(T, q)$ which is called a query to the ontology. The task of an *OBDA* system is to "rewrite" q and T in a new query q' in the data vocabulary so that, for any possible data A (in this vocabulary), the responses to q plus (T, A) are precisely the same as the responses to q' on A. Thus, the problem of querying data A (whose structure is not known to the user) in terms of ontology T (accessible to the user) is reduced to the problem of querying A directly (Kontchakov et al. 2013). We consider in this work the ontology languages supporting query rewriting, and we concentrate on description logics (*DLs*) as ontology formalisms and only provide the reader with references to languages of other types. For this, we assume that A is simply a *DL ABox* stored in a relational database.

The inclusion in the current *W3C* standard Web Ontology Language (*OWL* 2) of a special sub language (or profile) that is suitable for *OBDA* with databases and called *OWL* 2 *QL*. The *DLs* underlying *OWL* 2 *QL* belong to the so-called *DL* – *Lite* family (Artale et al. 2009; Calvanese et al. 2007). Below, we present *OWL* 2 *QL* in the *DL* parlance rather than the *OWL* 2 syntax. The *DL* – *Lite* language is defined as follows:

$$\begin{aligned} R &\rightarrow P \mid P^- \\ E &\rightarrow R \mid \neg R \\ B &\rightarrow A \mid \exists R \\ C &\rightarrow B \mid \neg B \end{aligned} \tag{1}$$

where A is an atomic concept, P is an atomic role, and P^- is the inverse of P. B (resp. C) is called basic (resp. complex) concept, and role R (resp. E) is called basic (resp. complex) role. We follow the description of *DL* – *Lite* used in (Calvanese et al. 2007); a $DL - Lite$ knowledge base K is a pair $K = \langle T, A \rangle$. $T = TBox$ is made up of a finite set of inclusion axioms between concepts of the form $B \subseteq C$ or $B \subseteq \neg C$. $A = ABox$ contains the finite set of assertions (facts) of atomic concepts and roles of the form $A(a)$ and $P(a, b)$. The $DL - Lite_F$ language extends $DL - Lite_{core}$ with the capability of functional specification on roles or their inverses of the form $(funct_R)$. The $DL - Lite_R$ language extends $DL - Lite_{core}$ with the ability to specify inclusion axioms between roles in $TBox$ of the form $R \subseteq E$.

Note that $DL - Lite$ language does not use connective and disjunctive operators. However, a logical transformation makes it possible to obtain conjunctions and disjunctions as follows:

- A conjunction of the form $B \subseteq C \cap D$ is equivalent to the pair of inclusion axioms: $B \subseteq C$ and $B \subseteq D$.
- A disjunction of the form $C \cup D \subseteq B$ is equivalent to the pair of inclusion axioms: $C \subseteq B$ and $D \subseteq B$.

3 Web Analytics

Web analytics is a method by which website consumption reports are obtained and compiled electronically. It is possible to use an analytics software as a platform to help you get to know who your customers are, where they come from, and how they use your website. Ultimately, providing access to your user records lets you make appropriate choices about your website, whether those choices refer to significant redesigns of your website or to continuing tweaks and small improvements that represent changes in customer usage or in your own existing programs and services. The beginnings of web analytics are from the creation of business websites, where tracking the attitudes and activities of users directly contributes to the buying behavior of customers. For a charity website, though, online analytics may be almost as useful as for an e-commerce platform. One strategy is to monetize the targets for the website, adding a dollar sum to such events such as visits to visitors or registrations of online programs.

Quantitatively, by statistics and ratios, numerical data makes sense. Stable trends of website traffic evolve over time. A sales page, for example, will be used more during holidays and sales. During emergencies, participating agencies will be accessed more regularly. After related news reports air, information, non-profit, and health organizations can be accessed, or after campaign messages are seen or heard. Being able to analyze analytical data efficiently needs knowledge of the participating company (Kent et al. 2011).

3.1 Web Analytics Business Components

The web analytics business process is a collection of business processes and sub-processes that define the ongoing operations that every company wants to master to be effective with web analytics when viewed collectively. The web analytics process can be categorized into five parts:

3.1.1 Measuring

The method by which web analytics is used is web calculation. While hits were common in the early days of the internet, over the years, analytics has become much more sophisticated. After deciding the result you want your website to obtain, simple metrics such as clicks, bounce rate, references, and conversions can be calculated.

3.1.2 Exit Rates and Bounce Rates

In assessing whether or not you have effectively captured the interest of your customers, bounce rates and departure rates are two metrics to evaluate. Exit ratings will inform you on a single tab how many users left the website, while bounce rates inform you how many people left the website without accessing a second page.

3.1.3 Referrers

A strong metric for understanding that certain tourists convert is where the tourist comes from. A referrer is a website which sends traffic to you. Referrers also take you traffic, in addition to search engines. To develop healthy, productive connections, check out others in your business or network, whether it be by social network interaction, blog discussions, or even writing articles in your niche.

3.1.4 Calling Clicks to Action

Your design should cater to your target group and motivate them to take action. Does the interface help them know like they are receiving benefit from the services and are they clicking on the call to action buttons, more importantly? To quickly understand what visitors do on your site and where they are clicking, Crazy Egg offers heat map reports. Heat map reports help you to see what's hot and what's not, so you can make conversion-enhancing adjustments.

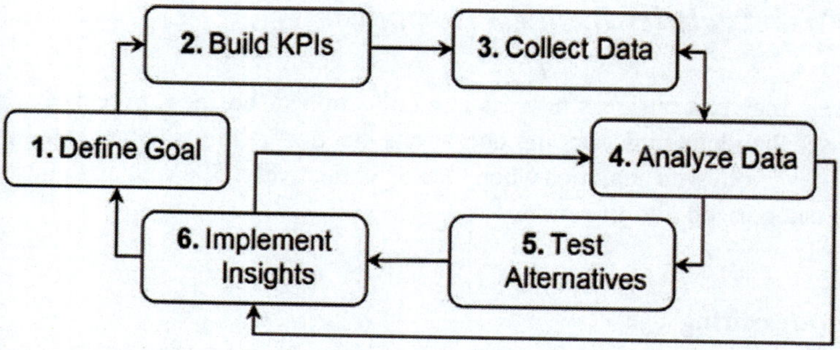

Fig. 2 Web analytics business process model. (Waisberg 2015)

3.1.5 The Transformations

If you've come this far, you know that most of your guests are pleased and have earned their attention. Having them to move is the key move now. A conversion is a term used by advertisers to explain the end product of a visit to the site. Don't thwart the guests from completing the planned action. This ensures that the online forms are designed for the customers and the form submission or checkout process is as user-friendly as possible.

3.2 Web Analytics Business Process

In an enterprise, web analytics can be much like a production cycle that begins from collecting specifications to validation. A visualization of an optimal method for online analytics is below. For tech organizations that have already established KPIs and daily/weekly/monthly updates of new features on their website, this process is more suitable (Järvinen and Karjaluoto 2015) (Fig. 2).

3.2.1 Define Goal

This is the beginning of the method of web analytics, and it deals with an analyst gathering stakeholder monitoring criteria. Similarly, this phase would also include checking new item product requirements that are part of a release cycle.

3.2.2 Build KPIs

The analyst will create a monitoring plan/analytics plan/solution concept document to identify variables for web analytics vendor tools (custom variables, page name variables, etc.) such as Omniture SiteCatalyst, WebTrends, Clicktracks, or Google Analytics, until all the specifications have been measured.

3.2.3 Collect Data

Web analytics isn't just about studying your landing page's page views and bounce rate. When you can start researching more concrete activities, such as pressing a button, filling out a field, submitting a message successfully on your contact form, etc., it gets really interesting. Indeed, by naming them names, groups, activity forms, etc., you need to manually execute these events on your website and send them to your web analytics providers when you wish to cause these events. To point out micro challenges that the client can face, or even to assess the accomplishment of clear targets, these particular details are very significant.

3.2.4 Analyze Data

In this phase, the analyst may normally collaborate with a developer to integrate the functionality on the website. The analyst also uses this move to support the developer with any concerns she has about the web analytics code or the monitoring strategy. This refers in particular to new developers who are not aware of the web analytics snippet.

3.2.5 Test Alternatives

This process deals with the web analytics data QA/testing that ends up in the web analytics tool. I have written a detailed article outlining the meaning of this phase, since this is a distinct method in itself.

3.2.6 Implement Insights

It's the task of the analyst to submit numbers arising from the functionality that went live during the previous release period after the data is found to be clean. In order to further strengthen the platform, the researcher will also provide insight (explaining the data or conversion, etc.) and potential recommendations/next steps.

4 OBDA Aware Business Analytics Web

One of the key purposes of this work is to collect, disinfect, organize, and incorporate data on business sites from numerous network monitoring sources. For

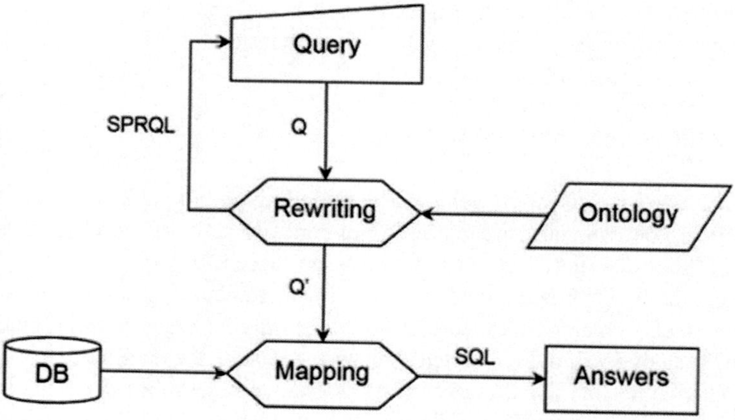

Fig. 3 Query/answering in OBDA approach

this purpose, we wanted to build a semantic sharing and reconciliation approach, using an agreed ontology model to store a shared interpretation of the domain in which the mechanism runs of six steps:

1. Determine the ontology's domain and context to restrict the complexity of ontology as the starting point.
2. Significant terms in ontology were derived from the minimal collection of variables expected from the specifications (Garıa-Nieto and Roldan 2014).
3. We describe classes and the class hierarchy from the list of concepts, and we got the ontology classes in the hierarchies (see Fig. 3).
4. Classification of objects and of data type properties based on the minimal collection of previously specified variables by specifying the properties of classes and slots in order to connect classes and identify attributes.
5. Description of limits on cardinality and constraints on value: in our ontology, value constraints are used to define the form of data for the value in each subset of the analytic parameters hierarchy.
6. Create instances (individuals in *OWL*) corresponding to the specific data obtained from a specific business.

4.1 Global Idea (Ontology)

The basic principle behind *OBDA* is to allow the user access to the data store through the use of a domain-specific class language. Via view descriptions, called mappings, this language is related to the database schema; thus, technical specifics of the database schema are concealed from end users (Kharlamov et al. 2017). In the case of web business analytics, the majority of earnings table names are not meant to be

Table 1 Sample ontology-based data access

Axiom	Description logic
hasSource	$\exists hasSource \subseteq Analytic - parameters$
hasBrowser	$\exists hasBrowser \subseteq Analytic - parameters \subseteq Device$
hasNetworkDomain	$\exists hasNetworkDomain \subseteq hasSource$
hasValue	$Value \subseteq hasContent$
hasTask	$Task \subseteq Content$
hasDevice	$Device \subseteq Price$
hasDuration	$Duration \subseteq Task$
hasReferer	$Referer \subseteq Price$
hasTotalRevenue	$\exists hasTotalRevenue \subseteq hasPrice \cap hasDevice$
hasTitle	$Title \subseteq Referer$
hasPrice	$\exists hasPrice \subseteq hasSource$
hasCategory	$Category \subseteq ID$
hasID	$\exists hasID \subseteq hasTitle$
hasLocation hasContent and *hasName*	$(Location, Content, Name) \subseteq Title$
hasContent	$hasContent \subseteq hasCategory$
hasPurpose	$\exists hasPurpose \subseteq hasCategory$

read by end employers, while the semantics of others are clearer such as the price, task, and value.

The employers formulate queries in terms of the classes and properties in an object-centric fashion. These queries over the domain vocabulary are then unfolded into queries over the database schemas and executed over the data (see Table 1). An important feature of the *OBDA* approach is that the answers are done via logical reasoning. For this aim, we proposed ontology, called "wbao. owl" (web business analytics ontology), to represent a business web filed.

4.2 Mappings Query Answering

The way ontological terms are linked to terms occurring in the relational schema can be described declaratively through mappings level. They are literally (read-only) view descriptions that describe how to represent classes with objects in *OWL* objects as follows:

$$\begin{gathered} Class\,(f(a)) \rightarrow \\ SQL(a),\, Property\,(f(a), f(b)) \rightarrow \\ SQL\,(a, b)\,,\, Property\,(f(a), g(y)) \rightarrow \\ SQL\,(a, b) \end{gathered} \tag{2}$$

where:

- *SQL*(a) and *SQL*(a, b) are *SQL* queries with, respectively, one and two output variables.
- *f*, *g* are functions that "cast" values returned by *SQL* into, respectively, objects and classes.
- $f(x)$ is computed from the values a returned by *SQL*(a).
- Properties can relate two objects.
- If it relates a table to a class or an attribute to a property, the mapping is direct.

We perform analyses regarding behavior and the product profile which entail a series of unsupervised procedures to classify products into different predefined types. Starting with a decision tree, rules are created to allocate samples in clusters to predefined forms. Finally, to allocate the new incoming data to each predefined sort, a classification process is used. *SQL* queries can be performed in the mapping descriptions because of a database and a collection of mappings over it, and groups and properties of the mappings can be filled, generating a collection of ontological truth. This approach is commonly referred to as the materialization of the ontological facts that the mappings describe (see Fig. 3).

We may also concentrate on a particular set of chronology in terms of time evolution and check if a consumer satisfaction plan may produce changes in the total employment over time or not.

Since answering questions in *OBDA* requires logical reasoning, it enables both overt and tacit question responses to be retrieved. The ontology and ontological facts that one would have if they were materialized from the database over which the mappings are described can be used as justification (Calvanese et al. 2007). For example, we have the following query:

$$\begin{array}{c} SELECT\ ?a\ WHERE \\ \left(?a : hasJob\ ?b.\right) \end{array} \tag{3}$$

This query selects all objects that have job in the database. For rewriting this query, it is gives the union:

$$\begin{array}{c} SELECT\ ?a\ WHERE \\ \left(?a : hasJob\ ?b.\right) \cup \\ \left(?a\ Investor.\right) \cup \\ \{?a\ Well - Investor.\} \cup \\ \{?a\ Small - Investor.\} \end{array} \tag{4}$$

Then, rewritten query with the sample mappings in the following *SQL* query:

$$\begin{array}{c} SELECT\ f\,(has.ID)\ As\ \ a\ FROM\ Well - Investor\ \cup \\ SELECT\ f\,(has.ID)\ As\ \ x\ FROM\ Small - Investor\ WHERE \\ a.ID = Job\ AND\ a.Name = "Investor" \end{array} \tag{5}$$

5 Conclusion

In this chapter, we proposed a semantic approach (*OBDA*) that uses an ontology for the representation and consolidation of the tracking data from web source's semantics. Semantic mappings between the source schema and the ontology can be easily queried by high-level algorithms. Finally, we expect that our work will open new doors for study in the fields of semantic access and semantic incorporation of federated and distributed relational databases into web data analytics, as it highlights the potential advantages of such an approach and poses significant practical problems that need to be tackled in order to ensure the viability of such a technology.

References

Artale, A., Calvanese, D., Kontchakov, R., et al.: The DL-Lite family and relations. J. Artif. Intell. Res. **36**, 1–69 (2009)

Calvanese, D., De Giacomo, G., Lembo, D., et al.: Tractable reasoning and efficient query answering in description logics: the DL-Lite family. J. Automated Reason. **39**(3), 385–429 (2007). https://doi.org/10.1007/s10817-007-9078-x

Calvanese, D., De Giacomo, G., Lembo, D., et al.: Ontologies and databases: the DL-lite approach. In: Tessaris, S., et al. (eds.) Reasoning Web Semantic Technologies for Information Systems Lecture notes in computer science, 5689, pp. 255–356. Springer, Berlin/Heidelberg (2009). https://doi.org/10.1007/978-3-642-03754-2_7

Calvanese, D., Cogrel, B., Komla-Ebri, S., et al.: Ontop: answering SPARQL queries over relational databases. Semantic Web. **8**(3), 471–487 (2017). https://doi.org/10.3233/SW-160217

Dean, M., Schreiber, G.: OWL Web ontology language reference, W3C recommendation 10 February 2004 (2004). Retrieved from: https://www.w3.org/TR/2004/REC-owl-ref-20040210/

Garıa-Nieto, J., Roldan, M.: D2.1 SME – E-compass requirements analysis. Public Deliverable (2014). Retrieved from: http://www.sme-ecompass.eu/public-deliverables.aspx

Järvinen, J., Karjaluoto, H.: The use of Web analytics for digital marketing performance measurement. Ind. Market. Manage. **50**, 117–127 (2015). https://doi.org/10.1016/j.indmarman.2015.04.009

Kent, M.L., Carr, B.J., Husted, R.A., et al.: Learning web analytics: a tool for strategic communication. Public Relations Rev. **37**(5), 536–543 (2011). https://doi.org/10.1016/j.pubrev.2011.09.011

Kharlamov, E., Hovland, D., Skjæveland, M.G., et al.: Ontology based data access in statoil. J. Web Seman. **44**, 3–36 (2017). https://doi.org/10.1016/j.websem.2017.05.005

Kogalovsky, M.R.: Ontology-based data access systems. Program. Comput. Soft. **38**(4), 167–182 (2012). https://doi.org/10.1134/S0361768812040032

Kontchakov, R., Rodríguez-Muro, M., Zakharyaschev, M.: Ontology-based data access with databases: a short course. In: Rudolph, S., et al. (eds.) Reasoning Web Semantic Technologies for Intelligent Data Access Lecture notes in computer science 8067, pp. 194–229. Springer, Berlin/Heidelberg (2013). https://doi.org/10.1007/978-3-642-39784-4_5

Waisberg, D.: Google Analytics Integrations. Wiley, New York (2015)

Web Analytics and Business Performance: Data Cleaning Does Matter

Aymen Salah Bennihi, Brahim Zirari, and Amina Fatima Zohra Medjahed

Abstract In the context of the relationship between web analytics tools and business performance, this chapter aims to explore the impact of data cleaning on data analysis accuracy and decision-making efficiency. The research is divided into two parts. The first part is a theoretical background about web analytics and its concept, tools, and data collection methods. Moreover, there is the data cleaning approach through outlier detection. The second part is a quantitative study, where we performed some statistical tests to detect outliers in the UK-registered retail dataset and showed its impact on the ARIMA estimation. The study revealed the effective impact of outliers' adjustment on the data collected from the web, which contributes to highlighting the massive need for data cleaning in web analytics to come up with good analysis as well as good decisions.

Keywords Web analytics · Business performance · Data cleaning · Outliers · ARIMA estimation

1 Introduction

Digitalization, technology, and the internet are a combination that has become a major concern in both academia and business. The necessity of the internet is increasing quickly, and it is followed by getting more users online. This revolution of the internet is a need in each company and organization because it facilitates communication and provides more information that helps in the decision-making process.

A. S. Bennihi (✉) · B. Zirari
Department of Economics, ITMAM Laboratory, University of Saida, Saida, Algeria
e-mail: bennihi.aymen@univ-saida.dz; brahim.zirari@univ-saida.dz

A. F. Z. Medjahed
Department of Informatics, University of Saida, Saida, Algeria

S. Sedkaoui et al. (eds.), *International Conference on Managing Business Through Web Analytics*, https://doi.org/10.1007/978-3-031-06971-0_4

The use of the internet has brought a new lifestyle and jobs and created a new branch of science. Data science, machine learning, artificial intelligence, and web analytics are the outcomes of internet implementation, and the more the development, the more they need for knowledge in these areas.

Web analytics is to educate customers' behavior, segmentation of the market, and analyze market trends (Kumar and Ayodeji 2020). Nowadays, the majority of companies are aware of the importance of owning a website to have a detailed report on their activities, which demands strong web analytics tools, data, and skills. In this information age, the capacity to collect and store new data grows rapidly, which leads to new challenges in the analysis process. Finding patterns and knowledge hidden in the data becomes a major problem for many analysts. According to Shilakes and Tylman (1998), the relevant market growth rate of data quality is about 17%, which is much higher than the 7% annual growth rate of the IT industry. For instance, approximately 30–80% of the time and costs are spent on data cleaning in data warehousing project development. The time series errors can be either timestamp errors or observed value errors.

Despite the consensus of the magic effect of web analytics, it is very difficult to perform any test in the absence of cleaned data, which is the key to good analysis, hence a good decision. Therefore, this chapter aims to answer the following question: **what is the impact of data cleaning on web analytics performance?**

To delve this problem, our study focused on delving the impact of data cleaning on the robustness of web analytics results through performing some statistical tests on a web dataset and analyzing their impact on the analysis.

The rest of the chapter is structured as follows: the first section is dedicated to the methodology, the second section to the related studies, the third section to the theoretical concepts about web analytics, and the fourth section to the applied data cleaning approach. The empirical analysis is presented in a fifth section followed by the conclusion.

2 Methodology

To demonstrate the impact of outliers on the estimation accuracy, this chapter uses the dataset of Online Retail II, which contains all the transactions occurring for a UK-based and registered, non-store online retail between 01/12/2009 and 09/12/2011. The company mainly sells unique all-occasion gift-ware. Many customers of the company are wholesalers. Due to the study purpose, the data were transformed into daily observations, and we focused on the sales made only in the UK. Figure 1 represents the sales made in the UK from 12/01/2010 to 12/09/2011.

In the first step, an auto-ARIMA model was used to forecast the sales on the horizon of 1 month using the original data. In the second step, the data were tested for the presence of outliers and cleaned using the tsoutliers package and then re-estimated the same horizon.

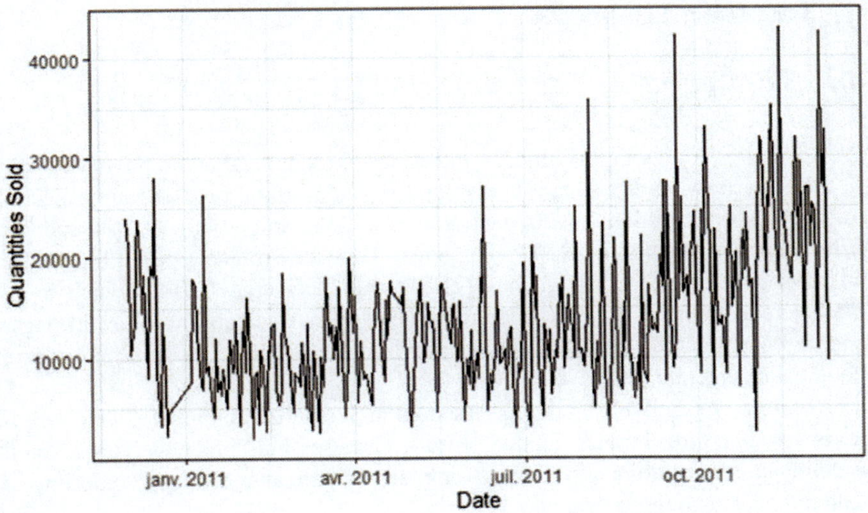

Fig. 1 Time series of retail sales made in the UK from 12/01/2010 to 12/09/2011

3 Literature Review

An empirical study was performed by Bekavac and Praničević (2015), where the author collected data of about 200 employees from a Croatian firm's IT and marketing branch to analyze the user's satisfaction from web analytics tools. The study revealed that web analytics tools are well accepted and applied in both IT and marketing. Moreover, web analytics tools have been used in business fields more than the marketing industry.

Padma Jyothi et al. (2017) conducted a study to discuss the benefits for companies in case they adopt web analytics in their websites. The analysis covered web analytics process and data collection methods. The output of the study highlighted the essential impact of web tools in dealing with problems at early stages.

Rana et al. (2016) focused on the impact of data cleaning tools in improving decision-making and its adverse effects in case being eliminated. Based on a comparative study about data cleaning tools and their features and benefits, the study found out that both RapidMiner and WinPure Clean & Match data cleaning frameworks have different features, but the same aim of ensuring unbiased analysis.

4 Theoretical Concepts

4.1 The Concept of Web Analytics

Web analytics is growing day by day to become more crucial and more developed. In the context of market size, it was valued at $2.63 billion in 2018 and projected to reach $10.73 billion by 2026 (Korad and Sinnarkar 2020). It has updated from a simple function of HTTP (Hypertext Transfer Protocol) traffic logging to an advanced method of using data (Zheng and Peltsverger 2014), which helps marketers to improve their strategic plan, developers to create more useful pages and improve their design, and finally leaders to make effective decisions and ensure good management.

According to the Web Analytics Association Standards Committee, web analytics is based on three metrics, unique visitors, visits/sessions, and page views, and it is defined as the combination of measuring, acquisition, analyzing, and reporting of data to understand the web experience.

The first phase of measuring data can be expressed in ratios, numbers, and KPIs (Bekavac and Praničević 2015). Secondly, Zheng and Peltsverger (2014) concluded that on-site and off-site are the two major web analytics technologies, where on-site refers to data collection on the actual traffic site and off-site or macro analysis refers to data collected from other sources as surveys.

Analyzing data depends on choosing good web analytics tools and qualified analysts in light of the SWOT analysis of the company. Finally, the reporting of data is conducted in many ways as dashboard presentations. However, the accuracy of data should be respected before the analysis as data cleaning and in the data visualization, where data-ink ratio and lie factor of data are the best-known diagnostic tests (Knaflic 2015).

$$\text{Lie factor of visualisation}: \frac{\text{the size of the effect in the shape}}{\text{the size of the effect in the data}}$$

$$\text{Data ink ratio}: \frac{\text{ink used in actual data}}{\text{ink used in the entire data}}$$

5 The Implementation of Web Analytics

Web analytics can be applied in several industries, which shows the importance of this tool; it can be used in e-commerce to develop the company and sell more products and in marketing to enhance the quality of campaigns, evaluate ads, reach more clients, and identify segments for improvement while ensuring an outstanding experience for the clients. It is also beneficial in web development to

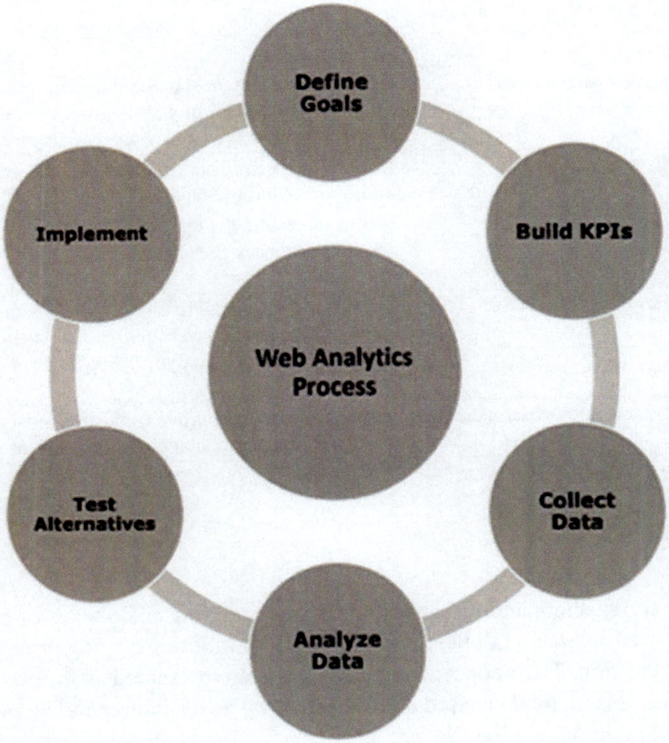

Fig. 2 Web analytics process

develop website design and performance, monitor web traffic/user flow, monitor the availability of the website, and protect it from malicious access.

In terms of the process, TutorialsPoint (2016) summarized web analytics as a process that brings more traffic to a website, thus increasing the inflow. Its process that starts from defining goals to change implementation is described in the figure below (Fig. 2).

6 Web Analytics Tools

Decision-making needs access to the data of the website through a set of tools that provide qualitative and quantitative data as summarized in Table 1.

Table 1 Types of web analytics tools

Web analytics tool	Concept
Traditional web analytics tools	Based on clickstream and internal resources data, it assesses the behavior of websites and visitors (Google Analytics, funnel analysis, Mixpanel)
Social media analytics tools	It tracks the performance of social media and improves it (Sprout Social, HubSpot, TapInfluence)
Visitors feedback tools	It aims to explore the reasons of the visitor's behavior and their feedback (Feedbackify, Qualaroo, Survicate)
Web analytics tools for mobile websites	These tools are crucial to give insights about visitors on their navigation to websites on their smartphones and tablets (Google Analytics, AppsFlyer, Adobe Analytics)
Web analytics tools for experimenting	These tools aim to improve the visitor's satisfaction through finding optimal technical or design solution

7 Data Collection

According to Waisberg and Kaushik (2009), there are four main ways of collecting data to be used in web analytics: web logs data, JavaScript tagging, web beacons, and packet sniffing. Their concept and advantages are presented as follows:

Log Files: this method is based on the server that hosts the website by registering the visitors' requests in a log file, and the extended log is the most common format that records several data about the customer after entering an URL. Log files are preferable for many reasons: owing data without needing a third party, availability of weblogs backward and weblogs saves helps to understand how search engines work (Fig. 3).

JavaScript Tagging: it works through inserting a small JavaScript code on each page, which will be activated each time the page is accessed by visitors, and their information will be saved in separate files. It is an effective way to count the number of users compared with log files; however, data need to be hosted with a third party, and sometimes companies won't share their data with others, and this might shed light to discuss the issue of data collection and privacy (Fig. 4).

Web Beacons: this third method is used with online advertising because of the type of ads (images) that demands tracking pixel to measure banner impressions and clicks path. Web beacons method is very beneficial to track customer behavior in different sites or different customers in the same site. It assesses how much banners ads perform across multiple websites.

Packet Sniffing: it collects data through the use of a packet sniffer that is usually positioned between the visitor's device and the webserver. This method differs from other methods in the context of tagging pages, where packet sniffing does not rely on tagging pages. Once the user clicks on the URL, the request goes to the webserver and passes through a packet sniffer that collects its attributes.

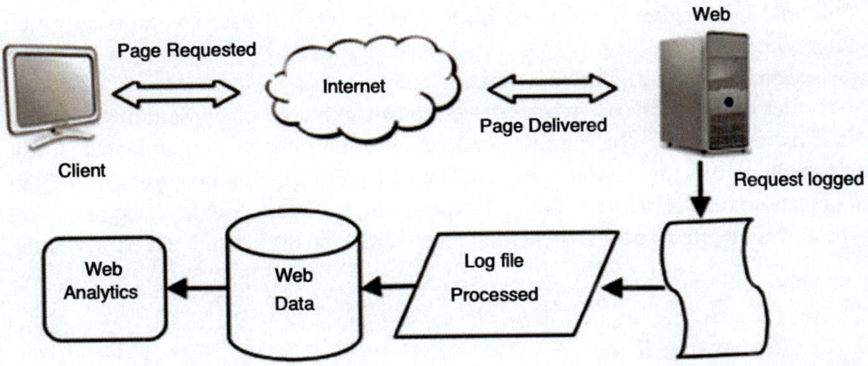

Fig. 3 Log files process

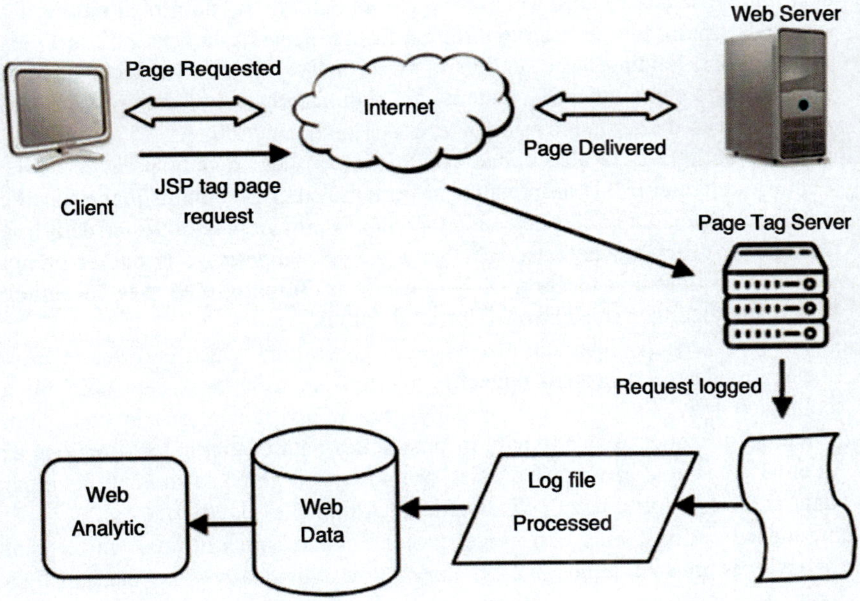

Fig. 4 Tagging files process

8 Data Cleaning (Outliers Approach)

Outlier detection has become a field of interest for many researchers and practitioners and is now one of the main tasks of time series data mining. According to Han et al. (2000), "An outlier is a data object that deviates significantly from the rest of the objects as if it were generated by a different mechanism." Another widely adopted definition in literature is proposed by Hawkins (1980): "outlier is an observation that deviates so significantly from other observations as to arouse suspicion that it was generated by a different mechanism."

Outlier observations are often referred to as anomalies, discordant observations, discords, exceptions, aberrations, surprises, peculiarities, or contaminants. In the first study on this topic, which was conducted by Fox (1972), two types of outliers in univariate time series were defined: type I, which affects a single observation, and type II, which affects both a particular observation and the subsequent observations.

9 Types of Outliers

We can distinguish the following:

Global outlier: It is also used to describe global outliers. It is a global outlier if a single data point is significantly different from the rest of the dataset. It's a very basic outlier. Finding these outliers is the primary goal of most techniques. If a computer's communication patterns are significantly out of the ordinary, it is considered an outlier in intrusion detection (Han et al. 2000).

Contextual outlier: It is also called conditional outlier. A data point is called contextual outlier if it is different or far from the other data points in the specific context. Contextual outlier detection techniques provide flexibility for detecting outliers in different contexts. Whether a 30 °C temperature is outlier or not depends on time and location. If it is winter in Toronto, then it is an outlier. If it is summer, then it is normal (Han et al. 2000).

Collective outliers: A collection of data point as a whole which is different from the entire dataset is called collective outlier. An individual data point in a collection may not be outlier. Usually, data points are related in collection. Finding subsequence as anomaly in time series dataset, finding sub-regions as anomaly in spatial imaginary dataset, or finding sub-graph as anomaly in graph dataset are examples of collective outliers (Athithan et al. 2019).

Subsequence outliers: This term refers to consecutive points in time whose joint behavior is unusual, although each observation individually is not necessarily a point outlier. Subsequence outliers can also be global or local (a review, etc.).

However, in this chapter, we will rely on the work of Chen and Liu (1993). The approach to dealing with these intervention events (or outliers) identifies the time locations and the type of outliers: the additive outlier (AO), level shift (LS), and transitory change (TC). Using the R package called tsoutliers constructed by, wish execute three basic stages globally and internally through the tso function (Automatic procedure for the detection of outliers). First, the outliers were located, given an ARIMA model adjusted to the data. Then, outliers were detected and located, checking the significance at all possible time points. This was done by means of the corresponding statistical τ, which is an approach to detect outliers and to examine the maximum value of the standardized statistics of the effect of the outliers. In the second instance, the procedure eliminated the outliers. Given a set of potential outliers, an ARIMA model is chosen and adjusted according to the statistics τ mentioned above. The importance of outliers was revaluated in the new adjusted model. The outliers that were not significant were eliminated from the set of possible outliers. And finally, the procedure performs iterations of the previous stages, first for the original series and then for the adjusted series (Cujia et al. 2019).

10 Outliers Identification Methods

According to Wang et al. (2019), these are the main identification methods of outliers:

Statistical-based methods: The fundamental idea of statistical-based techniques in labeling or identifying outliers depends on the relationship with the distribution model. These methods are usually classified into two main groups: the parametric and non-parametric methods.

Distance-based methods: The underlying principle of the distance-based detection algorithms focuses on the distance computation between observations. A point is viewed as an outlier if it is far away from its nearby neighbors.

Density-based methods: The core principle of these methods is that an outlier can be found in the low-density region, whereas inliers are in a dense neighborhood.

Clustering-based methods: The key idea for clustering-based techniques is the application of standard clustering techniques to detect outliers from given data. Outliers are considered as the observations that are not within or nearby any large or dense clusters.

Graph-based methods: Graph-based methods are based on the use of graph techniques to efficiently capture the interdependencies of interconnected entities to identify the outliers.

Ensemble-based methods: Ensemble methods focus on the idea of combining the results from dissimilar models to produce more robust models to detect outliers efficiently. They help to answer the question of whether an outlier should be linear model based, distance based, or another kind of model based.

Learning-based methods: In learning-based methods such as active learning and deep learning, the underlying idea is to learn different models through the application of these learning methods to detect outliers.

11 Some Identification Methods

kNN Anomaly Detection: One of the most widely used distance-based anomaly detection methods is the k-nearest-neighbor (kNN) algorithm. In most cases, it is a simple technique that works out of the box and accurately detects global anomalies.

In streaming data, the k-nearest neighbors of each data point must be discovered. The anomaly score is calculated based on these neighbors. Average distance to all k neighbors is used to calculate anomaly scores.

Local Outlier Factor (LOF): The LOF is a distance-based anomaly detection technique. It is used to identify local anomalies based on the density of the surrounding area. Streaming datasets must be analyzed to find the k-nearest neighbors for each data point. k-nearest neighbors are used to calculate the local reachability density of each data point in order to estimate its density (LRD). Lastly, the anomaly score is computed by comparing the LRD of a data point with all of its k neighbors.

Histogram-Based Outlier Score (HBOS): An unsupervised statistical anomaly detection method is employed here. Detecting anomalies in streaming data using histograms is the primary goal of this method. The first step is to create a histogram for each feature of the data. All features are then multiplied by their inverse height, based on the number of bins in which they reside. It is possible to create histograms with HBOS in two different ways: with static bin sizes and a predetermined number of items in each bin or with dynamic bin width and a variable number of items in each bin. In comparison to commonly used distance-based and clustering-based anomaly detection methods, this one is significantly less computationally expensive.

Isolation Forest (iForest): While distance and density measures are commonly used to detect anomalies, this method uses the concept of "isolation" instead. Anomalies are "isolated" from the rest of the dataset in this method. It is easier to isolate the data instances that are few in number and have attribute values that differ greatly from the rest of the data instances. The isolation tree (iTree) is a binary tree structure used to isolate these cases.

12 The Results

12.1 ARIMA Estimation Before Outlier Detection

At first, we conducted some unit root tests on our data in order to select the appropriate prevision model, which will be the reference to check the robustness of the estimation before and after the outlier's detection. The table below represents the unit root test using ADF and the result obtained from the estimation using the original data (Table 2).

As we see in the table above, MAPE (mean absolute percentage error) equals 44.11 meaning the estimation is accurate by 44.11% and it's very low. Now we clean our dataset from outliers and see if the estimation changes.

Table 2 Types of unit root tests and the selected model ARIMA (5,1,4)

Unit root test									
			ADF test			PP test			
I(0) p-value			0.01			0.01			
Decision			Stationary I(0) at 95% confidence			Stationary I(0) at 95% confidence			
ARIMA model									
	Ar1	Ar2	Ar3	Ar4	Ar5	Ma1	Ma2	Ma3	Ma4
coef	−0.983	−0.864	−0.606	−0.437	−0.396	0.1381	0.0033	−0.266	−0.238
s.e	0.1257	0.1357	0.1471	0.1042	0.0661	0.1332	0.1227	0.1173	0.1347
AIC = 6155.56					AICc = 6156.31		BIC = 6192.73		
Training set error measures			RMSE		MAPE		ME		
Training set			5808.307		44.11206		97.56072		

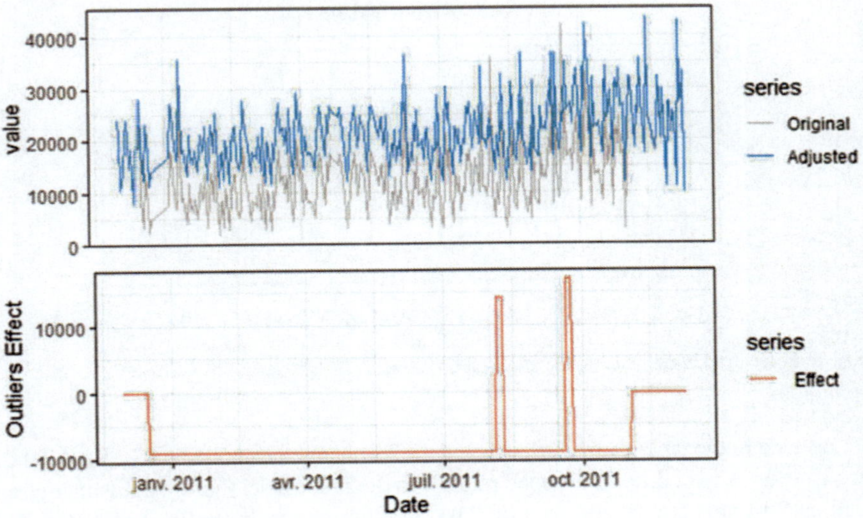

Fig. 5 Type of outliers in the UK time series

13 ARIMA Estimation After Outlier Detection

As seen from Fig. 5, there are two types of outliers: level shift (LS) and an additive outlier (AO). The LS outlier type is present at the end of 2010 and also at the end of 2011; this can be due to the reductions on the products at the end of the year. The AO outlier type is presented in two dates: the first in 08/04/2011 and the second in 09/20/2011. The clean time series of the original data that is adjusted for the impact of the outliers is presented in the figure below in blue color.

Now, we re-estimate the auto-ARIMA model using the adjusted time series; the results are presented in the table below (Table 3).

Table 3 ARIMA estimation after the adjustment of outliers

Unit root test					
		ADF test		PP test	
I(0) p-value		0.01		0.01	
Decision		Stationary I(0) at 95% confidence		Stationary I(0) at 95% confidence	
ARIMA model					
	Ar1	Ar2	Ar3	Ar4	Ar5
coef	−0.8615	−0.6979	−0.6391	−0.6304	−0.4226
s.e	0.0530	0.0642	0.0658	0.0639	0.0545
AIC = 6105.34			AICc = 6105.62	BIC = 6127.64	
Training set error measures		RMSE	MAPE	ME	
Training set		5421.533	20.8135	59.99746	

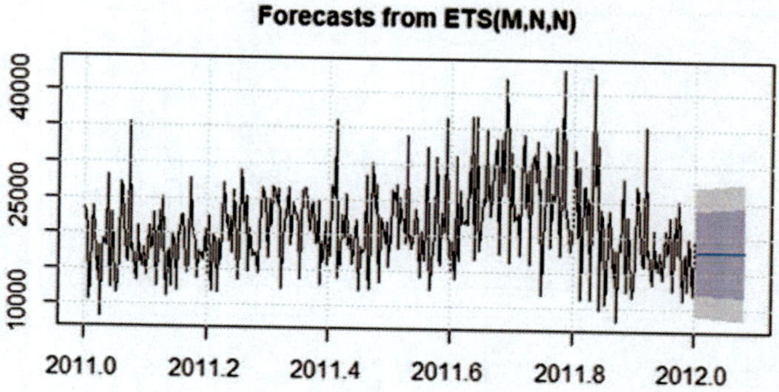

Fig. 6 Forecasted horizon using the adjusted time series

As seen in the table above, the selected model changed to ARIMA (5,1,0). All the error measurement criteria dropped relative to the first model (before adjustment for outliers). MAPE criteria dropped by 23.3; in other words, the estimation precision augmented from 55.89% to 79% after counting and adjusting the time series for outliers, and the figure below shows the forecasted horizon using the adjusted time series (Fig. 6).

14 Conclusion

This chapter aims to explore the impact of data cleaning (cleaning) on data analysis accuracy and decision-making efficiency. The research is divided into two parts. The first part is a theoretical background about web analytics and its concept, tools, and data collection methods. The second part is a quantitative study, where we performed some statistical tests to detect outliers based on the proposed and later developed method of Chen and Liu (1993) in the UK-registered retail dataset and showed its impact on the ARIMA estimation.

The empirical output based on the detection of outliers and their impact on the estimation performance shows the gap of biased results and bad decisions. The comparison between the two (before adjusting for outliers and after) values of the mean absolute percentage errors (MAPE) criteria showed that the presence of outliers affects the performance of the model. Thus, data cleaning is always needed for more accurate and precise results, Thus, data cleaning does matter.

References

Athithan, G., Murty, N., Suri, R.: Outlier Detection: Techniques and Applications: A Data Mining Perspective. Springer (2019)

Bekavac, I., Praničević, D.: Web analytics tools and web metrics tools: An overview and comparative analysis. Croat. Oper. Res. Rev. **6**, 373–386 (2015)

Chen, C., Liu, L.: Joint estimation of model parameters and outlier effects in time series. J. Am. Stat. Assoc. **88**(421), 284–297 (1993)

Cujia, A., Agudelo-Castañeda, D., Pacheco-Bustos, C., Calesso Teixeira, E.: Forecast of PM10 time-series data: a study case in Caribbean cities. Atmos. Pollut. Res. **10**(2019), 2053–2062 (2019)

Fox, A.J.: Outliers in time series. J. R. Stat. Soc. Series B (Methodological), 350–363 (1972)

Han, J., Kamber, M., Pei, J.: Data mining Concepts and Techniques. Third Edition, Morgan Kaufmann Series in Data management Systems (2000)

Hawkins, D.: Identification of Outliers, vol. 11. Springer (1980)

Knaflic, C.: Storytelling with Data: a Data Visualization Guide. Wiley, New Jersey (2015)

Korad, S., & Sinnarkar, M. (2020). Web Analytics Market. . Retrieved 09 21, 2020, from Applied Market Research: https://www.alliedmarketresearch.com/web-analytics-market-A05971#:~:text=The%20web%20analytics%20market%20size,19.3%25%20from%202019%20to%202026.

Kumar, V., Ayodeji, G.: Web Analytics for Knowledge Creation: A Systematic Review of Tools, Techniques, and Practices. Int. J. Cyber Behav. Psychol. Learn. **10**(1), 14 (2020)

Padma Jyothi, U., Bonthu, S., Prasanth, B.: A study on raise of web analytics and its benefits. Int. J. Comput. Sci. Eng. **05**(10), 59–64 (2017)

Rana, S., Gagan Prakesh Negi, E., Kapoor, K.: A Comparative Analysis of Data Cleaning Tools. Int. J. Adv. Res. Comput. Sci. Softw. Eng. **06**(04), 294–299 (2016)

Shilakes, C., Tylman, J.: Enterprise information portals. Enterprise. Enterprise Inf. Portals, 354–362 (1998)

TutorialsPoint: Web Analytics. TutorialsPoint, Telangana (2016). Retrieved from https://store.tutorialspoint.com/ebook_view_index.php?ebook=web_analytics_tutorial

Waisberg, D., Kaushik, A.: Web analytics 2.0: empowering customer centricity. SEMJ.org. **2**(1) (2009). Retrieved from http://www.semj.org/documents/webanalytics2.0_SEMJvol2.pdf

Wang, H., Jaward Bah, M., Hammad, M.: Progress in outlier detection techniques: a survey. IEEE Access. **2019**(7), 107964–108000 (2019)

Zheng, G., Peltsverger, S.: Web Analytics Overview. Encyclopedia of Information Science and Technology, USA (2014)

Web Analytics Tools for e-Commerce: An Overview and Comparative Analysis

Wassila Boufenneche, Mohamed Hebboul, and Omar Benabderrahmane

Abstract E-commerce had led to organizations spending considerable resources in online approaches to expand business processes on websites. Conventional approaches of determining web use disappoint the richness of information needed for the efficient assessment of such techniques. Web analytics has become one of the most vital tasks in E-commerce, because it enables business and E-merchants to track the actions of clients when browsing their website. There exists a collection of tools for web analytics that are utilized not just for monitoring as well as measuring website traffic but additionally for examining the business activity. Nevertheless, many of these tools concentrate on low-degree web features and metrics, making other advanced functionalities and analyses just offered for commercial.

Keywords Web analytics · Web analytics tools · E-commerce

1 Introduction

As more and more people get online, the world is becoming aware of how quickly the Internet is changing and growing. All organizations and businesses must have a web presence because the Internet provides numerous multimedia features that enable and change the way organizations communicate with their customers, suppliers, competitors, and employees. A user's perception of business success and the strategic importance of web context in modern business are directly influenced by their interactions with the Internet. It also shifts many business activities online, creating new business models known as web business models at the same time.

W. Boufenneche · M. Hebboul (✉)
University Center Abdelhafid Boussouf, Mila, Algeria
e-mail: w.boufeneche@centre-univ-mila.dz; m.hebboul@centre-univ-mila.dz

O. Benabderrahmane
International Islamic University – Malaysia, Gombak, Malaysia
e-mail: benabderrahmane.omar@live.iium.edu.my

S. Sedkaoui et al. (eds.), *International Conference on Managing Business Through Web Analytics*, https://doi.org/10.1007/978-3-031-06971-0_5

A critical enabler for any organization is its website. Yet websites, as major sales and marketing channels for organizations, need to have mechanisms to determine their success and weaknesses and improve their effectiveness.

2 Study Questions

The problem of the study is subdivided into the following two questions:

- How does web analytics work?
- What can we analyze through web analytics?

3 The Importance of the Study

The subject of the study is important because it seeks to identify the advantages and disadvantages of each tool, which will make it easier for experts and researchers to choose the best method that they use in analyzing their data.

4 Objectives of the Study

The aim of this study is to clarify how website analytics are defined, how they contribute to E-commerce, and the most useful tools used in data analysis nowadays.

5 Web Analytics

5.1 Web Analytics Definition

There are many definitions of web analytics:

- Web analytics is a technique which can be employed to collect, measure, report, and analyze website data.[1]
- Web analytics are techniques that assess quantitative data such as web traffic, surveys, sales transactions, and others to improve the performance of marketing activities.[2]

[1] Web Analytics, 2015, Tutorials Point (I) Pvt. ltd, p 1, on the site: Ltd.https://www.tutorialspoint.com/web_analytics/web_analytics_tutorial.pdf (consulted on 2/11/2020)

[2] Chaffey dave and all, 2009, Internet marketing strategy, implementation and practice. fourth ed. Pearson Education Inc., USA, p 12

- The term "web analytics" refers to the process of gathering, analyzing, and presenting information gleaned from the Internet in order to better understand and improve the user experience.

Measuring incorporates different metrics and is expressed in the form of numbers, ratios, and key performance indicators (KPIs).

Data acquisition activity is mainly done through one of the two most widely used methods: using log files that gather data from a server and using popular methods of tagging websites supported by JavaScript code.

Furthermore, the purpose of analyzing data is to transform data into information useful for a decision-making process. A company's specific characteristics and goals should be taken into consideration when selecting web analytics tools, as well as appointing employees who are capable of "discovering" useful information for supporting decisions based on large amounts of data collected.

As a final step, selected metric outputs are used to generate useful reports (d).[3]

5.2 *Importance of Web Analytics*

The importance of web analytics can be illustrated through:[4]

- Using data gathered from the Internet can provide insights into website traffic, transactions, server performance, and user information.
- Understanding of the web and website optimization provides a more adapted approach to a target audience to increase conversion rates, as well as customer loyalty.
- Website traffic analysis provides insight into the number of visitors and their geolocation, locations, and time spent on websites and other criteria.
- Web analytics provides other advantages such as increasing efficiency and cost reduction.
- Marketers can also find useful web analytics data to improve products/services and evaluate the success of a marketing campaign.
- Web designers and web developers use this data to improve the usability of the website and thus satisfy website user.
- Web analytics provides company management with insight into how to generate revenue from a website, how to create suitable user experience and improve its competitive advantage, as well as how to support continuous improvement and competitiveness.

[3]Ivan Bekavac and Daniela Garbin Praničević, 2015, Web analytics tools and web metrics tools: An overview and comparative analysis, Croatian Operational Research Review, 6(2), Croatian Operational Research Society, p 374. on the site: https://pdfs.semanticscholar.org/22ab/e974e3a7bc0ba418d5115241c10b847bc8bd.pdf (consulted on 2/11/2020).

[4]Ivan Bekavac and Daniela Garbin Praničević, 2015, p 375.

5.3 Pillars of Web Analytics

There are three basic pillars of web analytics that can contribute to enhancing a website's popularity. Data collection, storage, and evaluation are the three pillars of data management[5]:

– *Data Collection*: When deciding to analyze the site visitor traffic, then it must be clarified exactly which data is going to be analyzed. The proper implementation of web analytics helps to understand some basic questions such as "Where do online visitors come from?", "Which web pages did they come from?", and "How much time did they spend on the website?".
– *Data Storage*: As a result of this analysis, the storage of collected data will be examined, which is referred to as:
  • Internal storage: The most positive aspect of storing data internally on a server is the ownership of the data. Hardware, software, licenses, staff, energy, and security personnel, as well as infrastructure, are all potential cost factors in internal storage.
  • A subscription-based model for external storage and software: All collected data can also be stored externally by websites. Besides the lower costs, external storage offers additional positive aspects such as regular maintenances, software updates, software installation, technical assistance, etc.
– *Data Evaluation*: This is the last pillar of web analytics, and it is the assessment of collected and stored data. For web analytics users to get a quick overview of the most important information, proper metrics must be chosen. Evaluation is made easy with the use of a suitable key performance indicator (KPI), i.e., that is the basic metric against which the evaluation is done.

5.4 Dimensions of Web Analytics

There are three primary dimensions of web analytics:[6]

– *Web Content Analytics*: An important part of web content analytics is locating information or resources from a wide variety of sources. The agent-based approach and the database approach to web content analytics exist. Artificial intelligence systems that can act autonomously or semi-autonomously on behalf

[5]Lakhwinder Kumar and all, 2012, Web Analytics and Metrics: A Survey, paper in Proceedings of the International Conference on Advances in Computing, Communications and Informatics, pp. 966–967, on the site: https://www.csd.uoc.gr/~hy209/resources/web_analytics.pdf (consulted on 2/11/2020)

[6]Alghalith Nabil, 2015, Web Analytics: Enhancing Customer Relationship Management, Journal of Strategic Innovation and Sustainability, 10(2), pp. 13–14

of a specific user are part of the agent-based approach. Using a user's profile, some smart web agents can search for relevant information and then organize and interpret the information discovered. It is possible to organize and filter retrieved information using various information retrieval techniques and open hypertext documents. Using a database, you can create more structured and high-level collections out of the unstructured and semi-structured data on the Internet. Metadata or generalizations are then organized into structured groups so that they can be accessed and analyzed.

- *Web Usage Analytics*: Analysis of how people use a website is another aspect of web analytics. A user's habits (or patterns) can be discovered through daily access logs, which are automatically collected. Referrer logs, which keep track of the pages that users visit before and after registering, were recently added. In order to better structure a website, web usage analytics is essential. Users' behavior on the web is evolving at a rapid pace, necessitating the development of new types of user knowledge.
- *Web Structure Analytics*: In web structure analytics, links found on websites or web documents are analyzed for useful information. Search engines use it to determine the popularity of websites and the authority of portals.

5.5 The Web Analytics Business Process

The web analytics business process is a series of business processes and sub-processes that, when considered collectively, describe the ongoing activities that any organization needs to master to be successful with web analytics. These are management processes like "Define Business Objectives for the Web Site" and "Allocate Appropriate Resources" and operational processes like "Collect Data" and "Distribute Reports and Analysis."[7]

When it comes to web analytics, a thorough investigation must begin with the customer's goals and objectives and continue through the website's implementation. All of the various steps in the process, such as setting goals, creating effective metrics, gathering and studying data for better strategy, etc., are involved. It's actually a recursive process that helps with website optimization. The following are the stages of the web analytics process[8]:

[7] Web analytics demystified, 2007, the web analytics business process making the case for a process-driven approach to web site measurement and ten critical requirements for success, p 2, on the site: http://cdn2.hubspot.net/hub/74398/file-15425858-pdf/docs/web_analytics_demystified__the_web_analytics_business_process.pdf (consulted on 2/11/2020).

[8] Lakhwinder Kumar and all, 2012, pp. 967–968.

5.5.1 Defining Goals

To answer the question "Why should a website exist?", one must look at the goals. In order to measure the success of a website, it is important to have a clear idea of what the goals are for the site. The ability to measure the success of a website no matter what its goals are has been a major evolutionary trend in the last few years.

5.5.2 Defining Metrics

KPIs (key performance indicators) can be used to measure whether or not a website is meeting its goals. The proposed KPIs for a website should be linked to a specific action. A key feature of a KPI is that it can be easily adapted to different situations. A good metric should include the following:

- Un-complex: The metric can be used by people from different departments in a company to make decisions.
- Relevant: The metric must be relevant to the user, which means that it must accurately reflect the user's actions.
- Timely: Decision-makers need timely access to high-quality metrics so they can act quickly. Even the best metrics are of no use if they take a month to get data that changes every week in the industry.
- Instantly Useful: It's critical to know what a KPI is before looking at it, so that you can see the first glimmers of insight.

5.5.3 Collecting Data

Data must be saved for further investigation. In order to conduct further analysis, it is imperative that data be collected accurately and saved in a central database or on a local machine. Analytical results can only be achieved if data is gathered correctly. Gathering information about the user's behavior can be done in various ways:

- Cookies and IP: Cookies are small text files with a size of about 4 kilobytes. Cookies are small text files that a website transfers to a user's hard drive when they visit it. Generally speaking, there are two types of cookies: session cookies and persistent cookies. While session cookies get deleted once the visitor leaves the website, persistent cookies remain on the visitors' computer. Cookies are used to keep track of a user's visits. It is possible to distinguish between returning visitors and new visitors because each cookie contains a unique identifier. An IP address is a number that is assigned to a specific computer on the Internet. The user's location can be determined by looking up their IP address. Problems can arise when a user attempts to access the Internet via a router, such as one located in a business or public hotspot. IP address of the router is accessed instead of the IP address of the original user.

- Log File Analysis: The web server contains all of the files necessary to run the website. Each time a user accesses and interacts with the website, a record is kept in the log file. The IP address, browser type, operating system, accessed content, and date and time of access can all be included in a single entry. For every request for information from the site's server made by a visitor, a log file is kept of this information.
- Page Tagging: Every page of a website can be labeled with a small JavaScript tag. A visitor opens a page, and this script kicks in, saving the visitor's data and actions in a separate file.
- Web Beacons: Tracking customer behavior across multiple websites is a common use of web beacons. For example, it tells how well banner ads are doing on multiple websites. Advertisers can track anonymously the same visitor across multiple sites or different visitors to the same site because the same server collects the data, reads the cookies, and performs the tracking.
- Sniffing of Packets: However, despite its advanced technological capabilities, packet sniffing is still mostly used for multivariate testing. Because it doesn't need to tag pages, it has the greatest advantage of all packet sniffers.

5.5.4 Analyzing Data

Analysts need to take a few steps in order to gain insight into customer behavior from the data. After logging into any web analytics tool, you'll be presented with a summary report that includes the most important metrics.

- Visits: It is the number of visits to the website and the number of interactions with the website.
- Bounce Rate: The percentage of single page view and visits (this metric can also have different definitions, such as a visit that lasts less than 5 s).
- Page Views: The number of pages that were requested in all visits.
- Pages/Visit: The number of pages seen in each visit.
- Average Time: The time on which the users stay on the website.
- % New Visits: The count of the number of sessions which new users visited website first time.

5.6 Web Analytics Business Applications

Web analytics can be used for a variety of business purposes, including[9]:

- Examine the effectiveness of a company's marketing strategy.

[9] Alghalith Nabil, 2015, Web Analytics: Enhancing Customer Relationship Management, Journal of Strategic Innovation and Sustainability, 10(2), pp. 13–14.

- Offer users personalized content based on their browsing habits and preferences.
- Ads can be targeted based on user behavior.
- Come up with better marketing strategies that work.
- Predict the user's actions.
- Improve customer service and customer intimacy by providing better customer care.
- The maintenance and management of a customer base.

6 Web Analytics Tools

6.1 Categories Web Analytics Tools

As a foundation for making business decisions, a wide range of web analytics tools have been developed and are now available on the market. Five categories of web analytics tools are as follows[10]:

- Traditional web analytics tools relied primarily on data gathered by visitors, competitors, and internal company data. "What happens on websites?" and "How many conversions have been achieved on the website?" are two of the most common questions that clickstream data answers.
- Web analytics tools that track performance on social networks.
- "Why did the visitor behave or not behave in this manner?" is a question that web analytics tools are designed to answer.
- Web analytics tools for mobile websites that are becoming increasingly important as the number of people using mobile devices increases the turnover of websites. It is essential to have this type of tool to ensure that your website is compatible with mobile devices because it provides insight into the behavior of visitors on mobile devices.
- For testing and finding the best technical or design solutions to improve visitor satisfaction, web analytics tools are available.

6.2 Website Analytics Tools for e-Commerce

There are many website analytics tools for E-commerce that help understand customers and make better marketing decisions, including:

[10]Teixeira, J., 2011, Get Involved: 5 Types of Web Analytics tools to start using today!, on the site: https://www.morevisibility.com/blogs/analytics/get-involved-5-types-of-web-analytics-tools-to-start-using-today.html (consulted on 4/11/2020).

6.2.1 Google Analytics

Web analytics service Google Analytics is a freemium service that tracks visitor traffic to a website and generates reports on that data. After acquiring Urchin, Google launched the service in November 2005. As of this writing, Google Analytics is the most widely used web analytics service on the Internet.[11]

6.2.1.1 Advantages of Google Analytics

The advantages of Google Analytics are[12]:

– Cost: It is absolutely free. This is the coolest feature for most of the people involved.
– Usability: It is not just bound to the experts. Anyone even without having strong programming skill can use it. It has made analyzing very easy for both the specialists and the non-specialists.
– Availability of tutorials: Google provides an online tutorial to learn about Google Analytics. It has made learning the system extremely easy.
– Installation: It is easily installable. There are no programming skills required to install it unlike in other analytical tools.
– Visitor views by geography, time frame, and source: It allows getting and viewing the number of visitors to the site based on some time frame. It also allows one to find the geographical location of all the visitors and the source which helped them find it.
– Visitor's details: It also provides details of each visitor including the time of their stay on the website, the web pages they accessed, the number of links they clicked on, etc. There is also a concept of bounce rate that shows the number of visitors in and out of a web page and the visitors who stayed longer.

6.2.1.2 Disadvantages of Google Analytics

There are several disadvantages of Google Analytics, such as[13]:

– *Recording bot and spam traffic*: It's possible that some of the computers that access your site are not being operated by humans. Many bots are crawling

[11] https://ec.europa.eu/competition/information/digitisation_2018/contributions/at_internet.pdf (consulted on 2/11/2020).

[12] Deepika Verma and all, 2011, Google Analytics for Robust Website Analytics, p 10. on the site: https://www.sjsu.edu/people/rakesh.ranjan/courses/cmpe272/s2/Team-Matrix-Google-Analytics.pdf (consulted on 4/11/2020).

[13] Limitations of Google Analytics, 2019, on the site: https://masondigital.com/4-limitations-of-google-analytics/ (consulted on 4/11/2020).

websites all the time for a variety of reasons. Some are evil, while others are not. Bots have the potential to significantly skew data either way.

- *Time on site*: Google Analytics consistently underreports the amount of time spent on a page. Google's servers receive the precise time that a user requests a page to load. As soon as the next web page is loaded, the time is noted. Using this information, Google can estimate how long the user spent on the first page.
- *Measuring all users*: Each page of a website is loaded with a JavaScript code snippet from Google Analytics. Data is sent back to the Google servers for processing as soon as the page is loaded. There are some browsers that do not allow JavaScript to run. Additionally, Google Analytics makes use of cookies to keep tabs on user activity. Using web browsers and ad blockers, cookies can be disabled.
- *The need for customization*: There are a lot of webmasters and marketers who put Google Analytics code onto their website and then congratulate themselves because they're "crushing it." Unfortunately, this is just the beginning of the setup process. It is necessary to set up goal tracking and event tracking in order to track interactions with non-HTML content such as PDF downloads, video views, and form submissions in order to run the most important reports. In order to get the most out of Google Analytics, the tracking needs to be tailored to the user's specific requirements.

6.2.2 Piwik

Many of the features of Google Analytics can be found in Piwik, a free web analytics platform. Piwik requires users to download and install the software on their own servers, whereas Google Analytics can be accessed through your browser.[14]

6.2.2.1 Advantages of Piwik

The advantages of Piwik are[15]:

- Piwik provides a very simple install even for users with basic skills. It is user-friendly and relatively straightforward for website owners to place on their site and manage.

[14] **Stickler Rebecca, 2016,** 6 Ecommerce Analytics Tools for Digital Marketers, on the site: https://www.webfx.com/blog/marketing/6-ecommerce-analytics-tools-digital-marketers/ (consulted on 4/11/2020).

[15] Piwik PRO GmbH, p 4, Optimizing Piwik On-Premises for Top Performance and Security, on the site: https://www.dirk-buechner.de/content/6-erfolgsrezepte/2-kundenkontakte-in-krisenzeiten-klug gestalten/piwik_optimization_whitepaper.pdf (consulted on 4/11/2020).

- Piwik offers great privacy features. Many analytical or tracking tools send user behavior directly to advertisers, but with Piwik, your visitors can browse securely in the knowledge that their data and behavior are kept private.
- Piwik tracks user behavior and visitor activity via a JavaScript file. It is also possible to import your own web logs into Piwik to analyze them with all the enhanced features the software provides – another real plus.
- Piwik also comes with SDKs for both Android and iOS, so you can also measure your apps with it!
- The software also includes a wide range of additional features, such as analytics for any E-commerce activity that your site includes and a profile of your website's visitors (including real-time visitor analysis and mapping).
- Another big positive of Piwik is that it comes with its own mobile app, allowing you to access your reports at any time.

6.2.2.2 Disadvantages of Piwik

There are several disadvantages of Piwik, such as[16]:

- The UI isn't great. This is not the product if you're looking for slick reporting. The amount of information provided by the commerce sector is minimal. Piwik is a good reporting tool for basic metrics like the number of visitors and the URLs they came from.
- Search engine, keywords, etc. However, it doesn't appear to have a great deal of expertise in the field of business reporting. Because of this, this tool may not be ideal for a client who expects a lot of reports on conversion from visitor to customer and funnel analysis/dropout rate in a multi-step buying process.
- A good set of commerce variables can be found in GA. Filters' dropout and funnel analysis have been thoroughly tested and found to be effective. This is where commercial products come in handy because they can be customized a lot more than home-brewed solutions can.
- There is little to no external tool integration. AdSense and AdWords integration is simple because GA is a Google product. As a result, GA is a useful tool for tracking an entire marketing campaign (including email marketing and search engine marketing). This does not appear to be the case in Piwik. Will customers wait for development when both free products are available?
- Piwik requires PHP/MySQL and enough disk space on the server side in order to function properly. There is a monetary outlay associated with this. This is still a lot of money compared to the free storage on GA.
- In previous versions of this product, the reporting interval was fixed. As a result, Piwik may not be able to report on any custom interval (which is available in

[16]Ranganath Akshay, 2009, PiWik v/s Google Analytics, on the site: https://akshayranganath.wordpress.com/2009/01/27/piwik-vs-google-analytics/ (consulted on 4/11/2020).

GA). A feature missing from Piwik is the ability to compare data over two different time periods.

- Support for implementation. Piwik assumes that you are already familiar with analytics and only require a tool. No documentation on what analytics is and how to measure it is provided by this application.
- When it comes to implementation, GA's documentation is extensive, but its paid consultants are the most valuable resource. Piwik only provides PHP-based support for making the code work, but no guidance on how to actually use the product.

6.2.3 Kissmetrics

Kissmetrics is one of the most popular paid tools among E-commerce site owners, and it has several features tailored for increasing online conversions.[17]

6.2.3.1 Advantages of Kissmetrics

The advantages of Kissmetrics are[18]:

- Use smart campaigns: Automated, behavior-based email campaigns can increase sales and engagement.
- Customer engagement can be improved through the use of automated, behavior-based emails.
- Email campaigns can be automated so that they can be targeted to the right customers at the right time for individual and ongoing campaigns.
- Get a better understanding of what drives sales by analyzing how the campaign is performing beyond just the number of people who opened and clicked on it.
- By defining and tracking key customer segments based on their behavior, you can easily monitor their growth over time.
- It's time-saving: You don't have to sift through a mountain of data to find out what's going on in your business.
- Understand which users need your attention and begin new initiatives to ensure that your goals are met on time.
- Knowing what customers do, what works, and what doesn't is the key to understanding their behavior. To do this, you must watch the entire journey through devices.

[17]**Stickler Rebecca, 2016,** 6 Ecommerce Analytics Tools for Digital Marketers, on the site: https://www.webfx.com/blog/marketing/6-ecommerce-analytics-tools-digital-marketers/ (consulted on 4/11/2020).

[18]https://www.accuratereviews.com/marketing-analytics/kissmetrics-review/ (consulted on 5/11/2020).

- With the help of Surface's ready-to-use reports containing the necessary analyses and insights, you can make more informed marketing decisions.

6.2.3.2 Disadvantages of Kissmetrics

There are several disadvantages of Kissmetrics, such as[19]:

- Installation – The installation of this tool is quite an arduous job as it requires a lot of work and integrating with other tools is quite difficult.
- Cohort reports – You cannot demand the reports on the basis of week numbers; you have to schedule the reports if you want a weekly report.
- Confusing data at times – The data analysis reports can be a bit difficult sometimes. It means that data analytics can sometimes raise a few eyebrows.
- Usability testing is the only thing on the agenda here.
- Comparatively speaking, A/B testing is less effective than the other two methods.
- By default, demographic and geographic information is available.
- Kissmetrics has a firm grasp on what they are and are not. They're an excellent tool for comparing the number of unique users or visitors to overall traffic. They don't do a good job of analyzing traffic data at a finer level. Consequently, Kissmetrics allows customers to link their Google Analytics accounts with their own accounts so that they can pull data from GA into Kissmetrics.[20]

6.2.4 Mixpanel

- [21]Mixpanel is a web and mobile analytics platform that focuses on user behavior. Users' online actions, such as signing up for a service, watching a product demo, returning to the service after a break, and interacting with features of a product or service, are at the heart of this research strategy.

[19]https://www.semplaza.com/kissmetrics-review/

- https://www.mockingfish.com/blog/pros-and-cons-of-the-popular-ab-testing-tools-in-the-market/ (consulted on 5/11/2020).

[20]Zach williams, 8 Alternatives to Google Analytics, on the site: https://www.venveo.com/blog/8-alternatives-to-google-analytics (consulted on 5/11/2020).

[21]https://leadsbridge.com/blog/how-to-boost-productivity/what-is-mix-panel/ (consulted on 5/11/2020).

6.2.4.1 Advantages of Mixpanel

- As a web and mobile app analytics platform, Mixpanel has many advantages. In this section, we'll detail the benefits that users can expect from the system in particular[22]:

  - *User Engagement Stats*: Similar to Kissmetrics, this service provides a wide range of statistics that can help you learn more about your visitors.
  - *Rich Features*: Your website visitors can be segmented into groups based on their interests and demographics, as well as event tracking, retention reports, and a targeting system that groups users with similar profiles.
  - *Better Real-Time Data*: With this tool, you get better real-time data and a user interface that is easier to understand than Kissmetrics.
  - *Customer Retention*: You can use this tool to re-engage customers who have abandoned your site or are currently on a free trial by sending them notifications via email or text message. The tool allows you to schedule notifications.
  - *Email*: You can now send a pre-written email message to customers using the tool's new feature.
  - *Mobile App*: It has a mobile app.
  - *Easy-to-Digest Analytics*: For the average user, Mixpanel provides reliable data analytics and presents them in a comprehensive yet understandable manner. Information presented to users allows them to easily measure their actions regardless of whether it's an iOS-, Android-, or web-based app.
  - *Simple Interface*: The platform offers a simple interface that lets users dig in and unearth how people react to their app. To better understand how their app is being used, users can provide feedback, and this feedback helps the company make improvements to their app.
  - *Event Tracking*: Mixpanel is a tracking software that redefines the concept. Instead of making an evaluation or keeping track of page views and clicks, the app gauges the user's app based on events. With Mixpanel's help, companies can monitor events, discover new trends, and create sales funnel data.
  - *Targeted Data Collation*: Mixpanel has a narrower focus than other analytics platforms. The tool's output provides an infinite number of options. Customers' actions can be used to help users improve their products using the platform's billions of insights and events.

[22] https://www.woorank.com/en/blog/analytical-tools-other-than-google-analytics (consulted on 5/11/2020) https://comparecamp.com/mixpanel-review-pricing-pros-cons-features/ (consulted on 4/11/2020)

6.2.4.2 Disadvantages of Mixpanel

There are several disadvantages of Mixpanel, such as[23]:

- Setup is not easy. You need technical support or knowledge to fully integrate API tracking with this tool.
- You can track up to 1000 users per month for free with Mixpanel's free starter plan. This is ideal for startups in their initial stages. Mixpanel, on the other hand, gets pricier as you scale up your business.
- While Mixpanel does track traffic, it doesn't have the same level of attribution as Google Analytics.
- Some users of Mixpanel complain that it takes some time to get used to how it works because of the steep learning curve.

6.2.5 Klaviyo

Emails are a critical part of any e-commerce company's strategy for growing customer databases and providing them with useful content. For analyzing email data, Klaviyo is an e-commerce analytics tool. Because it recognizes the interdependence of e-commerce and email marketing, it provides numerous reports and automation tools to aid in the analysis of email insights and the formulation of strategic decisions.[24]

6.2.5.1 Advantages of Klaviyo

The advantages of Klaviyo are[25]:

- Automatic audience segmentation based on specific actions and parameters (purchase of a product, the opening of a message, purchase within a certain period of time), resulting in more personalized and faster messages.
- Welcome and auto-response email templates to make lead nurturing and fight against cart abandonment.
- Advanced reports that track openings, clicks, revenues, and all intermediate metrics.

[23] https://www.woorank.com/en/blog/analytical-tools-other-than-google-analytics. (consulted on 4/11/2020) https://practicoanalytics.com/segment-analytics-pros-cons/ (consulted on 4/11/2020)
[24] https://colorwhistle.com/7-alternatives-to-google-analytics/ (consulted on 4/11/2020)
[25] https://www.ecommerce-nation.com/best-analytical-tools-for-your-e-commerce/#Klaviyo (consulted on 4/11/2020)

6.2.5.2 Disadvantages of Klaviyo

There are several disadvantages of Klaviyo, such as[26]:

- There is no way to track leads. Klaviyo isn't the platform for you if you're looking for lead scoring functionality. Any automation that relies on a visitor's level of involvement with your site and marketing will require lead scoring. It's possible to have it integrated with a third-party app, but that's going to cost you money.
- Costs are going up. A small (very small) e-commerce business can benefit greatly from Klaviyo's free plan, which allows for up to 250 contacts and 500 email sends per month. However, if you want to grow your business, you'll have to pay a lot of money once you surpass the "freebie" numbers.
- Customer dissatisfaction. The platform's help center is excellent for finding quick answers to common questions. The hourly schedule of Klaviyo means that if you want a more personalized support experience, you'll have to work around it.

6.2.6 Woopra

Advanced analytics tools such as Woopra are used to track and analyze abandoned carts. It might be something related to price or a greater number of steps during the checkout process; Woopra can help to identify everything.[27]

6.2.6.1 Advantages of Woopra

The advantages of Woopra are[28]:

- Fully real-time data.
- Detailed tracking of customer activity.
- Reports allow for deep segmentation.
- Highly customizable reporting.
- Dynamic segment creation.
- Live dashboard with bespoke metrics.
- Make use of a variety of channels to interact with customers including email, chat, and more.
- "Appconnect" feature allows to utilize third-party features through Woopra.
- Retention analysis, funnel analytics, and segmented analysis.

[26] https://www.drip.com/blog/ecommerce/klaviyo-vs-mailchimp-vs-drip (consulted on 4/11/2020)

[27] Aarushi Ranjan, 2019, 7 Must-Have Analytics Tools for Your eCommerce Website, on the site: https://www.shiprocket.in/blog/ecommerce-analytics-tools/ (consulted on 4/11/2020)

[28] https://colorwhistle.com/7-alternatives-to-google-analytics/ (consulted on 7/11/2020)

6.2.6.2 Disadvantages of Woopra

There are several disadvantages of Woopra, such as[29]:

- The main disadvantage of Woopra is the pricing. While it is a great tool for small businesses, Woopra is not priced well for small businesses with equally small budgets.
- It has no traffic features.
- The free version of Woopra provides minimal solutions and features. There is a limitation on the number of days to view the data from.
- The new customizable dashboard may not be that great and requires a bit of improvement.

6.2.7 Hotjar

To be successful in e-commerce, it is essential that no decisions are made based on suppositions alone. Modern online merchants benefit greatly from Hotjar, a tool that aids in increasing their sites' click-through and conversion rates. With Hotjar, you can view heatmaps of the website along with real-time recordings of your visitors.[30]

6.2.7.1 Advantages of Hotjar

The advantages of Hotjar are[31]:

- Hotjar gives e-commerce and marketing managers all the tools they need to conduct in-depth analyses of their websites.
- Heatmaps, video recordings of user sessions, and conversion funnels are just a few of the features that allow you to track your visitors' activities on your website.
- More intriguing, however, is the ability to listen to and recognize a user's voice. You can embed a widget from Hotjar on any page of your website to collect visitor feedback. In this way, customers can rate your service and leave a message in just two simple steps. A variety of emotions, ranging from complete contentment to complete dissatisfaction, are represented by these emoticons.
- Ability to track the dynamics of the customer experience and quickly notice and fix bugs.

[29] https://www.canecto.com/best-25-google-analytics-alternatives-for-business-owners/ (consulted on 7/11/2020). https://growthbytes.co/google-analytics-alternatives/

[30] Aarushi Ranjan, 2019, 7 Must-Have Analytics Tools for Your eCommerce Website, on the site: https://www.shiprocket.in/blog/ecommerce-analytics-tools/ (consulted on 7/11/2020).

[31] https://www.promodo.com/blog/top-6-ecommerce-analytics-tools-for-online-stores-in-2019/#Hotjar (consulted on 7/11/2020).

6.2.7.2 Disadvantages of Hotjar

There are several disadvantages of Hotjar, such as[32]:

- Inability to exclude URLs.
- A separate site dashboard.
- Tracking codes slow down page speed.
- Poorly organized data.
- The Hotjar e-commerce analytics tool doesn't have a free trial and offers four plans depending on the number of page views per day. The cheapest one, which collects data from up to 20,000 page views per day, will cost you $89 per month. The most expensive plan limited by 400,000 per day costs $589 per month.

7 Comparative Analysis

A review of the various web analysis tools shows that each of these tools has similar features and follows similar goals but has entirely different focuses. This section of the study includes a comparison of these tools (Table 1).

8 Conclusion

Web analytics can be used to improve customer relationship management and to address ineffective search engines that produce incomplete indexing and the retrieval of irrelevant information. It's critical to have a system in place that makes it simple and quick for people to find the information they need on the Internet. Web analytics sifts through mountains of data on the Internet, but it also keeps tabs on and predicts the habits of individual users. Designers can use this data to construct and design a website with greater confidence. Users will appreciate the time and effort saved by web analytics technicians when creating user-friendly websites.

[32]https://www.promodo.com/blog/top-6-ecommerce-analytics-tools-for-online-stores-in-2019/#Hotjar (consulted on 7/11/2020).

Table 1 Comparative analysis between web analytics tools for e-commerce

	Google analytics	Piwik	Kissmetrics	Mixpanel	Hotjar	Woopra
Top features	Audience reports Conversion reports Site's search, speed, and flow reports Acquisition reports	A/B testing Funnel reporting Heatmap conversion optimization Fully customizable	Real-time data monitoring Funnel report A/B test report Cohort report Retention reports	Activity dashboards Targeted messaging campaigns Customer activity and behavior tracking and analysis Retention funnel	Heatmaps Visitor recordings Conversion funnels Form analysis	Journey reports Trend reports Cohort reports Retention reports Behavioral segmentation
Advanced features	Shopping and checkout funnels Real-time reporting	Search and marketing campaign Form analytics GDPP management	Customer profile management Email support management	Understanding the ratios of events Signal reports Scalability	Feedback polls Surveys	Triggered messaging Dedicated accounts manager
Pricing plan	Analytics – Free Analytics 360 – Monthly billing	On-premise free download Essential – 19( )/M Business – 29( )/M Enterprise custom plan	Start – $250/m Growth – $500/m, tracks up to 50,000 people monthly Power – $850/m, tracks up to 250,000 people monthly	Starter free Growth $799/year Enterprise custom price	20,000 page views/day 89 $/M 50,000 page views/day 189 $ /M 120,000 page views/day 289 $/M 400,000 page views/day 189 $/M Further custom quote according to page views	Core free up to certain actions Por: $ 999/M Enterprise contact sales team

(continued)

Table 1 (continued)

	Google analytics	Piwik	Kissmetrics	Mixpanel	Hotjar	Woopra
Real-time benefits	Analyze numbers of visits on pages Content activities reports Transactions activities reports with checkout funnels Audience and strategies marketing reports	High-level statistics Conversion optimization Form analytics for check-out forms reports Contribution traffic flow reports	Powerful user-based insights Company acquisition and retention rates full-proof guides Deep data tracking for businesses Solid retention reports	Increases marketing KPIs Increases customer loyalty High return on investment Significant increase in user engagement	Improves site performance using analysis and feedback tools Innovative screen recording functionality Tracks performance of feedbacks and surveys	Fully automated insights no SQL required Personalized customer experiences Follow-up triggers for re-visits One-click easy integration Customized tags for configure reports
Clientele	Twitter European Central Bank Nasaunicef	United Nations Red bult Huawei European Commission	Lucidchart Unbounce Mercy Corps SendGrid	Samsung Starz BMW Hinge Xpedia	HubSpot Reed.CO.UK InVision Homes.com Desbegar	Informatica Duke University Snapshot Ticket tailor Hubba

Source: https://magnetoitsolutions.com/blog/data-analytics-tools-for-ecommerce-businesses (consulted on 7/11/2020)

References

Aarushi Ranjan: 7 Must-Have Analytics Tools for Your eCommerce Website. (2019). On the site: https://www.shiprocket.in/blog/ecommerce-analytics-tools/ (consulted on 4/11/2020, 7/11/2020), https://colorwhistle.com/7-alternatives-to-google-analytics/ (consulted on 7/11/2020); https://www.canecto.com/best-25-google-analytics-alternatives-for-business-owners/ (consulted on 7/11/2020); https://growthbytes.co/google-analytics-alternatives/; https://www.promodo.com/blog/top-6-ecommerce-analytics-tools-for-online-stores-in-2019/#Hotjar (consulted on 7/11/2020)

Alghalith Nabil: Web analytics: enhancing customer relationship management. J. Strateg. Innov. Sustain. **10**(2), 13–14 (2015)

Bekavac, I., Praničević, D.G.: Web analytics tools and web metrics tools: an overview and comparative analysis. Croat. Oper. Res. Rev. **6**(2) (2015a) Croatian Operational Research Society, p 374. On the site: https://pdfs.semanticscholar.org/22ab/e974e3a7bc0ba418d5115241c10b847bc8bd.pdf (consulted on 2/11/2020)

Bekavac, I., Praničević, D.G.: (2015b), p 375

Chaffey dave and all: Internet Marketing Strategy, Implementation and Practice, 4th edn, p. 12. Pearson Education Inc, USA (2009)

Deepika Verma and all: Google Analytics for Robust Website Analytics, p 10. (2011). On the site: https://www.sjsu.edu/people/rakesh.ranjan/courses/cmpe272/s2/Team-Matrix-Google-Analytics.pdf (consulted on 4/11/2020)

Lakhwinder Kumar and all: Web analytics and metrics: a survey. In: Proceedings of the International Conference on Advances in Computing, Communications and Informatics, pp. 966–967 (2012a) On the site: https://www.csd.uoc.gr/~hy209/resources/web_analytics.pdf (consulted on 2/11/2020)

Lakhwinder Kumar and all: (2012b) pp 967–968

Limitations of Google Analytics: (2019). On the site: https://masondigital.com/4-limitations-of-google-analytics/ (consulted on 4/11/2020)

Piwik PRO GmbH, p 4, Optimizing Piwik On-Premises for Top Performance and Security, on the site: https://www.dirk-buechner.de/content/6-erfolgsrezepte/2-kundenkontakte-in-krisenzeiten-kluggestal-ten/piwik_optimization_whitepaper.pdf (consulted on 4/11/2020)

Ranganath Akshay: PiWik v/s Google Analytics. (2009). On the site: https://akshayrangananth.wordpress.com/2009/01/27/piwik-vs-google-analytics/ (consulted on 4/11/2020)

Stickler Rebecca: 6 Ecommerce Analytics Tools for Digital Marketers. (2016). On the site: https://www.webfx.com/blog/marketing/6-ecommerce-analytics-tools-digital-marketers/ (consulted on 4/11/2020); https://www.accuratereviews.com/marketing-analytics/kissmetrics-review/ (consulted on 5/11/2020); https://www.semplaza.com/kissmetrics-review/; https://www.mockingfish.com/blog/pros-and-cons-of-the-popular-ab-testing-tools-in-the-market/ (consulted on 5/11/2020)

Teixeira, J.: Get Involved: 5 Types of Web Analytics tools to start using today! (2011). On the site:https://www.morevisibility.com/blogs/analytics/get-involved-5-types-of-web-analytics-tools-to-start-using-today.html (consulted on 4/11/2020); https://ec.europa.eu/competition/information/digitisation_2018/contributions/at_internet.pdf (consulted on 2/11/2020)

W e b analytics de mystified: The web analytics business process making the case for a process-driven approach to web site measurement and ten critical requirements for success, p 2 (2007). On the site: http://cdn2.hubspot.net/hub/74398/file-15425858-pdf/docs/web_analytics_demystified__the_web_analytics_business_process.pdf (consulted on 2/11/2020)

Web Analytics: Tutorials Point (I) Pvt ltd, p 1, on the site: Ltd. (2015). https://www.tutorialspoint.com/web_analytics/web_analytics_tutorial.pdf (consulted on 2/11/2020)

Zach Williams: 8 Alternatives to Google Analytics, on the site: https://www.venveo.com/blog/8-alternatives-to-google-analytics (consulted on 5/11/2020); https://leadsbridge.com/blog/how-to-boost-productivity/what-is-mix-panel/ consulted on 5/11/2020); https://www.woorank.com/en/blog/analytical-tools-other-than-google-analytics (consulted on 5/11/2020); https://comparecamp.com/mixpanel-review-pricing-pros-cons-features/ (consulted on 4/11/2020); https://www.woorank.com/en/blog/analytical-tools-other-than-google-analytics (consulted on 4/11/2020); https://practicoanalytics.com/segment-analytics-pros-cons/ (consulted on 4/11/2020); https://colorwhistle.com/7-alternatives-to-google-analytics/ (consulted on 4/11/2020); https://www.ecommerce-nation.com/best-analytical-tools-for-your-e-commerce/#Klaviyo (consulted on 4/11/2020); https://www.drip.com/blog/ecommerce/klaviyo-vs-mailchimp-vs-drip (consulted on 4/11/2020)

A Qualitative Approach to Google Analytics to Boost E-Commerce Sales

Karima Yousfi and Ojo Johnson Adelakun

Abstract The valuation of data is not only important at the firm level but also essential at the national level. Since big data and analytics play an increasingly important role within the success of a business, Google Analytics may be a valuable tool to assist businesses to uncover vital insights and optimise for future growth. Google Analytics is one of the most comprehensive analytics solutions on the web and is a completely free service. Thus, it is expected that new or small business owners will resort to using their platform. Hence, this research investigated whether advanced web metrics, calculated using Google Analytics software, could be used to evaluate the general usability of e-commerce sites and to spice up e-commerce sales areas. Also, e-commerce tracking is effectively done via Google Analytics, which provides powerful reports. Our initial estimation shows that the worth of data generated is often tremendous. Moreover, online platform companies mostly capture the advantages of the information, because they create the worth that consumers lack the knowledge of. While the valuation of data will have important policy implications for investment, trade and growth, at present, data volume doubles every three years but the Internet of Things, the trend of 5G and the emerging online-to-offline transition is rapidly accelerating the buildup speed of data types and volume. Therefore, it is crucial to develop feasible methodologies to access the worth of data.

Keywords Web analytics · Google analytics · Usability · E-commerce web sites · Big data · Internet of Things

K. Yousfi (✉)
Economics Department, Abou Bakr Belkaid University, Tlemcen, Algeria
e-mail: karima.yousfi@univ-tlemcen.dz

O. J. Adelakun
Economics Department, National University of Lesotho, Roma, Lesotho

Federal University of Oye-Ekiti, Oye, Ekiti State, Nigeria
e-mail: Johnson.adelakun@fuoye.edu.ng

S. Sedkaoui et al. (eds.), *International Conference on Managing Business Through Web Analytics*, https://doi.org/10.1007/978-3-031-06971-0_6

1 Introduction

In the past decades, the advent of technology has advanced at a staggering rate, impacting several aspects of life, including economic and business. The internet and digital technology have become a crucial aspect in life, and it has changed the way people learn, find information, communicate, travel, govern and shop. Generally, the trending questions remain 'How big is the value of information possessed by online platform companies?' (Yin 2018; Frier 2018; Lee 2018; Bloomberg 2018). Nowadays, the activity of selling and buying, which was administered traditionally by the coming together of the seller and the buyer, can be done online. The vast change of technology within the field of communication, software application and computer hardware, browser technology and multimedia facilitate the knowledge search towards particular goods and services that people need. The use of electronic media or the internet for buying and selling is largely known as electronic commerce or e-commerce. E-commerce allows consumers to electronically exchange goods and services with no barriers of your time or distance (Lee 2018; SEC 2012, 2017; Hathaway and Muro 2017).

For example, transactions through an online e-commerce platform can generate an incredible amount of data. Whereas a transaction itself creates a standard economic benefit referred to as gains from trade. The information generated through the transaction also contains an economic value and the worth of such transaction of information has traditionally been accumulated within a firm as firm-specific knowledge on consumers, business partners and employees. The benefits of e-commerce could be for a short and long period. Hence, using e-commerce, the knowledge of the products might be spread on a bigger scale. This enables the consumers to choose the products with the best price, offers unlimited, fast communication and information access, in a limited time and cost-efficient manner, and also improve the quality of the service provided (Mayer-Schonberger and Cukier 2014; (Hartmans 2017). Apart from providing services to the customers, the main targets of e-commerce activities include the aspects of the business transaction, marketing and advertising. An in-depth review by Acquisti et al. (2016) concludes that the display of systematic and diffuse individual online price discrimination is currently scarce. According to Booking.com, the revenue of its corporate clients increased by 7% through a data-driven pricing strategy service provided to its third-party sellers.

E-commerce tracking provides several perspectives on website performance and publicities. It also shows what the website does in comparison to others and what it can improve with helpful insights. Whitmore (2018) is of the view that e-commerce helps in knowing the market trends and what people expect out of business. While there are several powerful tools within the market, Google Analytics (GA) is seen as the most popular and it is used to assist and to track website performance. According to GA for Mobile Apps (2019), the analytics platform was developed by Google in the year 2005 and is only for web analytics. Google Analytics allows for tracking essential metrics and observes the behaviour of customer about how customers are behaving on the websites. It also checks the pages customers visit the most, what proportion of time they spend on those pages, the geographical location of consumers, conversion rates, interaction per visit, etc. (Demunter 2018).

Because of improved programming capabilities and rapid price decline of information technology hardware and services, new business models have emerged, and many of them are built in several sorts of online platforms (Brynjolfsson et al. 2018a, b, c; Hartmans 2017).

The intellectual teasers of this study include the way to explore the influence of using GA to boost e-commerce sales and what the benefits are of using GA in e-commerce. This research examines the data, which demonstrates the power of GA in boosting e-commerce sales and explains the utilisation of GA from different perspectives within the scope of e-commerce. Therefore, the primary objectives of the paper are a qualitative analysis that critically looks at the influence of GA in boosting e-commerce sales and explores the benefits derivable from the utilisation of GA in e-commerce (Van de Ven 2018; Demunter 2018; Chen et al. 2018)

The authors thought that the economic outcome of this study would channel both theoretical and practical significance such as supplying information for the economic giants on the benefits of using GA to support e-commerce activities. It would also deliver information to the people that would like to boost e-commerce sales by using the technology of GA. Lastly, the result of this study is expected to be the source of useful information and reference for similar studies (Bloomberg 2018).

We have organised this paper into four main parts structured as follows: Section two is the literature review and section three is the methodology and case studies, followed by the discussion.

2 Literature Review

2.1 *Typologies of Google Analytics and E-Commerce: A Conceptual Review*

2.1.1 Online Platform

In this research, we adopt the European Commission's (2017) more all-encompassing definition of an online platform. Online platforms are defined as digital platforms that enable consumers to seek out information and businesses online and to take advantage of the benefits of e-commerce. Online platforms communicate key characteristics, including the utilisation of information and communication technologies to facilitate interactions between users; the aggregation and use of data about these interactions, and network effects, which make the utilisation of the platforms with the most users the most worthful to other users (the European Commission, 2017).

2.1.2 Big Data

The concept of big data describes the massive volume of data – both structured and unstructured – that overwhelms a business on a day-to-day basis. However, the quantity of data is irrelevant but what organisations do with the data is important. Big data are often analysed for insights that cause better decisions and strategic business moves. The combination of structured, semi-structured and unstructured data collected by organisations, which will be mined for information and utilised in machine learning projects, predictive modelling and other advanced analytics applications is known as Big data.

2.1.3 Importance of Big Data

The importance of big data does not revolve around what proportion of data you have, but what you are doing with it. Data can be analysed to seek out answers that enable: cost reductions, time reductions, new product development and optimised offerings, and smart decision-making. Once big data is combined with high-powered analytics, it will accomplish business-related tasks such as: determining root causes of failures, issues and defects in near-real time; generating coupons at the point-of-sale based the customer's buying habits; recalculating entire risk portfolios in minutes, and detecting fraudulent behaviour before it affects the organisation (Tefis Team 2017).

Companies use the accumulated data in their systems to enhance operations, provide better customer service, create personalised marketing campaigns from specific preferences from customer and, in the long run, increase profitability. Businesses that use big data hold a possible competitive advantage over those who do not because big data results in faster and more informed business decisions, provided they use the data effectively and efficiently. For instance, big data provides companies with valuable insights into their customers, helps in fine-tuning marketing campaigns and techniques to enhance customer engagement and conversion rates (Slotin 2018).

2.1.4 Google Analytics

Google Analytics (GA), a free online service, was introduced in 2005 by Google to provide data for all industries from small to enterprise-level businesses. Google Analytics is important for e-commerce business because it depends solely on data. In the e-commerce industry, correct data can help reach new heights and misinterpreted data can push business to the bottom.

Data play an important role in converting a lead into business or losing it. In addition, data helps in expanding sales, when compared to e-commerce stores or startups. Using GA drives business into a success and helps in initiating new techniques that will result in an increase in revenue.

As a free service provided by digital giant Google, it is one of the frontline analytics tools available online. *It is used in tracking visitors from search engines, social networking, direct access, and referring sites. According to* Big Data Made Simple (2017), Google Adwords is integrated with the most important paid search platform on the web. It is used for site traffic tracking and reporting and it provides information that helps grow a business in today's online world. It provides the business owner with an entire picture of the audience and their needs.

The data from GA is designed specifically for marketing and webmasters because of the quality of the traffic the website receives and the effectiveness of the user's marketing campaigns. According to Techopedia (2017), GA can provide feedback for subsequent marketing campaigns by tracking visitors from all referring sites and the number of visitors converted into customers or members from each page. In short, GA tracks the customer's journey as they interact with the content of the web site. Therefore, goals, tracking statistics, AdWords integration, campaigns management via URL builder and E-commerce tracking are the five main features that make Google analytics tools an interesting tool to use.

The most effective way of collecting and analysing website data is using GA. Enhanced Ecommerce gives site owners specific features designed for store optimisation, which is a useful platform for e-commerce A detailed report on customer behaviour in every step of the sales funnel are included metrics for shopping behaviour while the analytics for revenue and conversion rates, average order value and cart abandonment rates measure product performance. The tracking of internal and external marketing efforts, including affiliates and coupons, assesses marketing success rates and product attribution looks at the customer's 'Last Action' attribution.

2.1.5 The Objectives of Google Analytics

To optimise content within an on-site search, familiarisation with the buyer's journey, understand customer's interests, and measuring the duration by which your leads convert are several objectives derivable from using GA. GA further helps to annotate for better SEO management, map on-site engagement, realise the value of customers and decrypt ambiguous statistics to analyse data better. In the same vein, how long visitors are willing to spend on a particular page are determined and data catering are evaluated on e-commerce sites.

2.1.6 Advantages of Using Google Analytics for E-Commerce Sales

An advantage of using GA for e-commerce sales is that it is automatically free for collecting data. It has an advanced reporting feature and can integrate with other platforms easily. An internal site search which allows people to understand what they are looking for once they are on the website is measured by GA. While it can provide demographic details, it allows the business to know which social platforms they will target and it allows businesses to check if they are achieving their goals and helps them to redefine metrics.

2.2 Empirical Review

A study on blog Analyzer for Semantic Web Mining conducted by Fakhrun Jamal and Mamta Bansal (2020), the Semantic Web Mining observes and collects the information of visitors on the web site. It was found that entering of password and username is not require at any time because the Weblog analyser will remember the password and username, personalise information, locate memory and site understanding. It will also optimise your logs to trace what you would like to understand about visitors without complex filtering.

Analysing what appears interesting to the customers when buying a product, Thushara and Ramesh (2016) posited that to remain successful in e-commerce business it is important to know the customers better. To them, web usage mining allows the seller to examine and find patterns from composed information to make a primary statistical basis for decision-making. The precondition to use web usage mining properly as defined by the research method also allows for businesses to collect qualitative visitor data, and be able to know whether a visitor has followed a link to get to a seller's shop. It also checks whether a convincing product has placed a seller's site first in a search engine, etc. The research work is best suited for e-commerce business categories and development purposes.

An interpretive framework that delves into the definitional aspects, distinctive characteristics, types, business value and challenges of BDA within the e-commerce landscape was conducted by Akter et al. (2016). Also, broader discussions regarding future research challenges and opportunities in theory and practice were considered by the research work. In sum, the findings of the study synthesise diverse BDA concepts such as the definition of bag data, types, nature, business value and relevant theories, and provide deeper insights alongside cross-cutting analytics applications in e-commerce for small and medium enterprise development in Nigeria.

The utilisation of GA over a number of its competitors within the field of web analytics, owing to its open-source nature, simple use and natural integration with other renowned Google products such Google AdWords was advocated by Suraj Chande (2015). The study examined further a number of the distinct advantages of GA, such as high customisation as per the nature of the business and therefore the wide selection of reporting functions. A case study was used in making businesses conscious of the import of GA and also encouraged blending GA in sales and marketing activities.

A novel approach was offered and described by Neha Sharma and Pawan Makhija (2015) about session identification and personalisation of web services. The difference of internet sites and the improvement of web server performance are caused by web usage mining, and this had led to the growth of the present day

World Wide Web. The web usage mining applies data mining techniques to obtain web access patterns from log data. Weblog data should be rebuilt into sessions to reveal access patterns.

Vivek Dogne, Anurag Jain and Susheel Jain (2015) surveyed and described the abundance of information available on the World Wide Web (www), It also focused on how extracted knowledge from the web has gained significant attention among researchers in data mining and knowledge discovery. Web mining is applied to reflect the importance of webpages and is used to predict the visitations of users to the web. Also, using a survey of the rapidly rising area of web usage mining, showing the order of current technology, a standard overview of the web usage mining was produced.

Ravinder Singh and Bhumika Garg (2014) described web usage mining as discovering and analysing the patterns in clickstream and associated data that is self-controlled or generated as a result of user interaction with web browsers, on one or more websites. Web usage mining combines two approaches: web caching and web pre-fetching. The study presented an approach that integrates web caching and web pre-fetching to enhance the performance of the proxy server's cache. These hybrid approaches can be used together because the web caching technique exploits the temporal locality while the web pre-fetching technique utilises the spatial locality of web objects. The study concluded that the reaction time of a hit taken from the user cache is a lower compared to the data taken directly from the log file.

Karthik and Swathi (2013) studied the usage of web mining technology as it provides security for e-commerce websites using web structure mining, web content mining, decision analysis and security analysis.

Shanmuga Priya and Sakthivel (2013) proposed a new method of removing the noise associated in the web documents. This is done using web extraction in new architecture that is made to manage the web data. Unnecessary information was removed from the web information and patterns were created for that data using web content extraction. Three phases of data extraction were designed. Web documents are chosen in the first stage, documents are pre-processed in the second phase, and the results are presented to users in the final phase.

Similarly, Shanthi and Rajagopalan (2013) compared the different web mining algorithms. The context related to the web design of an e-commerce portal was identified and webpage collection, web-mining algorithm to manage time and space complexity was proposed.

Chhavi Rana (2012) researched online usage mining and focused on techniques that would predict user behaviour while the user interacts with the web. While behaviours and web surfer's session were also generated, the study supplies a summary of the state of the art web usage mining. The study further discusses the relevant tools available within the sphere and the niche requirements that the current tools lack. *It gives an outlook on the prevailing tools, their specialised focus for an applicative objective and therefore the need for a more comprehensive new entrant during this sphere within the light of the present scenario.*

Using time series, Mohammad Amin Omidvar, Vahid Reza Mirabi et al. (2011) developed a methodology to analyse the different variables on the dependent variables by. The study used time-series regression to find the relationship between the essential and primary index (page views per visit) on Google analytic. It also focused on using the most relevant data to gain a more accurate result. Since a linear regression analysis cannot model the impact of search engine visitors on page views, a variable referral visitor was added to the linear regression yielding a low impact. On the other hand, there was a significant impact on direct visitors on page views. The result further showed that higher connection does not lead to higher impact on page views and the content of the web page, the territory of visitors can help connection speed to describe user behaviour. However, returning visitors showed some similarities with direct visitors.

Hasan et al. (2010) developed a methodological framework for evaluating the usability of e-commerce websites. In designing the framework, user testing and heuristic evaluation methods were used alongside with GA software.

Similarly, Kazuo Nakatani and Ta-Tao Chuang (2005) designed a framework for the event of collaborative commerce applications. The framework used in the study integrates generic capabilities of collaborative technology and functional requirements of collaborative commerce. The framework supported previous research on the classification of e-commerce-related technologies, systems and collaborative functional requirements, and the usefulness of the framework was shown in the study.

3 Methodology and Case Studies

Fundamental business models used by online platform companies determine what data they collect, how data flow, and what value of data they create. However, the problems that there is no arms-length market for most intangibles and the majority of these intangibles are developed for a firm's use, have created a problem for the economist because intangibles are difficult to measure. Furthermore, Lev and Radhakrishnan (2005); Eisfeldt and Papanikolaou (2013); Brynjolfsson et al. (2018a, b, c) reported the use of sales, general and administrative (SG&A) expense as a proxy for a firm's investment in organisational capital. These expenses are reported in a firm's annual income statements to include employee training costs, brand enhancement activities, consulting fees, the installation and management costs of supply chains, etc., and these expenditures that generate organisational capital. Some items that are unrelated to improving a firm's organisational efficiency are included in the SG&A expenditures. However, the question remains whether SG&A expenditures is a valid measure of a firm's investment in organizational capital. Eisfeldt and Papanikolaou (2013) proposed five ways to validate the firm's investment in organisational capital and findings from their study revealed that four out of the five ways support this approach.

Li (2015), found that across U.S. high-tech industries, market leaders had a lower rate of depreciation than their followers. Li and Hall (2018) developed the R&D depreciation model, and it was adopted to estimate the depreciation rates of the organisational capital for four online platform companies, including Amazon, Booking Holdings, eBay and Google, using the available public data. A perpetual inventory method was propounded by Torrington and Hall (1998) and used to build the stocks of organisational capital and the associated growth rates for Amazon, Booking Holdings, eBay and Google (Wendy et al. 2018).

Some studies have attempted to explain the typology of online platforms (Demunter 2018; van de Ven 2018; Chen et al. 2018). In addition to the fundamental business models, Wendy et al. (2018) classify online platforms into eight major types. Type I includes e-commerce online platform, type II constitutes online resource sharing platform, type III is E-financial service online platform, type IV is made up of an online social network service platform, type V is the online auction or matching platform, type VI is the online competitive crowdsourcing platform, type VII constitutes the online noncompetitive crowdsourcing platform and type VIII includes the online search platform. Also, online platform companies use the existing business models ranging from one type of online platform to multiple types. Understanding the fundamental business models and the data activities involved is useful for businesses to establish a set of basic types of online platforms that can lead to business growth, though some complications may arise using the classification of the online platform.

Hence, the study qualitatively examines the benefits of GA to boost e-commerce focusing on type I: e-commerce online platform – Amazon Marketplace and type VIII: online search platform – Google Search.

3.1 *Type I: E-Commerce Online Platform*

E-commerce is the first online platform, using Amazon Marketplace as the case study (see Fig. 1). Amazon Marketplace is an online platform that facilitates sales between consumers and third-party sellers. Amazon Marketplace offers consumers a place to get a good range of products from more selections at lower prices. It also allows third-party sellers access to one of the world largest e-commerce markets in a cost-effective and time-efficient manner.

Amazon charges a commission of 30% for third-party sellers as commission for their sales according to the WSJ (2018). The commission covers not only the cost of accessing one among the world's large e-commerce markets but also the value of "basic" access to Amazon's consumer data. For instance, a consumer can purchase goods by taking advantage of an offline supermarket using cash; however, the supermarket and the third-party seller that gives the products do not obtain data about the consumer. If the customer decides to pay by a credit or debit credit, the supermarket will have some data about the buyer.

Considering the flow of data, Amazon collects data on clickstreams, purchases, reviews and locations from consumers. After collecting consumer information; it conducts analytics on those data to allow for data-targeting services to third-party sellers. For instance, the geolocation data of consumers and demand forecast

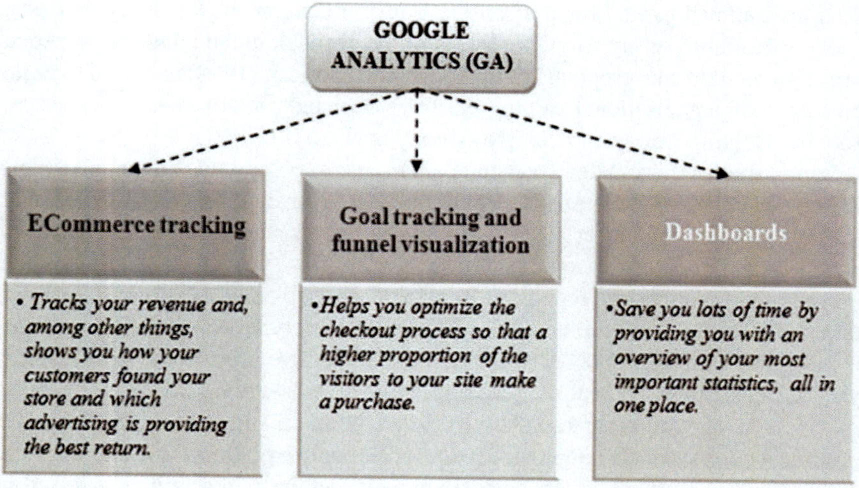

Fig. 1 Google Analytics goals for e-commerce. (Source: Adopted – Thomas Holmes 2013)

provides third-party sellers logistics consulting services such as where to site the warehouse. According to Bond (2018), Amazon offers corporate clients a premium data services, made-up of demand and trend forecasts, and the minimum of $100,000 per annum is paid as the cost for premium data service. Apart from this, Amazon gathers information on consumer price sensitivity by funding discounts on third-party products (Bond, 2018). Together with the price sensitivity data and other data collected, Amazon can conduct detailed profiling of every consumer and supply a data-driven pricing strategy service to third-party sellers.

According to Molla (2017), Amazon e-commerce accounted for 10% of U.S. retail sales in 2017, and 43% of the market share of the U.S. e-commerce market. In the same vein, Statista (2018) reported that 50.5% of its e-commerce sales are conducted through third-party sellers on Amazon Marketplace. Given the very fact that the 2017 sales for Amazon Marketplace are US $139.5 billion and Amazon charges third-party sellers a 30% commission on their sales, the Amazon 10K report reported an annual revenue from the commission is estimated at around US $41.8 billion for Amazon. With the accelerated growth rate, Amazon's annual data targeted advertising revenue amounted only to US $3 billion in 2017 covering 2.2% of its total revenue in 2017. Amazon does not depend on advertising revenue compared to Facebook and Google.

Note that online platform companies can also collect data from third-party sellers such as where they ship the products if they choose to fulfil the orders by themselves. When online platform companies provide data targeting services, they can incorporate the profile of their third-party sellers (Fig. 2).

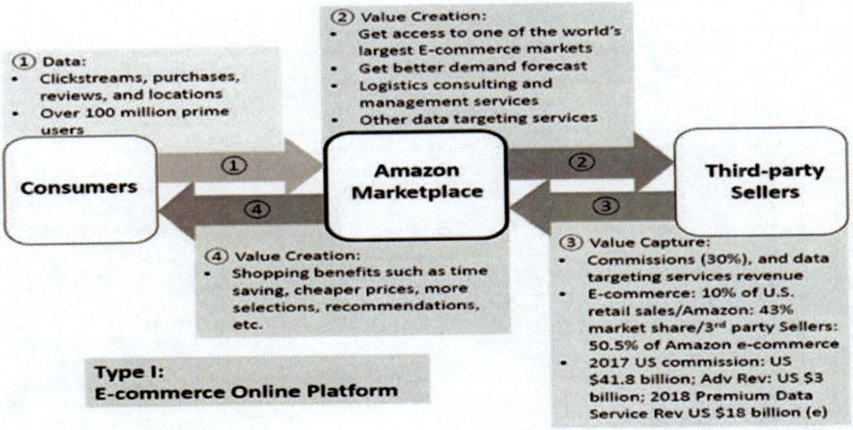

Fig. 2 Type I: E-commerce online platform. (Source: Adopted – Wendy et al. 2018)

3.2 *Type VIII: Online Search Platform*

Figure 3 reveals the type VIII online search platform, Google Search. Google Search is the most popular online platform in the world today. Google Search provides individuals with free, convenient, and relevant access to information immediately, and it allows advertisers and content providers access to a wide user base. In addition, advertisers' returns on investments (ROI) can be increased using data targeting ads. Google Search allows content providers to include search functionality on their webpages as it allows them to monetise their content. By integrating Google Play with other Google services, the content is directly linked to Google Play and can lead to an increase in the transaction.

Considering the flow of data, Google Search collects data on search terms, revealed preferences, browsing behaviours, locations, demographics, languages, etc., from users. After collecting the data, analytics are conducted on the data to provide its corporate clients data targeting services, such as targeted advertising, better demand forecast or marketing. Data targeting advertising revenue constitutes most of its revenues. Google's advertising revenue of 2017 was US $95.4 billion in which Booking Holdings paid Google US $3 billion for AdWords advertising.

Table 1 shows the estimated results based on this approach (see 4th column), annual commission or licensing revenue, and merger and acquisition prices associated with our case studies. For example, Amazon's estimated annual commission derived from the data is US $41.8 billion, and the estimated value of data derived from a data-driven business model is US $125 billion. These estimates are based on Amazon's financial statements.

The leading B2C e-commerce companies are based in China and the US (Table 2). In 2018, the world's top 10 B2C companies generated almost $2 trillion in gross merchandise value (GMV), consistent with the report. Alibaba (China) lead with a

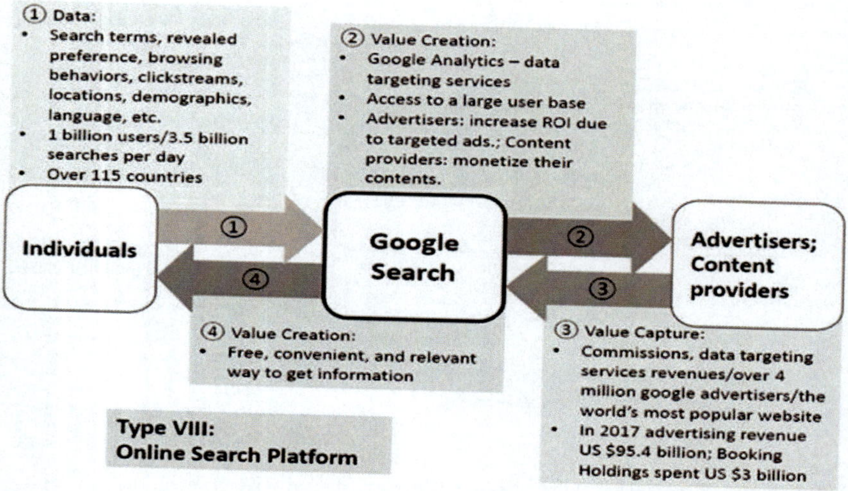

Fig. 3 Type VIII: Online search platform. (Source: Adopted – Wendy et al. 2018)

Table 1 Measurement of the value of data: Case studies

Type of online platform	Company	Annual commission or licensing access to data	Value-based on data-driven business model	Merger & acquisition price
E-commerce	Amazon	Commission revenue: US $41.8 billion (2017) Premium data service revenue: US $18 billion (2018)[a]	US $125 billion; Annual growth rate: 35%	
Search	Google	US $95.4 billion (2017)[c]	US $48.2 billion; Annual growth rate: 21.8%	

[a]Assume third-party sellers with annual sales over US$10 million order the premium data service. 19% of third-party sellers have sales over US $10 million per year

[b]Most of the revenue is from selling access to the data of its members to recruiters and sales professionals

[c]Data targeting service revenue: data targeted advertising revenue

GMV of $866 billion in 2018, followed by Amazon of the United States with $277 billion. In terms of revenue, JD.com of China and Amazon were ahead of Alibaba. Developing and transition economies accounted for about half of the highest 20 economies by B2C e-commerce sales. Hong Kong (China), China and the United Kingdom have the largest B2C e-commerce based on their GDP, while India, Brazil and Russia were ranked the least. However, the extent to which internet users engage in online purchases varies substantially among the top 20 economies. Records from

Table 2 E-commerce sales: Top ten economies in 2018

Rank	Economy	Total e-commerce sales ($ billion)	Share of total e-commerce sales in GDP (%)	B2B e-commerce sales ($ billion)	Share of B2B e-commerce sales in total e-commerce (%)	B2C e-commerce sales ($ billion)
1	United States	8.640	42	7.542	87	1.098
2	Japan	3.280	66	3.117	95	163
3	China	2.304	17	943	41	1.361
4	Korea (Rep)	1.364	84	1.263	93	102
5	United Kingdom	918	32	652	71	266
6	France	807	29	687	85	121
7	Germany	722	18	620	86	101
8	Italy	394	19	362	92	32
9	Australia	348	24	326	94	21
10	Spain	333	23	261	78	72
	10 above	**19.110**	**35**	**15.772**	**83**	**3.338**
	World	**25.648**	**30**	**21.258**		**4.390**

Source: UNCTAD ESTIMATES OF GLOBAL E-COMMERCE, UNCTAD Technical Notes on ICT for Development, N°15, UNITED NATIONS CONFERENCE ON TRADE & DEVELOPMENT (2019)

2018 revealed that 87% of Internet users in the United Kingdom shopped online, compared with only 14% in Thailand and 11% in India.

More than 1.4 billion people shopped online and more bought from abroad.

One-quarter of the world's population (1.45 billion people) are aged 15 and older and they made purchases online in 2018 according to UNCTAD. Moreover, the result revealed a 9% increase in 2017 and China constituted the highest number of internet buyers pegged at 610 million. Although some 330 million internet buyers made cross-border purchases in 2018, the majority of internet buyers bought from domestic suppliers.

3.3 Discussion

The beauty of GA is that it provides tremendous data that creates a way for an e-commerce business to realise more results. Also, it allows for unique and fresh content in a way that users like to read it and this increases the number of tourists to your website. Furthermore, visitors to your websites are given good customer service learning the buyer's interest and wishes. As a tremendous tool, GA affords customers with information and reports use that tends to extend sales, improve your revenue and provides your best customers services to all or any of your visitors.

However, many online platforms are providing digital goods or services to consumers at zero monetary cost in substitution for consumers' data. Since they store large data, online platform companies can monetise data by conducting analytics on the information to supply data targeting services and, for instance, transactions through an e-commerce online platform can generate an incredible amount of data. While on the contrary, a transaction itself creates a standard economic benefit referred to as gains from trade, the info generated through the transaction contains a value. The worth of such transaction data has traditionally been collected within a firm as firm-specific knowledge on consumers, business partners and employees. The second assumption looks at the condition of the identity of an observed consumer. The consumer identity in the transaction is not revealed or sometimes used by the service provider. In the case where anonymity is preserved, the incremental cost of the data provision is zero.

Thus, the household is considered as a "statistical" subject. As a set of data, it is probably going to possess a positive welfare effect. This is often the case where a household can access the knowledge generated by the data without disclosing its identity. An individual and a firm will share the increased value-added. The creative destruction process features a redistributive effect, but it does not necessarily imply welfare loss. For instance, a web platform company can accumulate an incredible amount of data that allows for an excellent competitive advantage and render obsolete the conventional business model of its competitors. As a result of this, the buyer surplus, income and rent generated by the traditional business decline while that of new business increases. Finally, our information goods act like knowledge.

Fundamentally, different business models explain online platforms, though they are generally asset-light but are often extremely profitable. Online platforms determine what sorts of data they collect, how data flow within online platform networks, how online platform companies monetise the information, and what consumers get from exchanging their data. Considering the underlying business model used in this research, we sort online platforms into eight major types. Therefore, to understand data activities in the different online platforms, case studies were conducted to analyse each type of online platforms depending on the size of business models, data flow, value creation for consumers, value creation for third-party sellers, and monetisation of data.

Furthermore, we found out that the degrees of vertical integration in the data value chain differ for online platform companies. This determines how they monetise their data. For instance, Twitter features a much lower degree of vertical combination within the data value chain compared to Google and Amazon. According to Bary (2018), 12.3% of Twitter revenue from licensing the utilisation of its user data is derived from data analytics firms. That is, online platform companies can monetise their data by licensing the utilisation of data and/or providing data targeting services, but what proportion economic benefits a web platform company can capture may depend upon its degree of vertical combination within the data value chain (Brynjolfsson et al. 2018a, b, c; Lee 2018). Furthermore, Lee (2018) posited that businesses often strengthen their virtuous cycle by compounding their learning relationship with data. That is, more data use can result in better digital goods and services, which successively attracts more users to their online platforms, generating even more data that further improve their digital goods and services.

As an intangible, data do not wear or tear while regular intangibles, such as R&D assets, depreciate owing to obsolescence. Through the aggregation and recombination of data, new values are produced. These unique features of data present significant challenges to firms and statistical agencies in measuring the worth of data.

The initial results indicate that the worth of data is often significant: for instance, Amazon's data can account for 16% of Amazon's market valuation and has an annual rate of growth of 35%. However, because some online platform companies are private, and due to the absence of data, to perform an equivalent estimation for all the firms was not viable in our case studies, which is a gap left for future work. Also, the present depreciation model assumes decreasing marginal returns to data-driven business model investments. Future research could modify the model to integrate the increasing marginal returns to the investments.

Currently, there is no definitive answer to the welfare implications of online platforms and data. For instance, we discovered from Booking.com that online platform companies offer data-driven pricing strategies for their corporate clients, like price discrimination strategies supported consumers' data on clickstreams and past transactions, to maximise their revenues. On the other hand, the households within the economy as an entity can receive the equivalent value of additional income. Also, price discrimination is often redistributive. However, if the firm's increased profits are distributed equally among households, the resulting distribution may become more egalitarian than before. Also, the increased profits accrue to only a couple of entrepreneurs, transfer through price discrimination does not necessarily cause an equal distribution. The resulting distribution depends on the ownership of an emerging business model.

Since online platforms are rapidly evolving, to look for growth, online platform companies may expand their existing business models from covering one sort of online platform to multiple types. This is often similar to where a firm conducts businesses in multiple industries. That is, online platform companies develop hybrid online platforms that cover several basic sorts of online platforms classified in this study. Also, the degree of hybrid platforms can vary across countries. Lee (2018) argued that a couple of Chinese online platform companies have developed super online platforms such as Tencent's WeChat, bundles and many online platform functionalities similar to Facebook, Uber, Expedia, PayPal, Amazon, LimeBike, and more combined. This outcome is often called "The App Constellation Model". On the other hand, most U.S. online platforms are less hybrid and specialise in original business models. More research is required to comprehend the impacts of the rapidly evolving trend of online platforms in areas, such as data collection, market competition and consumer welfare.

Lastly, data is like new oil. At the present, the pipelines of the latest oil are controlled by online platform companies. In the future, blockchain technology may allow each consumer to possess his or her pipeline, to require control of their ownership, and to make a decision whether and the way to sell personal data to AI companies, online platform companies or advertising companies. However, given the very fact that the creation of the worth of data takes place on the corporate side and depends on data analytics and business experts to create a business plan to monetise data, consumers are not in a position to know the worth of their data. Nevertheless, how briskly various industries can adopt blockchain technology may affect the long-term competition of online platform companies. Mayer-Schonberger and Cukier (2014) reported data volume doubles every three years, but the IoT, the trend of 5G and the emerging online-to-offline transition is rapidly accelerating the buildup speed of knowledge types and volume. Therefore, it is essential to develop feasible methodologies to determine the worth of data because the rating of data is not only important at the firm level but also the national level. For firms to derive important investment and outsourcing decisions on data, and to find a way to monetise them and gain a competitive edge through data, a proper valuation of data is necessary. The call for national accounts to incorporate this increasingly crucial new asset into the calculation of GDP and productivity growth is crucial at the national level.

Moreover, countries differ in terms of the ownership of personal data. Examples include Europe's use of data protection rule, the General Data Protection Regulation, and China's extreme openness of private data. Additionally, the U.S. allows foreign firms to gather personal data within the U.S. but China forbids it. The big questions, therefore, is how do the differences in data policy affect trade? Given the existence of the virtuous cycle between deep learning's relationship with data, the degree of openness of a country's data policy may affect relative competitiveness between domestic and foreign firms. Therefore, the valuation of data will provide important policy implications for trade and growth.

4 Conclusion

Big data analytics tools and techniques are rising in demand owing to the utilisation of big data in businesses. Organisations can find new opportunities and gain new insights to operate their business efficiently. These tools help in providing meaningful information for creating better business decisions. The import of big data analytics results in intense competition and increased demand for big data professionals. As an emerging field with huge potential, data science and analytics help in analysing the value chain of business and gain insights. To this end, the utilisation of Google Analytics can enhance the industry knowledge of the analysts. Data analytics experts provide the organisations with an opportunity to determine the opportunities for the business. There are huge requirements and significance for massive data analytics in several fields and industries. Hence, it becomes essential for professionals to stay abreast with the techniques. Simultaneously, a lot can be learnt by companies when these analytics tools are used correctly. As the de-facto standard web analytics tool, GA is offered free of charge at lower data volumes and provides tracking, analytics and reporting. It enables non-technical users to understand website performance by providing answers to questions such as: where are users coming from? Which pages have the very best conversion rates? Were the users undergoing friction and abandoning their shopping cart? As a simple website, GA offers a robust solution despite its free services, while GA provides detailed statistics of visitor's insight and engagement with the website. Consistent with the statistical usage, half the website population in the world uses GA, and it provides the geographical information of the visitors, site engagement, and their number of visits to the website. Furthermore, GA has some breathtaking features useful for the web merchant. With the information provided by GA and the insight the user gains from that data, the customer will be better ready to guide the web store to success and continue building thereon.

References

Acquisti, A., Taylor, C., Wagman, L.: The economics of privacy. J. Econ. Lit. **54**(2), 442–492 (2016)

Akter, S., Fosso, Wamba, S.: Big data analytics in e-commerce: A systematic review and agenda for future research. Int. J. Netw. Business, 1–53 (2016)

Bary, E.: Twitter earnings growth relied on selling user data. Market Watch, April 25th (2018). https://www.marketwatch.com/story/twitter-earnings-growth-relied-on-selling-user-data-2018-04-25

Bloomberg: Facebook Cambridge analytica scandal: 10 questions answered. Fortune, April 10th (2018) http://fortune.com/2018/04/10/facebook-cambridge-analytica- what-happened/

Bond, S.: Amazon's ever-increasing power unnerves vendors. Financial Times, September 21st (2018)

Brynjolfsson, E., Eggers, F., Gannamaneni, A.: New measures of the economy: Measuring welfare with massive online choice experiments: A brief introduction. AEA Papers Proc. **108**, 473–476 (2018a)

Brynjolfsson, E., Hui, X., Liu, M.: Does machine translation affect international trade? Evidence from a large digital platform, National Bureau of Economic Research Working Paper No. 24917, August (2018b)

Brynjolfsson, E., Rock, D., Syverson, C.: The productivity J-curve: How intangibles complement general purpose technologies, National Bureau of Economic Research Working Paper No. 25148, October (2018c)

Chande, S.: Google analytics – A case study, pp. 1–11 (2015, January)

Chen, Y., Dai, T., Korpeoglu, C., Korpeoglu, E., Sahin, O., Tang, C., Xiao, S.: Innovative Online Platforms: Research Opportunities, working paper (2018)

Demunter, C.: Towards a taxonomy of platforms in the collaborative economy: Outcomes of a workshop on measuring the collaborative economy. Presented at the 2018 OECD Workshop on Online Platforms, Cloud Computing, and Related Products, September 6th, OECD, Paris (2018)

Dogne, V., Jain, A., Jain, S.: Evolving trends and its application in web usage mining: A survey. Int. J. Soft Comput. Eng. (IJSCE). **4**(6), 98–101 (2015)

Eisfeldt, A., Papanikolaou, D.: Organizational capital and the cross-section of expected returns. J. Finance. **4**, 1365–1406 (2013)

Frier, S.: Is Apple your privacy hero? Bloomberg Businessweek, August, 8th (2018). https://www.bloomberg.com/news/articles/2018-08-08/is-apple-really-your-privacy-hero

Hartmans, A.: Airbnb has more listings worldwide than the top five hotel brands combined. Business Insider, August 10th (2017). https://www.businessinsider.com/airbnb-totalworldwide-listings-2017-8

Hasan, L., Probets, S., Morris, A.: Usability evaluation framework for e-commerce websites. 12th international conference on enterprise information systems technology, June 2013. Int. J. Comput. Sci. Mobile Comput. **2**(6), 145–150 (2010)

Hathaway, I., Muro, M.: Ridesharing Hits Hyper-Growth (2017). https://www.brookings.edu/blog/the-avenue/2017/06/01/ridesharing-hits-hyper-growth/

Jamal, F., Bansal, M.: Web log analyzer for semantic web mining. Int. J. Comput. Sci. Inf. Technol. **6**(3), 2658–2662 (2020)

Karthik, M., Swathi, S.: Secure web mining framework for e-commerce Websites. Int. J. Comput. Trends Technol. (IJCTT). **4**(5), 1042–1046 (2013)

Lee, K.F.: AI superpowers: China, Silicon Valley, and the New World Order. Houghton Mifflin Harcourt. September 25th (2018)

Lev, B., Radhakrishnan, S.: The valuation of organizational capital. In: Corrado, C., Haltiwanger, J., Sichel, D. (eds.) Measuring Capital in A New Economy, pp. 73–99. National Bureau of Economic Research and University of Chicago Press, Chicago (2005)

Li, W.C.Y.: Organisational Capital, R&D Assets, and Offshore Outsourcing. Working Paper. U.S. Bureau of Economic Analysis (2015)

Li, W.C.Y., Hall, B. H.: Depreciation of business R&D capital, the Review of Income Wealth (2018)

Mayer-Schonberger, V., Cukier, K.: Big Data. Mariner Books, Boston/New York (2014)

Molla, R.: Amazon could be responsible for nearly half of U.S. E-commerce Sales in 2017, Recode, October 24th (2017). https://www.recode.net/2017/10/24/16534100/amazon-marketshare-ebay-walmart-apple-ecommerce-sales-2017

Nakatani, K., Chuang, T.-T.: A framework for the development of collaborative commerce applications. Issues Inf. Syst. **VI**(2), 17 (2005)

Omidvar, M.A., Mirabi, V.R.: Analysing the impact of visitors on page views with Google analytics. Int. J. Web Semantic Technol. (IJWesT). **2**(1), 1–32 (2011)

Priya, V., Sakthivel, S.: An implementation of web personalization using web mining techniques. Comp. Sci. (2013)

Rana, C.: A study of web usage mining research tools. Int. J. Adv. Netw. Appl. **3**(6), 1422 (2012)

SEC: Facebook, Inc. 10-k annual report for the fiscal year ended December 31, 2012 (2012). https://www.sec.gov/Archives/edgar/data/1326801/000132680113000003/fb-12312012x10k.htm

SEC: Booking Holding Inc. (BKNG) SEC filing 10-K annual report for the fiscal year ended December 31, 2017 (2017). https://www.last10k.com/sec-filings/bkng

Shanthi, R., Rajagopalan, S.P.: An efficient web mining algorithm to my web log information. Int. J. Innov. Res. Comput. Commun. Eng. **1**(7), 1491–1500 (2013)

Sharma, N., Makhija, P.: Web usage mining: A novel approach for web user session construction. Glob. J. Comput. Sci. Technol. E Netw. Web Secur. **15**(3), 1–5 (2015)

Singh, R., Garg, B.: Hybrid approach for performance of web page response through web usage. Mining. **4, 2014**(7) (2014)

Slotin, J.: What do we know about the value of data? May 3rd (2018). http://www.data4sdgs.org/news/what-do-we-know-about-value-data

Statista: Percentage of paid units sold by third-party sellers on Amazon platform as of 2nd quarter 2018 (2018). https://www.statista.com/statistics/259782/third-party-seller-share-ofamazon-platform/

Tefis Team: Facebook's strong ad revenue growth to continue. Forbes, October 30th (2017). https://www.forbes.com/sites/greatspeculations/2017/10/30/facebooks-strong-adrevenue-growth-to-continue/#254871fe6fe7

Thushara, Y., Ramesh, V.: A study of web mining application on e-commerce using Google analytics tool. Int. J. Comput. Appl. (0975 – 8887). **149**(11), 21–26 (2016)

Torrington, D., Hall, L.: Human resource management. 4th Edition, illustrated, Prentice Hall Europe (1998). ISBN: 0136265324, 9780136265320

UNCTAD: Estimates of Global E-commerce, UNCTAD Technical Notes on ICT for Development, No 15. United Nations Conference on Trade & Development (2019)

Van de Ven, P.: Online platforms in a digital economy satellite account. Presented at the 2018 OECD Workshop on Online Platforms, Cloud Computing, and Related Products, September 6th, OECD, Paris (2018)

Wendy, C.Y. Li, Makoto, N., Kazufumi, Y: Value of Data: There's No Such Thing as a free Lunch in the Digital Economy. U.S Bureau of Economic Analysis (2018)

Whitmore, G.: How Google travel is changing the game with Google Flights, Google Trips, and more. Forbes, February 20th (2018). https://www.forbes.com/sites/geoffwhitmore/2018/02/20/how-google-travel-is-changingthe-game-with-google-flights-google-trips-more/#7683e1b34d98

Yin, Y.H.: The most successful internet company after Amazon – From Netherlands. Business Weekly, 1600, July, Taiwan (2018)

Ontology Alignment Systems to Contribute to the Interoperability of a Business Federation

Fatima Ardjani and Djelloul Bouchiha

Abstract This chapter proposes the use of ontology alignment to contribute to the interoperability of a business federation based on data interoperability. We proposed a system with a linguistic and syntactic matcher, called ABCMap. The ABCMap tool is based on an optimization method that relies on Artificial Bee Colonies (ABC). Experiments done using the implemented tool give the best results in terms of Recall and Precision.

Keywords Business federation · Interoperability · Ontology alignment · ABCMap · Syntactic matcher · Linguistic matcher · Structural matcher

1 Introduction

The Web of Data was designed to extend the Web through shared structured data. Its basic idea, expressed by Tim Berners-Lee in 2001 in (Berners-Lee et al. 2001), is inspired by the structure of Web pages linked together by hypertext links, to offer a new standardized representation of data, which can be used by humans as well as by machines. The Web of Data is based on the RDF (Resource Description Framework) language that represents data as triples. A triplet is made up of three elements: subject, predicate, and object. These triples relate RDF resources which are resources of the Web. Each RDF resource has a unique URI (Uniform Resource Identifier) of a Web page related to the resource. The RDFS (RDF Shema) and OWL (Ontology Web Language) languages are used to organize RDF resources into hierarchical classes and to define the relationships that can bind the resources. These languages also make it possible inferences from RDF data based on description

F. Ardjani (✉)
Ctr Univ El Bayadh Algeria, El Bayadh, Algeria
e-mail: fatima.ardjani@univ-sba.dz

D. Bouchiha
EEDIS, Lab Ctr Univ Naama Algeria, Naama, Algeria
e-mail: djelloul.bouchiha@univ-sba.dz

S. Sedkaoui et al. (eds.), *International Conference on Managing Business Through Web Analytics*, https://doi.org/10.1007/978-3-031-06971-0_7

logics. Using RDFS and OWL to structure data and organize it into triples to form data graphs makes RDF databases simpler than relational databases. RDF databases can be queried using SPARQL language. The most tangible form of the Web of Data is Linked Data, which appeared in 2008 at the same time as SPARQL. It is a collection of bases that concern various fields and that follow common rules of structuring. These bases are linked together by equivalence relationships between RDF resources representing the same elements in different bases. The RDF resources in the Linked Data databases are all associated with user-readable Web pages, and all provide a means of accessing their content. The Linked Data initiative aims to publish structured and Linked Data on the Web using Semantic Web technologies. These technologies offer different languages for expressing data in the form of RDF graphs and querying them in SPARQL. The Web of Data enables the creation of RDF databases and services based on RDF data. RDF is a simple data model for knowledge representation on the Web. RDF is used to describe RDF resources (Heath and Bizer 2011). Each RDF resource is unique throughout the Linked Data. An RDF triplet expresses a relationship between a subject and an object, that is, a triplet describes a property of the subject having the value of the object of the triplet. The subject is a resource, explicitly identified by a URI. A URI (Uniform Resource Identifier) is a unique identifier on the Web. The relationship is always an identified resource (URI); the object is either a resource (identified or not) or raw data, also called a literal. The database must contain links to other Linked Data databases. Binding describes the relationship between two resources and consists of three URI references. URIs, in the subject and subject of the link, identify the linked resources. The predicate URI defines the type of relationship between the resources. There are two types of RDF link, internal RDF links that is to say the links described in the same database (intra-Base links) and external RDF links that is to say the links described between a set of bases (inter-Base links). Internal links connect resources in a single Linked Data source. Thus, subject and object URIs are in the same namespace. External links connect resources in different Linked Data sources. The URIs of the subjects and objects of external links are in different namespaces. An external link is a collection of RDF external links between two sets of data. This is a set of RDF triples where all subjects are in one dataset and all objects are in another dataset.

In Linked Data, ontologies provide the framework for the data structure of an RDF database. The ontology defines the set of classes and relationships that are used in the RDF database. Their definitions are made so as to frame their uses by constraints to keep the consistency of the data in the database.

Linked Data enables the implementation of applications that reuse data distributed over the Web. To facilitate the interoperability of the company, data from different suppliers must be linked. This means that the same entity in different datasets must be identified. One of the main challenges of Linked Data is to deal with this heterogeneity by detecting links between sets (Heath and Bizer 2011).

Ontology alignment is an automatic or semi-automatic process that consists in identifying the semantic correspondences between ontology entities to align. Alignment is the solution that was previously used to align databases, schemas, etc.

It represents a kind of artificial intelligence trying to reconstruct human interaction with computer systems and knowledge of the relationships between the various concepts in the world (Grau et al. 2013).

The process of ontology alignment consists of several steps (Grau et al. 2013) that can be summarized as follows: the first step is to extract the entities from the two ontologies to be aligned; then, the second step is to calculate the similarity between these entities using different methods, such as linguistic, syntactic, structural, etc.; the third step is to combine the similarity values calculated by the different matchers using various strategies; finally, the last step consists in extracting the semantic matches.

In this work, we propose the use of ontology alignment system called ABCMap to contribute to the interoperability of a business federation based on data interoperability.

The interoperability of data refers to make different data models and query languages working together, and sharing information from heterogeneous data sources, which can moreover reside on different machines under different operating systems and database management systems. The proposed work has the following properties:

- It focuses on addressing data interoperability issues among multiple enterprise information systems.
- Ontology is brought to this issue as a major component.
- Ontology alignment is used as a federated approach and a way to make data interoperable and sharable.
- The hypothesis of the research is that the ontologies from enterprises already exist.

In the remainder of this chapter, we present our proposed ABCMap system. Then, we present experiments that show how to get the best ABCMap alignment. Finally, the conclusion section provides concluding remarks and perspectives.

2 General Presentation of the ABCMap System

ABCMap relies on Artificial Bee Colonies (ABC) to create mappings between ontologies. So, it is an automatic ontology alignment system designed to solve the problem of ontology matching. Thus, it builds ontological alignment at the terminological and linguistic level. The ABCMap system uses different terminology and linguistic matching methods with a local filter to find matches between two ontologies to be aligned.

To obtain the mapping, ABCMap system compares each entity of the source ontology with all entities of the target ontology. Then, the system creates a similarity matrix, which contains a vector for each pair of entities. The vector is composed of three similarity values (syntactic similarity-1 "Jaro-Winkler," syntactic similarity-2

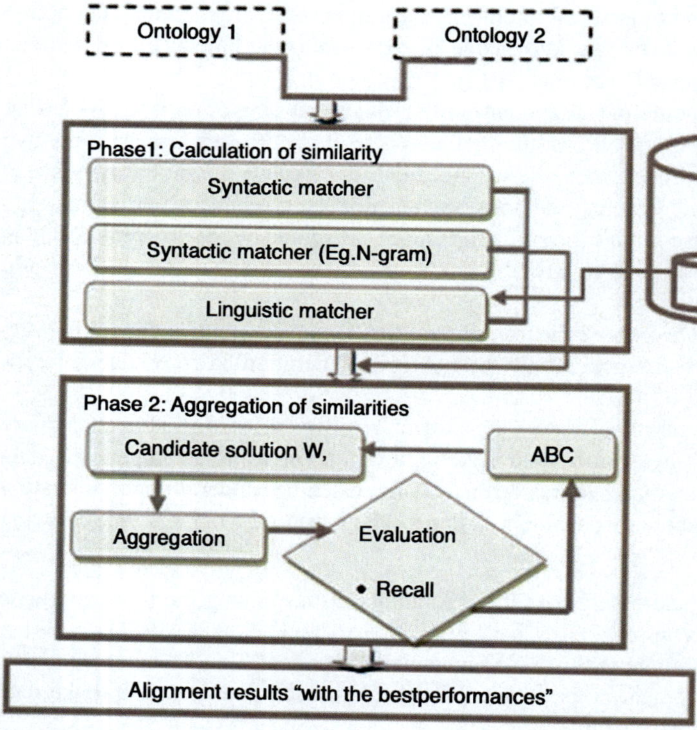

Fig. 1 Architecture of ABCMap alignment system

"N-gram," and linguistic similarity). Figure 1 represents the ABCMap alignment system architecture.

ABCMap associates with each similarity value a weighting coefficient "weight," and it computes the sum of the weighted similarities to achieve a new and unique aggregate similarity. To choose optimal weights for optimal alignment, we chose optimization by the Artificial Bee Colonies (Ardjani and Bouchiha 2019).

2.1 Syntactic Matcher

We use the distance of Jaro-Winkler and N-gram to calculate the similarity between the entities.

Distance Jaro-Winkler (Winkler 1999) Let s and t be two strings. Let P be the length of the longest common prefix of s and t. Let n be a positive number. The distance Jaro-Winkler is a function of DSJaro-Winkler dissimilarity: SXS → [0,1], such that:

$$\overline{\mathrm{DS}_{\text{Jaro-Winkler}}(s,t)} = \overline{\mathrm{DS}_{\text{Jaro}}(s,t)} - \frac{\max(P,n)}{10}\overline{\mathrm{DS}_{\text{Jaro}}(s,t)}$$

N-gram Distance The N-gram distance calculates the ratio of the number of common N-grams to the total number of N-grams between two character sequences. Typically, let N-gram (s, n) be the set of all substrings of s of length n. The N-gram distance between two chains s and t is defined by the following dissimilarity function (Grau et al. 2013).

$$\mathrm{DS}_{\text{n-grams}}(s,t) = \mid \mathrm{ngram}(s,n) \bigcap \mathrm{ngram}(t,n) \mid / n * \mathrm{Min}(|s|;|t|)$$

2.2 *Linguistic Matcher*

The similarity between two entities represented by terms can be also deduced by analyzing these terms using linguistic methods. These methods exploit essentially expressive and productive properties of the natural language (Berners-Lee et al. 2001). The exploited information can be intrinsic ones (internal linguistic properties of terms, such as morphological or syntactical properties) or extrinsic ones (using external resources, such as vocabularies or dictionaries).

In our system ABCMap, the linguistic matcher uses the WordNet dictionary (Fellbaum and Miller 1998). WordNet is a lexical database of the English language, which groups the terms (nouns, verbs, adverbs, and adjectives) into a set of synonyms called synsets.

3 ABCMap Evaluation

Since 2004, researchers in ontology alignment, with the increasing number of automatic alignment systems, organize annual evaluation campaigns called OAEI "Ontology Alignment Evaluation Initiative" whose objective is:

- Highlight the strengths and weaknesses of these systems
- Compare the performances of the different techniques
- Promote communication between the developers of these systems
- In general, advance research in the field of ontology alignment

We have chosen the evaluation campaign of alignment systems OAEI.

This campaign proposes a set of categories to evaluate the alignment systems according to several criteria. We present the results obtained from our ABCMap ontology alignment system by testing different series (OAEI 2012[1]) with:

[1] http://oaei.ontologymatching.org

Table 1 Brief description of the reference tests

ID	Description
101–104	The aligned ontologies are identical. The first ontology is the OWL-Lite restriction of the second ontology
201–210	The same structure for the two aligned ontologies, but the lexical and linguistic functionalities are different
301–304	Aligned ontologies are real cases

Table 2 Results from our ABCMap alignment system

ID	ABCMap recall	ABCMap precision
101	0,93	0,98
103	0,81	0,90
104	0,86	0,93
201	0,91	0,97
206	0,70	0,92
302	0,61	0,89

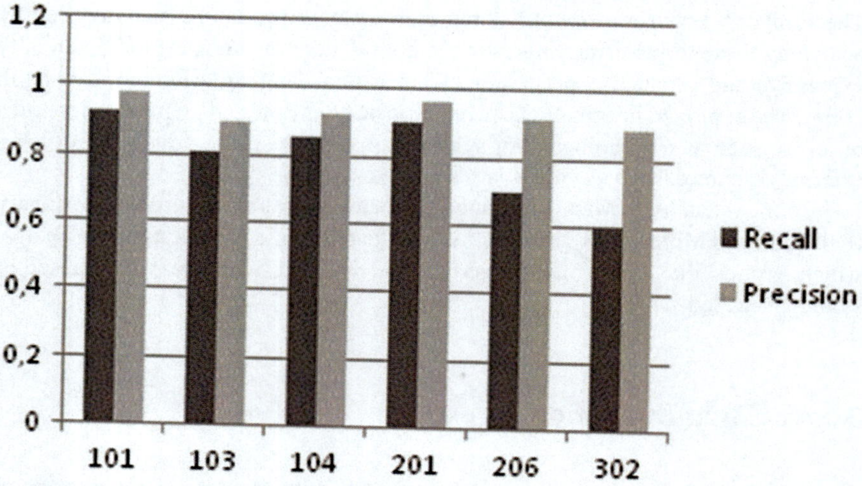

Fig. 2 Results from our ABCMap alignment system

- Number of iterations equal to 10
- Number of active bees equal to 3
- Number of inactive bees equal to 1
- Threshold equal to 0.7

We take the pairs of entities that have a similarity greater than the threshold, and the similarities below the threshold are rejected.

The WordNet API was used to calculate the linguistic similarity (Van Rijsbergen 1977) (Table 1).

Table 2 and Fig. 2 show the best results obtained by ABCMap. Precision and Recall are metrics used to evaluate the quality of our ontology alignment systems.

Precision is the ratio of the number of relevant pairs found "Ncorrect," reported to the total number of pairs not found "Not Found" by our ABCMap alignment system.

The precision function is defined as follows:

$$\text{Precision} = \frac{\text{Ncorrect}}{\text{Nfound}}$$

The results obtained for the ABCMap system show that the Precision value has reached a rate of 98% for the test of "101," 90% for the test of "103," 93% for the test of "104," 97% for the test of "201," 92% for the test of "206," and 89% for the test of "302."

The Recall is the ratio of the number of relevant pairs found "Ncorrect" by our ABCMap alignment system, relative to the total number of relevant "Nexpected" pairs.

The recall function is defined by:

$$\text{Precision} = \frac{\text{Ncorrect}}{\text{Nexpected}}$$

After experiments, the results obtained for the ABCMap system show that the Recall value reached a rate of 93% for the test of "101," 81% for the test of "103," 86% for the test of "104," 91% for the test of "201," 70% for the test of "206," and 61% for the test of "302." This shows that our ABCMap alignment system did not find all the correct matches established by the domain expert (correspondences of the reference alignment).

4 Conclusion

In order to develop the interoperability of business data, this chapter is focused on the adoption of the ontology alignment technique to contribute to interoperability approach from federated enterprises.

In this chapter, we have described a system with a linguistic and syntactic matcher (ABCMap). The analysis of the experimental results gives slightly better results in terms of Recall and Precision.

This work may be improved in the future by considering other semantic matchers in the ontology alignment process and by using other similarity measures. So, we will use other optimization methods to discover new matches.

References

Ardjani, F., Bouchiha, D.: A new approach based on the bee optimization algorithm for ontology alignment: ABCMap+. Int. J. Infor. Retrieval Res. (IJIRR). **9**(4), 13–22 (2019)

Berners-Lee, T., Hendler, J., Lassila, O.: The semantic web. Sci. Am. **284**(5), 34–43 (2001)

Fellbaum, C., Miller, G.: Wordnet: an electronic lexical database (language, speech, and communication) (1998)

Grau, B.C., Dragisic, Z., Eckert, K., Euzenat, J., Ferrara, A., Granada, R., et al.: Results of the ontology alignment evaluation initiative. **2013** (2013, October)

Heath, T., Bizer, C.: Linked data: evolving the web into a global data space. Synth. Lect. Semant. Web Theory Technol. **1**(1) 1–136 (2011)

Van Rijsbergen, C.J.: A theoretical basis for the use of cooccurrence data in information retrieval. Journal of documentation (1977)

Winkler, W.E.: The State of Record Linkage and Current Research Problems. In Statistical Research Division, US Census Bureau (1999)

Feature Selection Based on Term Frequency for Arabic Text Classification Using Multilayer Perceptron

Ouahab Abdelwhab

Abstract Text classification (TC) is an important field of the text mining and information retrieval. The TC is the operation of attributing to a group of texts an appropriate category from predefined classes depending on its contents. In this work, an Arabic text classification approach has been realized using multilayer neural network (MLP). A new feature selection approach based on term frequency is proposed. We have used a dataset called Waten-2004 for learning and testing. The used dataset contains three categories (Economic, Sports, and Culture). We have used precision, recall, and accuracy to evaluate the proposed approach which gives an acceptable accuracy with 95%. The results obtained revealed that the proposed approach gives good results.

Keywords Arabic text · Classification · Multilayer neural network · Feature selection

1 Introduction

Text classification or categorization (TC) is one of the most important fields of the text mining and information retrieval. With the huge amount of the information on the internet and in the digital libraries, the TC has become an active research. The TC is the operation of attributing to a group of texts an appropriate category from predefined classes depending on its contents (Alsaleem 2011).

Generally, the TC has three essential steps: text preprocessing, text modeling, and classification (Alhaj et al. 2019). The text processing is the work that transforms the text into a suitable format for the classification methods. There are many preprocessing operation that can be performed such as stemming, tokenization, and stop word suppression (Alhaj et al. 2019). The text modeling and representation is

O. Abdelwhab (✉)
Department of Mathematics and Computer Science, African University Ahmed Draia, Adrar, Algeria

S. Sedkaoui et al. (eds.), *International Conference on Managing Business Through Web Analytics*, https://doi.org/10.1007/978-3-031-06971-0_8

the operation that extracts features from the text and transforms them into a vector. The text classification step is the operation of constructing the model, testing, and evaluation.

TC has been used in different applications such as information retrieval systems, medical information systems, automatic indexing, web site classification, analyzing customer feedback, opinion detection, and spam filtering (Alhaj et al. 2019; Al-Radaideh 2020).

Arabic language is one of the most popular languages. About 6.5% of the world's inhabitants speak this language (Al Qadi et al. 2019). The statistics provided by the Internet World Stats reports that the Arabic language is ranked fourth in the world in terms of the Internet users about 5.2% Arabic Internet users (April, 2019) (Boukil et al. 2018).

The number of applications that uses Arabic language is increasing rapidly. Therefore, using the TC is necessary to organize these text resources into classes for better applications manipulation.

Many classification methods have been applied on Arabic texts such as the Naive Bayes probabilistic, K-nearest neighbor, neural networks, and support vector machines (Boukil et al. 2018).

Recently, deep learning models have been used for Arabic text classification (Elnagar et al. 2020).

Among the challenges that we faced when classifying texts are the complexity of the texts, multidimensionality, or including a large number of the features (Zareapoor 2015). These challenges can reduce the classification performance. The feature extraction can be used to reduce this problem (Zareapoor 2015). The feature extraction is to choose the most representative and meaningful features that can enhance the classification performance.

In this chapter, we propose a new strategy for feature extraction which is based on the term frequency. The features extracted are trained by multilayer perceptron (MLP) to show the performance of the proposed approach.

2 Proposed Approach

The proposed approach for Arabic text classification contains three essential steps which are the preprocessing step, feature extraction step, and classification step as shown in Fig. 1.

2.1 *Preprocessing Step*

The text processing is the work that transforms the text into a suitable format for the classification methods. In this step, stop words, punctuations, digits, and non-Arabic words were deleted. The text preprocessing includes:

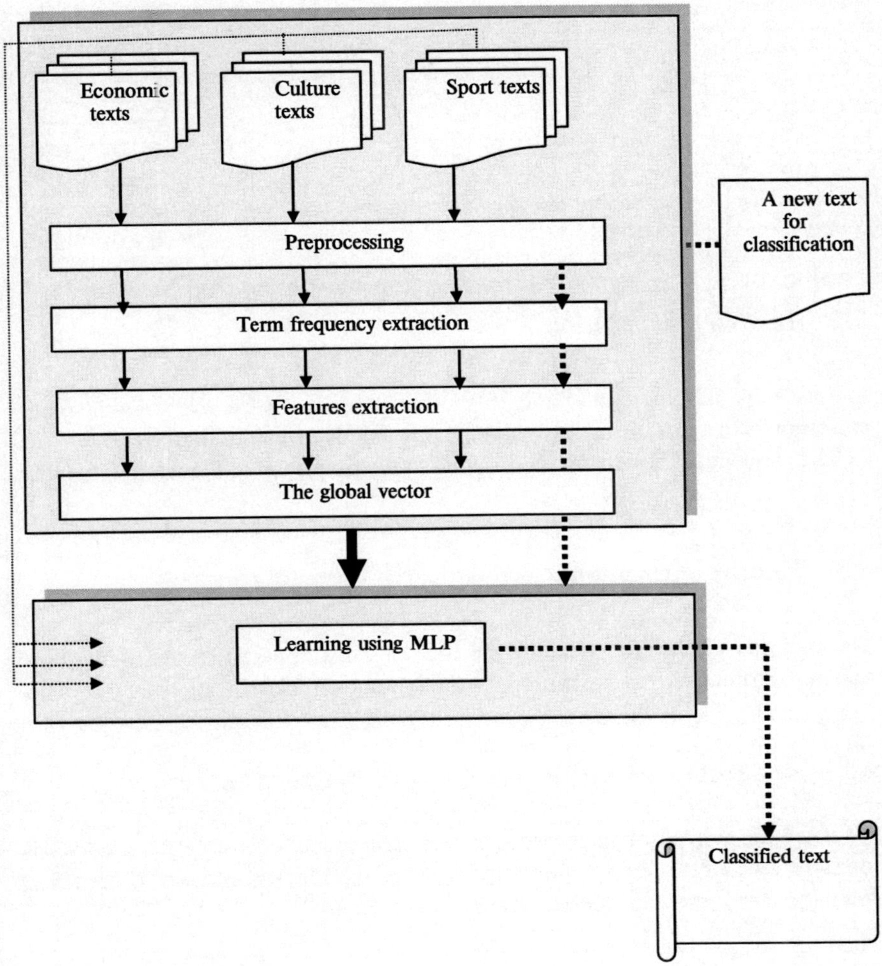

Fig. 1 Architecture of our approach

2.1.1 Tokenization or Segmentation

It is the operation of separating out the words from text. We use blanks, periods, commas, quotes, and semicolons to separate out words. The text is represented in a vector. Each element of this vector contains a word.

2.1.2 Stemming

Stemming is the operation of reducing inflected word to its base, stem, or root as shown in Table 1.

Table 1 Stemming example

Word	Root
الشاعر	شعر
مؤخرا	أخر
العلوم	علم
افتتاح	فتح
عمليات	عمل
حوادث	حدث
الشعب	شعب

2.2 *Text Representation*

In this step, all texts are represented in a matrix. Every row of this matrix corresponds to a text, and every column corresponds to a word in the text.

The elements of this matrix correspond to the frequency of the word in the text.

2.3 *Feature Extraction*

This step is essential for reducing the time of execution and removing the noisy features. Feature selection can enhance the accuracy of classification.

2.3.1 The Text Representative of a Category (A Global Vector)

We count the number of occurrences of each term in the texts of each category. The terms representative of a category are the terms with high number of occurrences. We name this vector "the global vector of the category."

2.3.2 Removing Repeated Words

In this step, we remove the terms which are repeated in the global vectors of any two categories. The reason of this operation is to reduce the confusion into classes when classifying the texts.

2.3.3 The Global Vector of the System

The global vector of the system is constructed from the global vectors of all categories.

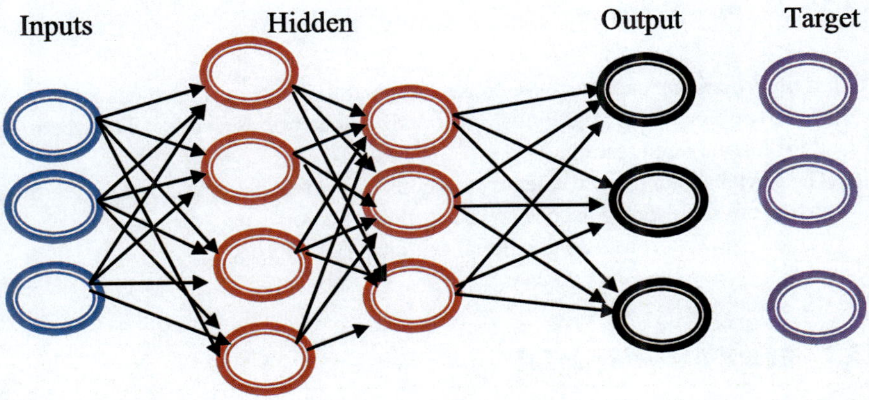

Fig. 2 Architecture of MLP

2.4 Classification Using Neural Networks

Neural networks (NNs) have been widely used in text classification by many researchers. It can manipulate nonlinear and linear problems for text classification. Utilizing neural networks can accomplish a good result (Harrag and Al-Qawasmah 2010).

Multilayer perception neural network (MLP) is composed of an input layer and at least one hidden layer and an output layer. Each layer is completely associated with the next layer with a random weight as shown in Fig. 2.

MPL uses backpropagation for training and learning. The training phase is based on the minimization of a cost function between the target and the outputs by updating the weights (Zheng et al. 2016).

3 Experiments

3.1 Dataset

We used a corpus named Waten-2004. This corpus contains 20,291 texts of 6 categories (Religion, Economic, Culture, Sports, Local News, and International News). Each category contains the average of 3382 texts (Einea et al. 2019). In our approach, we used only three categories (Economic, Sports, and Culture). Each category includes 1300 texts. 1000 texts were used for training and 300 texts were used for testing.

3.2 Feature Selection

After preprocessing each text, we construct the global vector of each category.

Next, we concatenate the global vectors of all categories in ones. This vector is used for learning and testing.

The global vector of each category contains 100 words. All texts used for learning or testing are represented according to the global vector.

Table 2 shows a subset of the global vector of our approach.

3.3 Learning Using MLP

Each text is represented in a vector format that contains the number of occurrences of each term appearing in the global vector. This vector is considered as input of MLP.

Table 2 A subset of extracted features

شعر
ثقف
سرح
فنن
فكر
بدع
شخص
أرخ
أدب
غني
أنس
غرب
مهرج
قرأ
قصص
سلسل
شعب
نبأ
جرب
عصر
روج

3.4 Evaluation Criteria (Mayy et al. 2015)

Recall, precision, and accuracy are computed to evaluate the proposed approach.

3.4.1 Precision

Precision represents the proportion of the texts classified into a category that are truly including to this category.

3.4.2 Recall

Recall represents the proportion of the texts including to a category and are classified into this category. Precision of a category is calculated as:

$$\text{Precision}_i = \frac{\text{TP}_i}{\text{TP}_i + \text{FP}_i} \tag{1}$$

$\text{Precision}_i = \frac{\text{TP}_i}{\text{TP}_i + \text{FP}_i}$ and recall of a category I is calculated as:

$$\text{Recall}_i = \frac{\text{TP}_i}{\text{TP}_i + \text{FN}_i} \tag{2}$$

where FN_i, TP_i, and FP_i represent falsely negative, truly positive, and falsely positive.

3.4.3 Accuracy

Accuracy is computed using the following equation:

$$\text{Accuracy} = \sum_{i=1}^{i=\text{Classes number}} \frac{\text{TP}_i + \text{TN}_i}{\text{TP}_i + \text{TN}_i + \text{FP}_i + \text{FN}_i} \tag{3}$$

3.4.4 Confusion Matrix

A confusion matrix illustrates how the system classified each text and when it might do mistake. The rows of this matrix represent the actual class, and the columns represent the system's results as shown in Table 3. A confusion matrix allows users

Table 3 A confusion matrix

		Predicted class	
		Positive	Negative
Actual class	Positive	TP	FN
	Negative	FP	TN

Table 4 The values of evaluation metrics

	Precision	Recall
Culture	0.93	0.90
Sports	0.99	0.97
Economic	0.90	0.96
Average	0.94	0.94
Accuracy	0.95	

Table 5 A confusion matrix of our approach

		Predicted		
		Culture	Sports	Economic
Current class	Culture	90.66%	0.66	8.666
	Sports	2%	97%	1%
	Economic	3.66%	0%	96.33

to know which category the system may be incapable to classify correctly. It is used for calculating other metrics like accuracy, precision, and recall.

3.5 *Results and Discussion*

From Table 4, it can be seen that the mean values of precision and recall equal 0.94 and 0.94, respectively. The accuracy equals 0.95. These results are very acceptable.

These results demonstrated that using MLP for training gives good results.

Based on the confusion matrix in Table 5, we deduce that the sports and economic texts are good classified with precision equal to 97% and 96%. There is a remarkable confusion between the economic class and the culture class, as 3.66% of economic texts predicted as culture texts and 8.66% of culture texts predicted as economic texts.

4 Conclusion

In this work, we presented a new feature selection strategy for Arabic text classification. In this strategy, we select terms with high frequency, and we remove the terms that appear in more than one class. The used dataset contains three categories (Economic, Sports, and Culture). We used MLP for learning and testing. The results obtained revealed that the proposed approach gives good results.

Our future work is to add more categories for classification and to compare our approach with existing methods.

References

Al Qadi, L., El Rifai, H., Obaid, S., Elnagar, A.: Arabic text classification of news articles using classical supervised classifiers. *University of Sharjah.* 2nd International Conference on new Trends in Computing Sciences (ICTCS), (2019)

Alhaj, Y.A., Xiang, J., Zhao, D., Al-Qaness, A.A., Al-Qaness, Abdelaziz, M., Dahou, A.: A study of the effects of stemming strategies on Arabic document classification. IEEE Access. **7**, 32664–32671 (2019). https://doi.org/10.1109/ACCESS.2019.2903331

Al-Radaideh, Q.: Applications of mining Arabic text: a review chapter. In: Sadollah, A., Sinha, T.S. (eds.) Recent Trends in Computational Intelligence. IntechOpen (2020). https://doi.org/10.5772/intechopen.91275

Alsaleem, S.: Automated Arabic text categorization using SVM and NB. Int. Arab J. e-Technol. **2**(2) (2011)

Boukil, S., El Adnani, F., El Moutaouakkil, A., Cherrat, L., Ezziyyani, M.: Arabic stemming techniques as feature extraction applied in Arabic text classification. Adv. Inf. Technol. Serv. Syst. Lect. Note Netw. Syst. **25** (2018). https://doi.org/10.1007/978-3-319-69137-4_31

Einea, O., Elnagar, A., Al Debsi, R.: SANAD: single-label Arabic news articles dataset for automatic text categorization. Data Brief. **25**, 104076 (2019). https://doi.org/10.1016/j.dib.2019.104076

Elnagar, A., Al-Debsi, R., Einea, O.: Arabic text classification using deep learning models. Inf. Process. Manag. **57**(1), 102121 (2020). https://doi.org/10.1016/j.ipm.2019.102121

Harrag, F., Al-Qawasmah, E.: Improving Arabic text categorization using neural network with SVD. J. Digit. Inf. Manag. **8**(4) (2010)

Mayy, M., Al-Tahrawi, A., Sumaya, N., Al-Khatib, B.: Arabic text classification using polynomial networks. J. King Saud Univ. Comput. Info. Sci. **27**, 437–449 (2015)

Zareapoor, M.: Feature extraction or feature selection for text classification: a case study on phishing email detection. Int. J. Info. Eng. Electr. Bus. **2**, 60–65 (2015)

Zheng, L., Duffner, S., Idrissi, K., et al.: Siamese multi-layer perceptrons for dimensionality reduction and face identification. Multimed. Tools Appl. **75**, 5055–5073 (2016). https://doi.org/10.1007/s11042-015-2847-3

Part II
Social Network Analysis and Graph Algorithms

Social Network Mapping Software: An Approach to Human Resource Systems

Rabia Ahmed Benyahia and Smail Benamara

Abstract Network analysis has developed for several decades. It's popular in every kind of academy social science. In this chapter, we tried to highlight the critical role of social network analysis software in the human resource system. We have provided an introduction to social network analysis and review some of the software available to human resource practitioners in the visualization and analysis of network; finally, we took the case of Bitrix24 as a modern example in the field of social network analysis software.

Keywords Social network analysis · Software · Human resource

1 Introduction

At the beginning of the twenty-first century, the technological revolution has contributed to the worldwide spread of Internet use, which has reinforced individual's tendency for using this tool to benefit its various services. The social networking service attracts a significant segment of surfers and allows them to stay in touch and interact with their knowledge, whether they are friends, colleagues, or family members.

This new situation has encouraged the use of social networks by individuals and organizations in different areas: economic, education, or business. One of the important areas that we will focus on is the human resources of the company, where it is essential for the company to analyze social networks to better understand and communicate with employees.

From the above, many social network mapping software appeared in the field of human resources, and these programs are seen as virtual business space to provide services for managers and employees to achieve the company targets and maximize the work effectiveness.

R. Ahmed Benyahia (✉) · S. Benamara
University of Djilali Bounaama, Ain Defla, Algeria

S. Sedkaoui et al. (eds.), *International Conference on Managing Business Through Web Analytics*, https://doi.org/10.1007/978-3-031-06971-0_9

Based on the above, we can raise the following key issues:

What is the importance of social network mapping software in the development of human resource systems?

2 Research Structure

In order to answer the previous question, we divided this chapter into four main sections:

- Background about social network
- Social network software
- Social network and HR system
- Case of Bitrix24

3 Background About Social Network

3.1 The Concept of Social Network and Its Evolution

Before addressing the definition of the social network, we will give a glimpse history of its emergence and evolution.

3.1.1 Evolution of Social Network as a Concept

Social network is a term that appeared to exist in 1954, coined by anthropology J. A. Barnes; this term has been used to refer to applications that link individuals or friends over the Internet, where SNS (social networking service) users can communicate with each other using blogs, chat rooms, email, or instant messaging.[1]

The expansion of Internet usage across the world, the proliferation of computers, and their access to homes in the 1990s helped to create chat sites that allowed individuals to communicate with each other in audio and video, such as Microsoft Windows Live Messenger,[2] but there were some imperfections in these sites, such as when it was not possible to see other users' profiles then, it spread and became more interacted across the world through popular websites such as MySpace and Facebook. In 2003, it became the most popular and attractive for website users.[3]

[1]Celia Romm-Livermore, Kristina Setzekorn, "Social Networking Communities and E-Dating Services: Concepts and E-Dating services: concepts and Implications", Information Science Reference, New York, USA, 2009, p230.

[2]Alvin Chin, Daqing Zhang, "Mobile Social Networking, Springer", New York, USA, 2014, p 2.

[3]Anil Kumar Jharotia, 'Use of Social Media in Marketing of Library and Information Services in Digital Era', Web seminar "Use of social Networking in Knowledge Sharing", Modern Rohini Education Society, 26th September 2015, New Delhi, India, p 3.

Currently, having an account on a network site is common, especially with the evolution of using smartphones and tablets. The Internet coverage growth at both home, workplace, and university areas has expanded, with the number of accounts active across. The number of accounts active across these sites reached 3.8 billion in (January 2020),[4] an increase of 9% over the past year, with 321 million new users in 2019.

3.1.2 Social Network Definition

There are many definitions related to this concept, as follows:

According to Vala Ali Rohani and Siew Hock Ow, "social networking sites are web sites that allow people to stay connected with other people in online communities.[5]"

Charles Steinfield, Nicole Ellison, Cliff Lampe, and Jessica Vitak have defined it as "a term social used to refer to Web sites that enable users to articulate a network of connections of people with whom they wish to share access to profile information, news, status updates, comments, photos, or other forms of content.[6]"

Roughly, we can define a social network as a package of relationships and interactions between individuals or groups, which has for objective to spread information, ideas, and influence among its members.

3.1.3 Social Network Classification

Existing social networking sites available to users whether individual or organization vary according to their interests and needs, allowing them different options for interacting online. Social networking sites can generally be classified as follows[7]: social connections, multimedia sharing, professional, and academic.

Social Connections These sites are intended to keep the communication going on permanently between family members, between friends, or between companies and their customers, and the most frequently used sites are:

[4]https://datareportal.com/reports/digital-2020-global-digital-overview, 09/08/2020, 14h30.

[5]Vala Ali Rohani, Siew hock Ow, "On Social Network Web Sites: Definition, Features, Architectures and Analysis Tools", January 2009, p 43. https://www.researchgate.net/publication/233927861_On_Social_Network_Web_Sites_Definition_Features_Architectures_and_Analysis_Tools, 11/08/2020, 13h00.

[6]Charles Steinfield, Nicole Ellison, Cliff Lampe, and Jessica Vitak, "Online Social Network Sites and the Concept of Social Capital", the Internet Turning 40 Conference School of Journalism and Mass Communication Chinese University of Hong Kong, 2012, p 2.

[7]Mary Gormandy White, "What Types of Social NetworksExist?",https://socialnetworking.lovetoknow.com/What_Types_of_Social_Networks_Exist, 06/08/2020, 21h00.

- *Facebook*: the most used online social networking website with 2.5 billion users[8] (January 2020) that enables users to create profiles that allow them to share posts, photos, and videos together and interact with other users' posts through comments or likes.
- *Instagram*: a Facebook-owned app since 2012, allowing users to share and interact with photos and short videos, which celebrities often use to interact with their followers, with a billion of users[9] around the world (January 2020).
- *Twitter*: a free social networking service with 340 million users[10] around the world, allowing only account-owning users to publish tweets, and retweets, while unregistered users are allowed to read only; tweets cannot exceed 140 characters, while voice and video tweets are limited in 140 s for most accounts.

Multimedia Sharing Social networking is an effective support for sharing videos, photos, and audio online, making them contribute to multimedia sharing, such as:

- *YouTube*: a video sharing service that allows users to share and interact with videos from each other via comments and alike and unlike features (2 billion users in January 2020[11]).
- *Flickr*: an application that allows hosting high-quality images and videos from professionals and amateurs, created by Ludicorp in 2004, and then has been owned by Samsung since April 20, 2018[12].

Academic A space dedicated to academic researchers to share their scientific research with their counterparts and review it in an interactive academic framework, the most important of which are:

- *Academia.edu*: a platform for academics to share their research, to share it and study its effects, as well as to follow developments in their scientific specialization. More than 132 million academics have signed up for this site, 25 million papers have been published, and more than 52 million monthly visitors visit the site.[13]
- *ResearchGate*: a professional network dedicated to scientists and researchers, with 17 million members worldwide sharing and discussing research.[14]

[8] https://datareportal.com/reports/digital-2020-global-digital-overview, 08/08/2020, 15h00.
[9] https://datareportal.com/reports/digital-2020-global-digital-overview, 08/08/2020, 15h25.
[10] https://datareportal.com/reports/digital-2020-global-digital-overview, 08/08/2020, 15h50.
[11] https://datareportal.com/reports/digital-2020-global-digital-overview, 08/08/2020, 16h10.
[12] https://en.wikipedia.org/wiki/Flickr, 08/08/2020, 18h50.
[13] https://www.academia.edu/about, 08/08/2020, 23h50.
[14] https://www.researchgate.net/about, 09/08/2020, 00h05.

Professional Designed specifically for the Professional category, it allows them to search for new jobs or share experiences among themselves, and some sites focus on specific occupations, some examples of which are:

- *LinkedIn*: a professional website specializing in online work and employment, allowing its members to create professional links and exchange ideas, and allowing organizations to select competent employees according to their needs. The active account holders reached 660 million users (November 2019[15]).
- *Classroom* 2.0: a specially targeted social network to facilitate communication and assistance between teachers in matters related to their profession.

3.1.4 Advantages and Disadvantages of Social Networking

The widespread use of social networks has increased its impact on its users, leading to a debate about their positive and negative effects, where they can be divided into two parts: advantages and disadvantages.

Advantages there are many advantages, the most important of which can be mentioned:

- *Worldwide connectivity*: whatever the purpose of the search, whether looking for a former classmate, first teacher, or a friend, there's no easier or faster way to connect from the social network. Facebook, Twitter, and LinkedIn are the most popular social networking communities, where people create new friendships or business relationships or expand their personal relationships by connecting and interacting with friends. They can also[16]:
  - Make a new friend
  - Seek a new job
  - Exchange views on goods and services
  - Make or receive advice on career or personal issues
- *The ability for the content reader to become a publisher*: content readers cannot be considered just like consumers, as social networks allow them to become content publishers. This way, social networks enable content readers to share it with their followers' network by posting or reposting the message in their way.[17]
- *An effective way to run a business at low cost*: using social networks by companies reduces costs and increases income, as it is a very-low-cost advertising tool that requires only Internet access, electronic equipment, and electricity.

[15] https://www.businessofapps.com/data/linkedin-statistics/, 08/08/2020, 20h55.

[16] RakhiTyagi, "Social Networking: Advantages And Disadvantages", Web seminar "Use of social Networking in Knowledge Sharing", Modern Rohini Education Society, 26th September 2015, New Delhi, India, p 11.

[17] "Social media marketing", SEOP, p 7.

Compared to other advertising formats where the organization must pay a high cost to broadcast its advertising live on TV or radio channels, everything on the social networks is free. Targeting the right audience also saves the organization high costs that it would spend on the marketing budget, agent salaries, and other associated high costs.[13]

– *Recruiters and employers are more visible*: internal recruiters in companies or recruiters in staffing agencies are flocking to the social network sites in the hopes of finding competencies to fill positions. In return, by having a well-written profile on LinkedIn or other virtual communities, job seekers are looking for a job that suits their ambitions.[19]

Disadvantages the advantages of social networks cannot cover risks to their use, which can be summarized in the following points:

– *Endangering the privacy of network users*: by violating their personal information and selling it to different companies. For example, when you comment on Facebook, you can be aware of ads related to a user's post, Facebook won 3.8 billion $ in 2011 through those ads, and this justifies the free site. Different social network sites generate revenue by selling targeted ads in particular. In general, the social network site is not the commodity, but its users are commodities. These sites provide user-friendly advertisements by using searched keywords and other information stored on your computer or profile.[20]
– *Addiction to social networks*: which can have many negative effects, including: More time wasted, as a recent study in 2019 showed that users spent an average of 153 min per day on social networking; in other words, it represents nearly 6 years and 8 months in their lifetime[21] (average of 72 years lifespan in base of WHO estimation).

  Social linkages collapse, as a result of excessive social networking use by creating a fiction of belonging a member of a virtual community, that appear in isolation from the real world of each one family member in his world.
– *Social networking can spread false or unreliable information quickly*: studies showed that 78% of traditional media correspondents used social networks as sources for finding breaking news; the trend is that there is a lack of fact verification before the sharing process. People tend to get the news that is

---

[18]Caroline Mutuku, "Advantages and Disadvantages of Using Social Networks in Business", Grin verlag, München, Germany, 2017, p 1.

[19]Diane Crompton, Ellen Sautter, "Find a Job Through Social Networking"- second edition, JIST Publishing, USA, 2011, p 22.

[20]FatemahAziziRostam, "Investigating the advantages and disadvantages of social networks on social media",2020,https://www.academia.edu/42998165/INVESTIGATING_THE_ADVANTAGES_AND_DISADVANTAGES_OF_SOCIAL_NETWORKS_ON_SOCIAL_MEDIA, 06/08/2020, 17h00.

[21]"Daily time spent on social networking", https://www.broadbandsearch.net/blog/average-daily-time-on-social-media, 11/08/2020, 18h20.

compatible with their beliefs even if it's fake. For example, news items that contain inaccuracies spread six times faster on Twitter than articles that contain the truth.[22]

3.2 *Social Network Analysis*

After we have addressed the general concepts related to social networking, we will briefly address the analysis of social networks and their concept and importance.

3.2.1 Social Network Analysis Meaning

Social network analysis as a concept has many definitions, from which we can mention it:

> SNA is described as the measurement and mapping of various aspects of relationships between people, organizations, and groups. It also includes mapping these groups of people to computers, sites that they visit and other types of information sources.[23]

> Social network analysis is a large and growing body of research on the measurement and analysis of relational structure.[24]

Based on the above definitions, we can define social network analysis as a thorough study of the relationships that arise from interactions among the influent in the social network environment, in order to measure and analyze them.

3.2.2 Fundamental Concepts in Network Analysis

When discussing social network analysis topic, we must address the key concepts that form the focus of the study; these concepts are[25] actor, relational tie, dyad, triad, subgroup, group, relation, and social network.

[22] Keith Miller,"21 Advantages and Disadvantages of Social Networking", https://futureofworking.com/10-advantages-and-disadvantages-of-social-networking/, 11/08/2020, 19h10.

[23] *Elizabeth Wamicha,* "What is Social Network Analysis?",https://study.com/academy/lesson/what-is-social-network-analysis.html, 14/08/2020, 20h10.

[24] Carter T. Butts, "Social network analysis: A methodological introduction", Asian Journal of Social Psychology, 2008, p 13.

[25] Stanley Wasserman and Katherine Faust, "Social Network Analysis: Methods and applications", Cambridge University Press, USA, 1994, P 17.

3.2.3 The Importance of Social Networking Analysis

Social network analysis is important by the nature of their level and purpose, where we can identify two levels of focus: microanalysis and macroanalysis.[26]

- *Microanalysis*: the use of the analysis at this limited level can be done with own data or a subset of the social data that one may have access to. This may provide answers to recommendations based on own tastes or the friend's tastes or at the understanding level for each of the communities found in a large social network, how those communities influence other communities, or the interaction way of people in each community and becoming influencers of their peers.
- *Macroanalysis*: it requires the analysis of large networks, where interesting measures like centrality or using algorithms such as community search are important. In this case, the importance of specific algorithms like platforms like Apache Graph can be very useful to obtain interesting metrics or algorithms executed for large graphs.

4 Social Network Software

In analyzing the social network number of graph structures utilized to define the general connectivity of a network, first, we have the nodal degree which means the number of ties between nodes or actors. For no directional ties, the number of connections is determined as either present 1 or not present 0. For directional ties, the quantity of the connection depends on the value related to the level of the relationship (i.e., 5 = I speak with this individual every day).[27]

There are numerous commercial and freely computer packages that give the capacity to perform social network analysis.

Some programs were originally developed for network visualization, and other programs were particularly created to integrate network analysis and visualization.

The age of the program was not a measure for choice, although the release dates of the last versions of the majority of the reviewed software were within 2003 or 2004.

UCINET 6 is the most notable program; it offers the ability to compute network measures through its joined visualization computer program NetDraw, which is included with the package.

Table 2 represents the main objective or characteristic of several programs.

The data format: distinguishes three aspects:

[26] Josep Lluis Larriba Pey, "Why is social network analysis important?", https://www.quora.com/Why-is-social-network-analysis-important, 15/08/2020, 14h00.

[27] John-Paul Hatala, (2006), Social Network Analysis in Human Resource Development: A New Methodology, Human Resource Development Review 5: 45 pp51–52.

Table 1 Social network key concept

Concept	Definition
Actor	Social entities who may be either discrete individual, corporate, or collective social units
Relational tie	Linkage established between a pair of actors
Dyad	A pair of actors and the possibility or possibilities between them
Triad	A subset of three actors and the possibility or possibilities among them
Subgroup	A subset of actors and all ties among them
Group	The collection of all actors on which ties are to be measured
Relation	Collection of ties of a given kind measured on pairs of actors from a specified actor set
Social network	Consists of a finite set or sets of actors and the relation or relations defined on them

Source: Stanley Wasserman e Katherine Faust, Social Network Analysis: Methods and Applications, Cambridge, Cambridge University Press, 1994, pp. 17–21

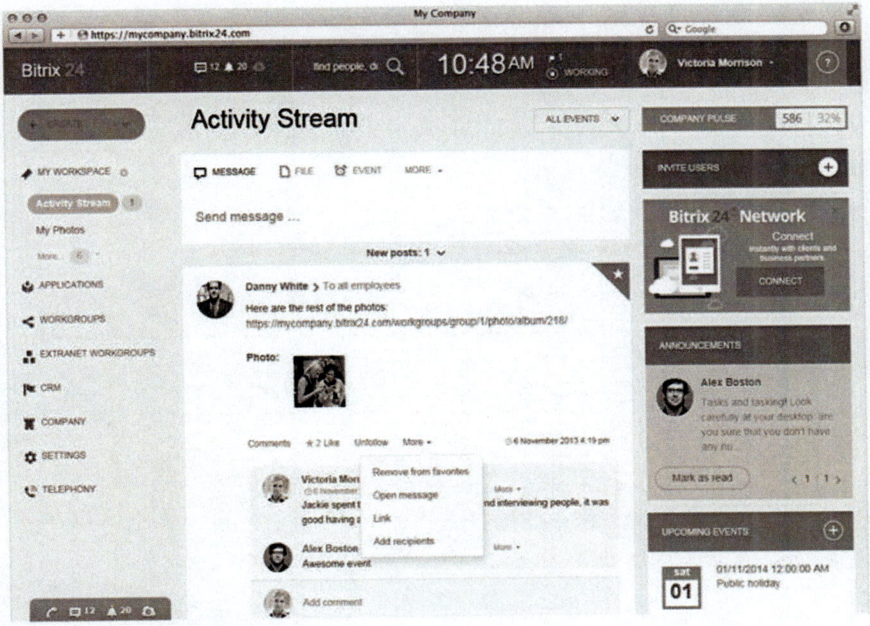

Fig. 1 Bitrix24 platform. (Source: https://www.dreamsite.ca/en/bitrix24/ about-corporate-portal/ (25/07/2020))

1. Type of data the program can handle.
2. Input format.
3. Whether there is an option to indicate missing value codes for network relations.

The functionality: indicates whether the software contains (network) visualization options and the kind of analyses it can perform.

Table 2 Characteristics of several program

Program	Version	Objective	Data			Functionality		Support		
			Type[a]	Input[b]	Miss.	Visual.	Analyses[c]	Avail.[D]	Manual	Help
Social network Analysis fu	2.0	General Macsf	C	M, ln	No	Yes	D, Sl	Free[d]	No	No
StOCNET	1.5	Statistical analysis	C	M	Yes	No	D, Dt, S	Free	Yes	Yes
UCINET	6.55	Comprehensive	C, E, A	M, ln	Yes	Yes[e]	D, Sl, Rp, Dt, S	Com[f]	Yes	Yes
Visone	1.1	Visual exploration	C, E	M, ln	No	Yes	D, Sl	Free	No	No
Agna	2.1.1	General	C	M	No	Yes	D, Sl, sequential	Free	Yes	Yes
Fat cat	4.2e	Contextual analysis	C	Ln	Yes	Yes	D, Sl	Free[g]	No	Yes
GRADAP	2.0e	Graph analysis	C	Ln	Yes	No	D, Sl, Dt	Com[g]	Yes	No

Source: Peter J. Carrington, John Scott, Stanley Wasserman (2005), Models and Methods in Social Network Analysis, Cambridge University Press, pp. 271–276

[a]c, complete; e, ego centered; a, affiliation; l, large networks
[b]*m* matrix, *ln* link/node, *n* node
[c]*d* descriptive, *sl* structure and location, *rp* roles and positions, *dt* dyadic and triadic methods, *s* statistical
[d]Open-source software
[e]No graph drawing routines
[f]An evaluation/demonstration version is available
[g]DOS program which is no longer updated
[h]com, commercial product; free, freeware/shareware

The support: distinguishes the availability of the program (free or commercial), presence and availability of a manual, and presence of online help during the execution of the program.

5 Social Network and HR System

The theory of social networks emphasizes that human decisions are functions of people-to- people ties. Individuals get support, knowledge, and power from the network structure and from their position in the network.[28]

Analysis of the social network concerned the relationship patterns of frontline workers and their interaction with each other and the communication flow between managers and their employees. Identifying these relational patterns will help develop training initiatives and employees to meet organizational and individual needs.

Collecting relational data can help to identify particular similarities in individual groupings. For example, it is necessary to understand who a new employee is conversing with in order to predict future performance.

If we can identify who the new employee frequently communicates with, we may be able to determine that individual's performance direction based on the profiles of their main contacts. If the contact performance levels are in line with the standards of the organization, the supervisor can encourage the employee to continue to communicate with those individuals. If the profiles are negative, the supervisor may wish to interfere and direct the new employee to higher-level workers.[29]

Using the social network analysis can help alleviate the resistance to change often associated with organizational reconfigurations such as downsizing, layoffs, or restructuring.

Analysis of social networks can assess the effects of a social environment on learning participation and performance improvement through the identification of cultural influences.[30]

There are several ways in which an HR analytics function can use the analysis of social networks to help improve operational effectiveness, some of which are summarized below:

Optimizing communication and collaboration: By evaluating workplace interdependencies. Several occasions have shown that this decreases operational

[28]Teresa Torres-Coronas, Mario Arias-Oliva (2005), e-Human Resources Management: Managing Knowledge People, IDEA GROUP PUBLISHING, London, pp69–70.

[29]John-Paul Hatala, (2006),Social Network Analysis in Human Resource Development: A New Methodology, Human Resource Development Review 5: 45 pp51–52.

[30]John-Paul Hatala, (2006),Social Network Analysis in Human Resource Development: A New Methodology, Human Resource Development Review 5: 45 pp65–67.

costs and increases the pace of communication, thereby enhancing operational performance.

Enhanced knowledge sharing: The position of internal expertise and the optimization of knowledge sharing are major concerns for most organizations. Data sources and dashboards can help visualize who deals with which subject in the organization and what competencies can be linked to it.

With these interactive connections, any employee may find it easier to quickly detect the right person they need to contact.

Using resources more efficiently: Social network analysis will help recognize the actual work each employee is doing and help redefine their position. Through this way, you can theoretically spot talents doing the "wrong" job, and more flexibility can be provided to participants through more value-added activities.

Leverage connections: An employee holding together a network (a central node) that can help to enhance efficient communication by exploiting its connections and can be used as an informal leader. Apart from his job duties, it is necessary to acknowledge and award the importance of such an employee.

Succession planning: These important connectors can also play a major role in the preparation of successions. Organizations can use social network analysis to evaluate the social networks of workers who are soon to retire and their role in them.

Mergers and acquisitions: It is important to understand the informal networks existing in the acquired companies to better integrate employees and facilitate communication and collaboration between them.

Analysis of social networks can be used to define the key nodes which can help shift faster.

Employee absenteeism and turnover predictions: In addition to conventional employee characteristics such as age or performance, employee data may also be improved by social network analysis. Analysis of linkages between various groups of employees may help to identify more complex absenteeism or turnover trends of employees.[31]

6 Case of Bitrix24

Bitrix24 is a unified workspace that brings together a complete set of enterprise management tools in one intuitive interface. Bitrix24 is composed of seven corporate portal options. They differ in their functionality and capabilities. There are

[31] Saskia Menke, (2017), The value of Social Network Analysis in HR analytics, on site https://www.linkedin.com/pulse/value-social-network-analysis-hr-analytics-saskia-menke?fbclid=IwAR3teIM2vdYhs6yVvCCAd-FUZkLNd9CoMk2-9n6y45qMeUMlhgSHaKO5SNU (01/07/2020).

four cloud service rates, "Free," "Plus," "Standard," and "Professional," and three versions of the product in the box: "CRM," "Business," and "Enterprise."

6.1 Corporate Portal

Corporate portal: is the internal information management system of the company designed for joint activities, projects and documents, and efficient internal communications.

6.1.1 Internal and External Communications

Work pleasurably on the corporate platform: as on the social network. Collaborate on projects and debate all in real time. Use familiar tools to communicate and manage tasks and documents.

The corporate platform invites colleagues to business community and call, chat, write, and edit the same document at the same time, through voice and video assistance. Call colleague who is not logging in: call him directly from the portal on the mobile phone. Call clients to landline phone from CRM with integrated telephony support. Work on extranet with partners can provide a stable and neutral work environment.

6.1.2 Project Management and Tasks

Projects are handled using useful tools. Using practical features for collaborating with partners (on extranet): calendar integration, collaboration process, checklist method, task templates, filters, and constructor tasks.

The fulfilment of control activities in divisions: Estimate the time and other costs during project creation.

Manage the progress of project activities using Gantt charts: total number of tasks, how many have been completed and how many are in progress, and which tasks are overdue and which ones have no deadline. Evaluate work with obtained information about people, organizations, and programs. At the end of the month, make conclusions and summarize results.

6.1.3 Joint Work with Documents

Connect "Bitrix24. Disk" and access your work files both from your computer and from the portal. Attach the company files and group drives to your Disk directories, allow colleagues access, and work together on files. Share documents with colleagues, address them on "Operating Unit," and get them external connections.

Function with papers, sometimes without apps on the desk. You can open, view, and edit files of all common online formats via external web services Google Docs and Microsoft Office Online.

Edit some Bitrix24 documents on your computer with software helping: MS Office and even Adobe Photoshop. The Updated file will be transferred directly to Bitrix24 right away.

6.1.4 Scheduling and Time Tracking

Employees will mark the start and finish of the day, breaks, and absences and also schedule everyday activities. According to this data, a report will be generated about the use of working time.

Project calendar events. Download it to your cell phone or tablet to have it always with you, even on the road. Invite your colleagues directly to a meeting on "Operating band." Internal meeting service can help: invite participants quickly, send out the agenda and discussion results automatically, create events in personal calendars, and set tasks based on conclusions.

6.1.5 CRM: Customers and Sales

CRM (customer relationship management) system. Build leads, contacts, and company database; record all events connected with the business (e.g., phone calls, emails, meetings, transactions); invoice clients; schedule your operations; and build "sales funnel," charts, and graphs.

6.1.6 Open Channels

Open channels connect the most popular social networks and messengers with Bitrix24 CRM. Client messages from Facebook, Telegram, Skype, and other platforms are distributed among Bitrix24 CRM according to the specified guidelines. Even if your employees use Bitrix24 to communicate with customers in real time, customers will see all responses within the social network or the messenger they initiated the contact with.

6.1.7 Company Structure

Bitrix24 Company Structure/Organizational Chart tool comes with a Visual Builder (Drag'n'Drop) that makes it easy to create an interactive diagram with departments, sub-departments, department heads, and subordinates representing your organization's hierarchy. This chart can be used to assess which employee is the most appropriate for a particular task or problem. Importantly, the information found in the organizational map Bitrix24 is integrated into several other features in the corporate portal Bitrix24: monitoring, task delegation, workflow, and more. Department heads, for example, will see all of their subordinates' activities.

6.1.8 Business Process Automation

Business processes let you work with virtually any sort of portal details.

The use of "business processes" allows a range of business processes to be managed: from basic to more complex. Start with news publishing on the portal, and then proceed to more serious processes: sending orders and networks of interaction with partners or clients.

6.1.9 Desktop Application for the Portal

Desktop application for the portal will replace all traditional rapporteurs.

You'll still be aware of the latest happenings. In "Operating band," even if you are not connected to the network, you can receive essential updates about events and their opinions about assigned tasks and their course of implementation.

6.1.10 Mobile Application for the Portal

Install a mobile application on your tablet or smartphone (iOS, Android) and work with the portal: read and comment "Operating band," and manage documents, tasks, and files. Manage CRM client base, grant calendar meetings for friends, and confirm your attendance at new events. Send the pictures straight from your phone to the band. Push messages help you keep up with company activities and keep in contact with colleagues.

6.1.11 Integration with Microsoft, Google, and Apple

The product of Corporate Portal can be integrated with numerous leading software applications. Edit documents directly to the portal through the browser with MS Office, Open Office, and LibreOffice. Synchronize your portal calendars, contacts, and tasks with applications from Microsoft, Google, and Apple.

6.1.12 Security and Reliability

"Bitrix24" offers maximum protection against various threats. The protection of this network is checked by thousands of websites that operate online and serve a wide variety of companies: from popular brands to the largest online stores and information portals. Certified companies audit this product, thus ensuring confidential information security is credible.

7 Conclusion

Organizations have a lot to gain from using social network software. Specifically, defining social structures within an organizational context can improve our understanding of why individuals are behaving and responding to various inputs based on that we can recognize common stresses, through looking at the individual's behavior within a group context, and it can help us define the ability of the individual to perform effectively.

Human resource professionals must comprehend the functions of social network software in order to transfer the method of working and provide value to the field of human resources by evaluating the risks and opportunities they may present to an organization.

References

Butts, C.T.: Social network analysis: a methodological introduction. Asian J. Soc. Psychol., 13 (2008)

Chin, A., Zhang, D.: Mobile Social Networking, p. 2. Springer, New York (2014)

Crompton, D., Sautter, E.: Find a Job Through Social Networking, 2nd edn, p. 22. JIST Publishing, USA (2011)

Daily time spent on social networking. https://www.broadbandsearch.net/blog/average-daily-time-on-social-media. 11/08/2020, 18h20

Elizabeth Wamicha: What is Social Network Analysis? https://study.com/academy/lesson/what-is-social-network-analysis.html. 14/08/2020, 20h10

FatemahAziziRostam: Investigating the advantages and disadvantages of social networks on social media. (2020). https://www.academia.edu/42998165/INVESTIGATING_THE_ADVANTAGES_AND_DISADVANTAGES_OF_SOCIAL_NETWORKS_ON_SOCIAL_MEDIA. 06/08/2020, 17h00

Hatala, J.-P.: Social network analysis in human resource development: a new methodology. Hum. Resour. Dev. Rev. 5(45), 51–52 (2006a)

Hatala, J.-P.: Social network analysis in human resource development: a new methodology. Hum. Resour. Dev. Rev. 5(45), 65–67 (2006b)

Jharotia, A.K.: 'Use of Social Media in Marketing of Library and Information Services in Digital Era', Web seminar "Use of social Networking in Knowledge Sharing", Modern Rohini Education Society, 26th September 2015, New Delhi, India, p 3; https://datareportal.com/reports/digital-2020-global-digital-overview, 09/08/2020, 14h30

Josep Lluis Larriba Pey: Why is social network analysis important? https://www.quora.com/Why-is-social-network-analysis-important. 15/08/2020, 14h00

Keith Miller: 21 Advantages and Disadvantages of Social Networking. https://futureofworking.com/10-advantages-and-disadvantages-of-social-networking/. 11/08/2020, 19h10

Mary Gormandy White: What Types of Social Networks Exist? https://socialnetworking.lovetoknow.com/What_Types_of_Social_Networks_Exist, 06/08/2020, 21h00; https://datareportal.com/reports/digital-2020-global-digital-overview. 08/08/2020, 15h00; https://datareportal.com/reports/digital-2020-global-digital-overview. 08/08/2020, 15h25; https://datareportal.com/reports/digital-2020-global-digital-overview. 08/08/2020, 15h50; https://datareportal.com/reports/digital-2020-global-digital-overview. 08/08/2020, 16h10; https://en.wikipedia.org/wiki/Flickr. 08/08/2020, 18h50; https://www.academia.edu/about. 08/08/2020, 23h50; https://www.researchgate.net/about. 09/08/2020, 00h05; https://www.businessofapps.com/data/linkedin-statistics/. 08/08/2020, 20h55

Menke, S.: The value of Social Network Analysis in HR analytics. (2017). On site https://www.linkedin.com/pulse/value-social-network-analysis-hr-analytics-saskia-menke?fbclid=IwAR3teIM2vdYhs6yVwCCAd-FUZkLNd9CoMk2-9n6y45qMeUMlhgSHaKO5SNU

Mutuku, C.: Advantages and Disadvantages of Using Social Networks in Business, p. 1. Grin Verlag, München, Germany (2017)

RakhiTyagi: “Social Networking: Advantages And Disadvantages”, Web seminar “Use of social Networking in Knowledge Sharing”, Modern Rohini Education Society, 26th September 2015, New Delhi, India, p 11

Vala Ali Rohani, Siew Hock Ow: “On Social Network Web Sites: Definition, Features, Architectures and Analysis Tools”, January 2009, p 43. https://www.researchgate.net/publication/233927861_On_Social_Network_Web_Sites_Definition_Features_Architectures_and_Analysis_Tools. 11/08/2020, 13h00

Romm-Livermore, C., Setzekorn, K.: Social Networking Communities and E-Dating Services: Concepts and E-Dating Services: Concepts and Implications, p. 230. Information Science Reference, New York, USA (2009)

Social media marketing. SEOP, p 7

Steinfield, C., Ellison, N., Lampe, C., Vitak, J.: Online Social Network Sites and the Concept of Social Capital, p. 2. The Internet Turning 40 Conference School of Journalism and Mass Communication Chinese University of Hong Kong (2012)

Torres-Coronas, T., Arias-Oliva, M.: e-Human Resources Management: Managing Knowledge People, pp. 69–70. Idea Group Publishing, London (2005)

Wasserman, S., Faust, K.: Social Network Analysis: Methods and Applications, p. 17. Cambridge University Press, USA (1994)

Toward a New Recursive Model to Measure Influence in Subscription Social Networks: A Case Study Using Twitter

Hemza Loucif and Samir Akhrouf

Abstract This chapter presents a new version of one of the models that we have proposed to measure the influence of web users in social networks like Facebook. The new version enhanced the previous one through the incorporation of two substantial modules, namely, the global impression that the postings of the potential influencer get from his followers and the entropy which manifests the quantity of information carried by those postings. The comparison of our model with the precedent version and the PageRank benchmark has shown the effectiveness of our updates and the importance of incorporating the entropy and the global impression factors in its formulation.

Keywords Influence · Information diffusion · Social graph · Twitter · Entropy · Subscription networks

1 Introduction

Nowadays, social media has become a powerful tool for businesses to reconfigure their marketing strategies. The latest statistics[1] reveal that more than 3.5 billion people (about 45% of the population) are active in social media platforms. An average of 3 h is the amount of time spent by every person on social media per day. 91% of web users access social media platforms through mobile devices, and 71% of them recommend new appreciated items – products or services, for example – after

[1] https://www.oberlo.com/blog/social-media-marketing-statistics

H. Loucif (✉)
Computer Science Department, M'sila University, M'sila, Algeria
e-mail: hemza.loucif@univ-msila.dz

S. Akhrouf
Computer Science and its Applications Laboratory of M'sila (LIAM), M'sila, Algeria
e-mail: samir.akhrouf@univ-msila.dz

S. Sedkaoui et al. (eds.), *International Conference on Managing Business Through Web Analytics*, https://doi.org/10.1007/978-3-031-06971-0_10

to their friends and followers. Such statistics are sufficient for businesses to keep straggling for reserving a place in those digital communities. To help businesses in their mission, social network analysts agreed that the focus should be directed toward the way information circulates in social networks (Loucif et al. 2015; Mengting et al. 2017; Qiang et al. 2019; Khalid 2019). More particularly, businesses need to know how information can be diffused largely and low costly within digital communities using social media platforms. Researchers in this domain have studied the diffusion process and proposed multiple approaches to help marketers in taking the greatest advantage of those platforms. As social network analysis (SNA) researchers, we propose in this chapter an approach to identify the users who can ensure the largest diffusion of information across social networks. Due to the restriction on the length of the chapter, the rest of this chapter is organized as follows. We present the motivation behind the proposition of the new model in Sect. 2, and then we present the formulation of the different parts of the model in Sect. 3, and we reserve the fourth section to show the conducted experiments; finally, we end with a conclusion and a perspective.

2 Motivation

The work presented in this chapter is the compliment of a previous work we have participated with in an international conference on Software Engineering and New Technologies (Loucif et al. 2014). In the first work, we have proposed a new recursive simplistic model that allows us to evaluate the social influence a web user can exert on his community (Samanta et al. 2020; Essaidi et al. 2020; Zheng et al. 2020). The model was very simplistic since it covers just a small structural aspect of the whole social graph. In other words, the model misses the factors which are related to one of the fundamental pillars of social networks, namely, the interactional ones. This is exactly what motivated us to enhance the model in the way that both structural and interactional features will collaborate to gage influence in social networks.

Interested in content diffusion on social networks (Afrasiabi-Rad and Benyoucef 2011), Rad and BenYoucef have conducted a deep experimental analysis work over a real-world database from YouTube. This work has motivated us to concentrate our attention on one of the fundamental types of social communities (i.e., groups) that can be found in social networks. This type corresponds to subscription egocentric networks in which the focus is on the web user who receives subscription links from other web users who are called followers in the social network jargon.

The database on which the authors have worked comprises two kinds of networks, a friendship network on the one side and followers (i.e., subscribers) network on the other side. Data in those networks concern only the comments that are posted on videos by users who have a link (direct or indirect) via a subscription or friendship to the uploader.

Firstly, the authors have focused on the magnitude (i.e., extent), which represents the largest number of hops that can be reached by the content. They found that five-hop networks give significant results in both networks. As a pre-processing phase, all the interactions that were not established among friends/subscribers were filtered out from millions (a total of 16.4 million in friendship vs 44.7 million in subscription networks) of interactions that are crawled in the dataset. The same process was applied to remove all the links between channels that do not maintain a friendship path to the uploader user. Surprisingly, it has been found after the pre-processing stage that the significant fraction of interactions happens between those who do not have a path through subscription.

Secondly, the authors were interested equally in the discovery of the relationship that binds the popularity of the content (i.e., video) and the extent that is expected to be reached in friendship/subscription networks.

The number of views and ratings are the only criteria that were selected to measure the popularity of a given video.

According to the empirical results, the global findings of this work can be outlined as follows:

- The effect on the propagation of people who are not in either a friendship network or a subscription network is higher than that of friends or subscribers.
- A relatively small number of highly active YouTube users are responsible for the wide propagation of content. This fact is manifested empirically by the power law distribution of commenting feature in the friendship network.
- Despite the high connectedness of subscription networks in comparison to friendship ones, subscribers have less impact on the extent of propagation than friends. It is found that the aforementioned fact is the result of the lower personal connection between subscribers than that existing among friends.
- The analysis revealed that subscription networks don't provide significant propagation rates since it is found that major content is diffused at most to two layers of subscribers.
- Videos are propagated at most to three layers (i.e., hopes) of friends, where only a tiny fraction of the videos is moved to the second and third hopes.
- In friendship open networks, the number of connections among strangers (i.e., don't maintain a friendship interaction) accounts for a high percentage of the total connections.
- The popularity of videos is not affected by the propagation of friends, and vice versa. Conversely, the extent of propagation is influenced directly by the popularity of the video in subscription networks, so that the most popular is the highly propagated.

As three substantial findings for businesses who are looking for increasing vastly and low-costly the promotion of their services and products, Rad and BenYoucef have suggested that:

The low propagation rate within friendship and subscription networks suggests that open social networks are not generally well suited for them that need to spread the word in communities. Alternatively, businesses are invited to launch their advertising campaigns within open social communities to make a product/service well known.

Due to the fact since the propagation rate of videos is not affected by their popularity, businesses are advised to ensure the high quality of their messages within subscription networks.

3 The Propagation Process Is Orchestrated Mainly by a Small Subset of Active Subscribers

This work provided the SNA community with a great opportunity for analyzing the behavioral and interactional aspects of friendship and fellowship networks and most importantly their influence on content diffusion.

We started from the second and third findings to elaborate a new model to deal with subscription networks since there are not many literature studies on it.

4 Formulation of the Model

First, we start by defining the formulation of the social network. The formulation is represented in a 4-uplet Net = (*SG*, *T*, *R*, *RT*) whose elements are defined as follows:

- *SG* is the social graph which represents the topology (i.e., structure) of the social network. The graph $SG < N, L >$ has a set of N nodes representing Twittrers and L the set of links representing unidirectional follow relationships between the Twittrers. The direction is so important in such context, where A follows B in Twitter, which doesn't necessary mean the reciprocity of the relation. For this reason, we represent each link starting from a node x (the follower) toward y (the followed) by an ordered subset $l = <x, y>$.

  We define here a function Followers (x) that returns the set of nodes following a given node.
- T is the collection of tweets which have been published by a Twittrer. We define equally a function tweets (x) that returns the set of tweets that have been published by a given node.
- R is the set of reactions that have been added to a tweet. We define reactions_x (t) as the function that returns the set of nodes who have reacted positively with the tweet t that have been published by the node x.
- RT is the set of retweets that have been received by the tweet. We define retweets_x (t) as the set of nodes that have retweeted the tweet t which have been published initially by x.

Based on reactions$_x$ (t) and retweets$_x$ (t), we can define the function impression$_x$ (t) which tells us whether the nodes have appreciated or depreciated the tweet t that have been published by the node x. It is worth mentioning here that Twitter platform doesn't provide any option (dislike button, e.g.) to react negatively against the content carried by a given tweet. This means more formally:

$$\text{impression}_x\ (t) = \begin{cases} 1\ (*) \\ -1, \text{Otherwise} \end{cases}$$

$$(*)\ \text{if}\ \alpha\ \text{reactions}_x\ (t) + \beta\ \text{retweets}_x\ (t) > \theta$$

Here, α and β represent the weights that can be adjusted to be assigned to retweeting or reacting factors to show the importance of one over another. In our case, we chose 0.5 and 0.75 for α and β, respectively.

For θ, it is the parameter that stands for the threshold that we require to be exceeded to confirm the impression that can be obtained by the tweet.

To measure the global impression G – impression$_x$ () for all the tweets that have been posted by a given Twittrer, we proceed simply as follows:

$$G - \text{impression}(x) = \frac{\sum_t^T \text{impression}_x\ (t)}{T}$$

We model the collection of the tweets the node x has published in the form of a matrix M_{r*c} (x) in which the rows represent the tweet, and the columns represent the topics to which the tweets are related.

Concerning the topics, it is very smart to take benefit of the hashtags (#) which are created specially to bring tweets together under the same idea.

The elements m_{ij} in Table 1 denote the probability of membership of the ith tweet to the jth topic. For each topic, we have attributed a set of most popular hashtags that can be grouped semantically under the same lexical field.

To facilitate our work, we change the probabilities m_{ij} as follows:

$$m_{ij} = \begin{cases} 1, \text{if}\ m_{ij} \geq 0.5 \\ 0, \text{Otherwise} \end{cases}$$

Table 1 Topical distribution of the tweets

	Topic 01	Topic 02	Topic 03	...
Tweet 01	m_{11}	m_{12}	m_{13}	...
Tweet 02	m_{21}	m_{22}	m_{23}	...
...	...	...	...	...

The reason behind this step is to facilitate the calculation of the entropy that will be defined in the next paragraph.

Based on the topical distribution of the tweets summed up in Table 1 and the approach, we define another function to assess the value (i.e., importance) of the content which is carried by the tweets a certain Twittrer has published. This value will be considered as the criterion through which the followers of the Twittrer (i.e., the publisher of the tweets) would decide to retweet (i.e., forward) or not his tweets. Mathematically, this function is simply the direct application of Shannon entropy such that the inputs are obtained from the topical distribution matrix we have seen in the precedent paragraph.

The entropy of Shannon accepts as a source of information a discrete random variable with n symbols. In our case, the random variable X corresponds to the occurrence of each topic in the tweets of the given Twittrer, since each topic t_i has a probability p_i to occur.

The entropy E is calculated simply as follows:

$$E(X) = \sum\nolimits_{j=1}^{N} p_j \log_b \left(\frac{1}{p_j}\right)$$

where p_i is calculated as follows:

$$p_j = \frac{1}{M}\sum\nolimits_{i=1}^{M} m_{ij}$$

For simplicity, we make b equals to 2 since we have restricted ourselves to just four topics, namely, #economy, #sport, #gaming, and #politics.

After the presentation of the new components, we present briefly the first version of the model as follows:

$$\text{Frank}(v) = \frac{\sum_{u \in F(v)} [\text{Int}_{(u,v)}.\text{Att}_{u \to v}.\text{Frank}(u)]}{|\text{Followers}(v)|}$$

- Followers(v) stands for the set of v's followers, and u is one of them. |Followers (v)| is its size. In the structural SNA, this parameter can be substituted by the in-degree of v.
- $\text{Int}_{(u,v)}$ stands for to the function we dedicate to measure the "degree of interestingness" received by the content published by v from his friend u. For doing so, we take the proportion of communities (i.e., groups) to which both u and v are affiliating as an approximate measure. Formally:

$$\text{Int}_{(u,v)} = {|C_u \cap C_v|}\Big/{|C_u \cup C_v|}$$

where C_u and C_v are the sets of circles u and v are members in, respectively.

- $\text{Att}_{u \to v}$ denotes the index that reflects the degree of attention that v is receiving from his friend u. This parameter can be formulated simply as:

$$\text{Att}_{u\to v} = {}^{1}\!\Big/_{|\text{Followers}(v)|}$$

The new version called TFrank of the recursive model Frank can be formulated as follows:

$$\text{TFrank}(v) = (1 - G - \text{impression}(x)) * \frac{\sum_{u\in F(v)} \left[\text{Int}_{(u,v)}.\text{Att}_{u\to v}.\text{Frank}(u)\right]}{|\text{Followers}(v)|} + E(X)$$

5 Experiments and Results

To check its accuracy, the model has been tested using a dataset provided by the famous Stanford Large Network Dataset Collection.[2] The statistics of the dataset are summarized in Table 2.

As shown by the statistics, we can remark the hugeness of both amounts of nodes and edges. Therefore, we were verily in need of restricting ourselves to just a small subset of 250 nodes with 1307 edges to facilitate carrying out our experiments in a reasonable period of time. In the experiments, we have conducted a

Table 2 Statistics of the database

Dataset statistics	
Nodes	81,306
Edges	1,768,149
Nodes in largest WCC	81,306 (1.000)
Edges in largest WCC	1,768,149 (1.000)
Nodes in largest SCC	68,413 (0.841)
Edges in largest SCC	1,685,163 (0.953)
Average clustering coefficient	0.5653
Number of triangles	13,082,506
Fraction of closed triangles	0.06415
Diameter (longest shortest path)	7
90 percentile effective diameter	4.5

[2]https://snap.stanford.edu/data/

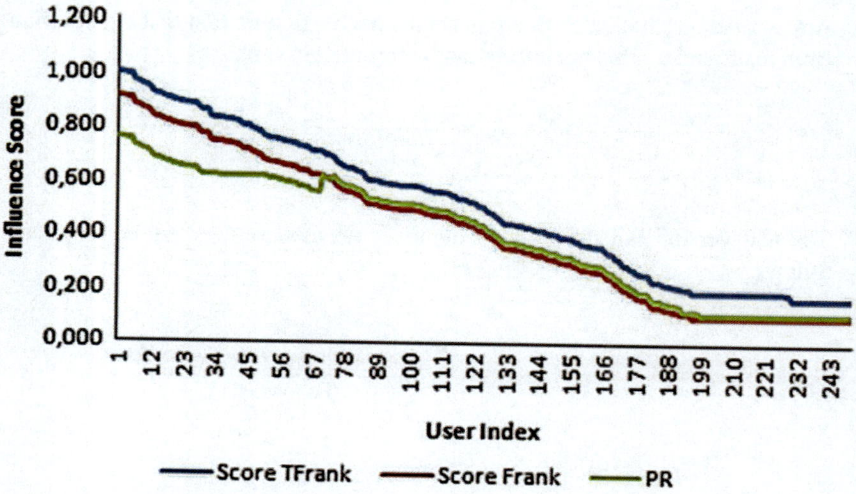

Fig. 1 TFrank vs Frank and PR scores

comparison between TFrank on the one side and Frank and PageRank[3] on the other side (Riquelme and González-Cantergiani 2016; Zambuk et al. 2019) (Fig. 1).

The chart shows the scores of each user which are measured by the three algorithms. At first glance, we remark that the slopes of the three curves are almost equal. This means that the three approaches regardless of the values themselves of the scores gave the same ranking for almost all the participated nodes. This result enables us to witness the accuracy of both TFrank and Frank in the measurement of influence which can be obtained with high precision using the PR benchmark.

However, to reveal how TFrank can give the most realistic influence scores, we have measured the correlation between the scores provided by TFrank and the sub-scores issued from the three principal components, namely:

- In-degree (i.e., number of followers).
- G-impression.
- Entropy.

It should be mentioned that measuring correlation allows finding out how strongly are related two numerical items to each other. The closest the correlation value is to +1, the strongest the relation between the variables.

Figure 2 reveals the clear correlation between TFrank scores and the sub-scores issued from entropy.

Similarly, we notice the high degree of correlation between TFrank scores and G-impression sub-scores (Figs. 3 and 4).

On the contrary, we remark the poor correlation between TFrank and in-degree.

[3] https://en.wikipedia.org/wiki/PageRank

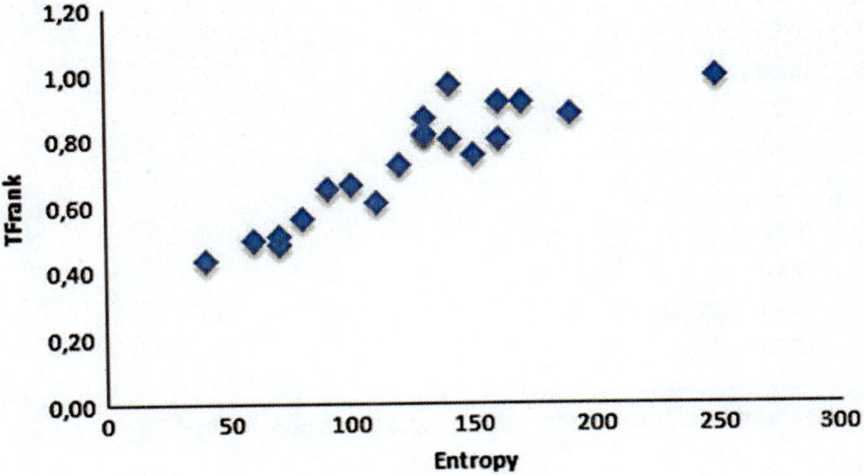

Fig. 2 Correlation between TFrank scores and entropy

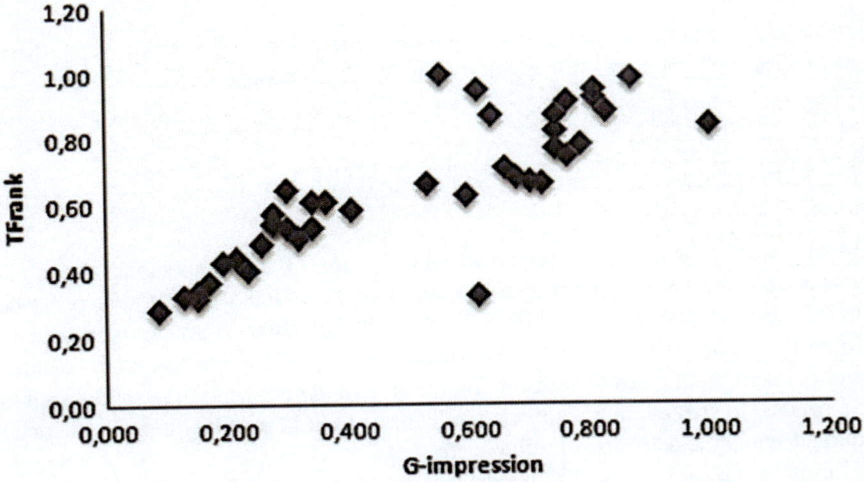

Fig. 3 Correlation between TFrank and G-impression

From the aforementioned observations, we can find out the strong points of TFrank in comparison to its predecessor. Firstly, the results confirm that it is the quantity of information carried by the postings that cause the largest broadcast of information across the social network. Secondly, the interaction and the reactions of followers contribute positively to the propagation of information, particularly through the forwarding process. Thirdly, the in-degree (i.e., number of followers) as a structural feature is not a good index to judge the popularity of the content which

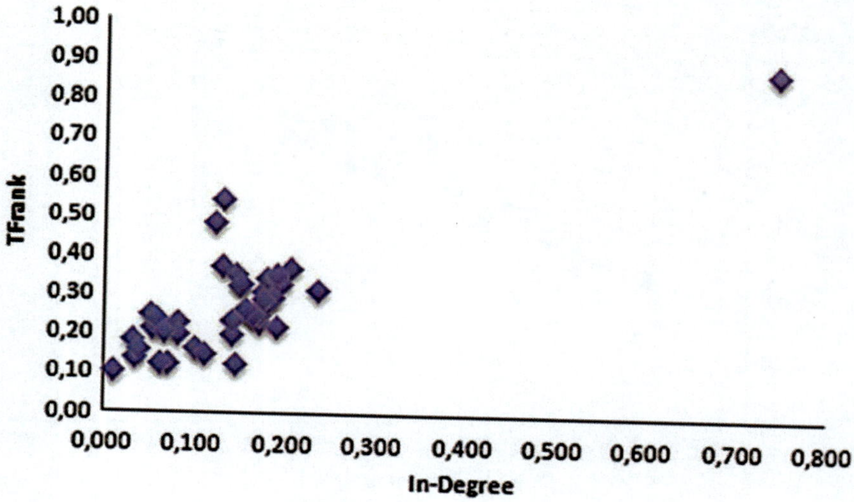

Fig. 4 Correlation between TFrank scores and in-degree

is supposed to be broadcasted to a large extent due to the popularity of its publisher as previously demonstrated in (Loucif et al. 2015; Meeyoung Cha et al. 2020).

6 Conclusion

In this work, we have contributed to tackling one of the challenging issues researchers in the social network analysis community are facing today. Our experience in this domain allowed us to better understand the diffusion process and then to focus strictly on the factors that maximize it. The integration of interactional features and the incorporation of the entropy besides the in-degree within the same formula allowed us to create a more realistic model to be used in the measurement of social influence and the detection of influencers.

Finally, the evaluation of influence requires the collaboration of both structural and interactional aspects of the social network as it has been proven in previous works (Loucif et al. 2015; Loucif et al. 2014).

As future work, we have started in the elaboration of a new model which makes use of one of the powerful tools of machine learning, especially the famous deep learning.

References

Afrasiabi-Rad, A., Benyoucef, M.: Towards detecting influential users in social networks. In: Revised Selected Papers of the 5th International Conference on E-Technologies: Transformation in a Connected World, Les Diablerets, Switzerland, pp. 227–240 (2011)

Essaidi, A., Zaidouni, D., Bellafkih, M.: New method to measure the influence of Twitter users. In: 2020 Fourth International Conference On Intelligent Computing in Data Sciences (ICDS), pp. 1–5 (2020). https://doi.org/10.1109/ICDS50568.2020.9268726

Khalid, H.: Systematic literature review on social network analysis. In: 2019 International Conference on Innovative Computing (ICIC), pp. 1–7 (2019). https://doi.org/10.1109/ICIC48496.2019.8966673

Loucif, H., Boubetra, A., Akrouf, S.: New recursive model for ranking web users in Facebook based on their social influence. In: Proceedings of the 3rd International Conference on Software Engineering and New Technologies ICSENT – 2014, Hammamet, Tunisia, pp. 20–22 (2014)

Loucif, H., Boubetra, A., Akrouf, S.: A simplistic model for identifying prominent web users in directed multiplex social networks: a case study using Twitter networks. New Rev Hypermed Multimed J. **22**, 4 (2015)

Meeyoung Cha, C., Hamed, H., Fabricio, B., Gummadi, P.: Measuring user influence in Twitter: the million follower fallacy. In: Fourth International AAAI Conference on Weblogs and Social Media (2020)

L. Mengting, W. Xiang, G. Kai, and Z. Shanshan, "A Survey on Information Diffusion in Online Social Networks: Models and Methods" J. Info., https://doi.org/10.3390/info8040118. 2017. Published: 29 September 2017

Qiang, Z., Pasiliao, E.L., Zheng, Q.P.: Model-based learning of information diffusion in social media networks. Appl. Netw. Sci. **4**, 111 (2019). https://doi.org/10.1007/s41109-019-0215-3

Riquelme, F., González-Cantergiani, P.: Measuring user influence on twitter: a survey. Inf. Process. Manag. (2016). https://doi.org/10.1016/j.ipm.2016.04.003

Samanta, S., Dubey, V.K., Sarkar, B.: Measure of influences in social networks. Appl. Soft Comp. J. (2020). https://doi.org/10.1016/j.asoc.2020.106858

Zambuk, F.U., Gital, A.Y.u., Boukary, S., Jauro, F., Chiroma, H.: Evaluation of Iterative Pagerank Algorithm for Web Page Ranking. In: 2019 4th international conference on electrical, electronics, communication, Computer Technologies and Optimization Techniques (ICEECCOT), pp. 365–370 (2019). https://doi.org/10.1109/ICEECCOT46775.2019.9114728

Zheng, C., Zhang, Q., Long, G., Zhang, C., Young, S.D., Wang, W.: Measuring time-sensitive and topic-specific influence in social networks with LSTM and self-attention. IEEE Access. **8**, 82481–82492 (2020). https://doi.org/10.1109/ACCESS.2020.2991683

Social Influence Analysis in Online Social Networks for Viral Marketing: A Survey

Halima Baabcha, Meriem Laifa, and Samir Akhrouf

Abstract One of the most exciting developments of the last few years has been the rise of online social networks. The richness of this network's content provides unprecedented opportunities for data analytics, which can be taken advantage of. One of the most important areas of social network analysis is the study of social influence. It can be used in a variety of ways, from viral marketing to advertising. In addition to identifying influential nodes in a social network, the modeling of influence diffusion and influence maximization in social networks is an important challenge in this area. There has been a lot of research done on the influence of online social networks, particularly in viral marketing contexts, for various applications. Methods for influence modeling, maximization, and identifying influential nodes are discussed in this chapter. Using cutting-edge research on viral marketing's impact on social influence, we hope to serve as a resource for aspiring researchers.

Keywords Online social networks · Influence analysis · Viral marketing · Influence maximization · Influential spreaders

1 Introduction

The recent rapid increasing popularity of online social networks (OSN) and new communication technologies availability (smartphone, tablet, etc.) provided the world population with a great facility to share and to communicate with others throughout the whole world and to share and exchange information, opinions, products, ideas, and services. The rich content of OSN that can be in diverse

H. Baabcha (✉) · M. Laifa
Department of Computer Science, University Mohammed El Bachir El Ibrahimi, Bordj Bou Arreridj, Algeria
e-mail: halima.baabcha@univ-bba.dz; meriem.laifa@univ-bba.dz

S. Akhrouf
Department of Computer Science, University Mohamed Boudiaf, M'sila, Algeria
e-mail: samir.akhrouf@univ-msila.dz

S. Sedkaoui et al. (eds.), *International Conference on Managing Business Through Web Analytics*, https://doi.org/10.1007/978-3-031-06971-0_11

formats (text, image, video, audio, etc.) has been attracting researchers to study and analyze large-scale social structures and users' behaviors in order to – among other purposes – understand the flow of information through social networks. A lot of real-world applications like public opinion monitoring, recommender systems, and political campaigns often make use of OSNs for influence diffusion (Fu et al. 2016; Can and Alatas 2019; Saxena and Saxena 2020). Many existing models for influence diffusion have been proposed for various applications. In this chapter, we look into influence diffusion in the setting of viral marketing.

In essence, viral marketing is an efficient solution to advertisement for commercial companies through OSN where companies try to promote their products and services through word-of-mouth propagation among friends or followers. One of the fundamental objective of viral marketing is to find a set of users with the maximum influence in the network where the output is called K-seed set users with K as the optimal number of users chosen for influencing other users in the network.

The goals of this chapter are to highlight some of the most used and most recent work that has been done in social influence diffusion and influence maximization model for viral marketing and to call attention to the topic of deep learning for social influence. To the best of our knowledge, this work is a literature review of influence analysis for viral marketing in online social networks. Firstly, we give a better understanding of the preliminary knowledge concerning social influence analysis, and we illustrate the categories of relevant research works on influence analysis in the context of viral marketing. After that, we categorize and compare a number of relevant research works on influence maximization algorithms in social networks and the identification of influential spreaders.

In the following section, we describe the main concepts that will be addressed throughout the chapter, namely, online social networks, social network analysis, social influence analysis, viral marketing, and the definition problem of influence maximization. Section 3 presents related work to diffusion influence modeling, identification of influential spreaders, and influence maximization. Section 4 exhibits the current methodology using deep learning for influence spreading modeling. Finally, we conclude the chapter in Sect. 5.

2 Background

In this section, we begin with an overview of the basic terminology.

2.1 *Online Social Networks*

OSNs can be defined within the context of systems, but in general, they can be defined as a network of interactions or relationships, where nodes consist of actors or persons and the links consist of the relationships or the interactions between the actors through the network. According to (Kemi 2016), an OSN alternatively referred to as a virtual community or profile site, a social network is a website on the Internet that brings people together in a central location to talk, share ideas and interests, or make new friends. With the emergence of the World Wide Web (WWW), OSNs have dramatically expanded in popularity around the world. For further details about online social networks and their features, we recommend readers to check (Fu et al. 2016; Boyd and Ellison 2007).

2.2 Social Network Analysis

Social network analysis (SNA) or social network mining is the research of social relations between nodes or people. The study of social network mining technologies focuses on the level of individuals, groups, organizations, and whole networks. It is the drawing and determining associations and drifts among users and other interlinked information entities using networks and graph theory (Otte and Rousseau 2002). Many research applications benefit from SNA techniques such as community or group detection (Cai et al. 2016; Dabaghi-Zarandi and Rafsanjani 2019), expert finding (Yuan et al. 2020), link prediction (Daud et al. 2020), recommender systems (Pourhojjati-Sabet and Rabiee 2020), predicting trust and distrust among individuals (Towhidi et al. 2020; Girdhar et al. 2019), influence propagation (Ortiz-Gaona et al. 2020; Abd Al-Azim et al. 2020; Singh et al. 2019; Gulati and Eirinaki 2018), etc.

2.3 Social Influence Analysis

According to (Sun and Tang 2011), influence is usually reflected in changes in social action patterns (i.e., user behavior) in a social network. Typically, it refers to the phenomenon that an individual's emotions, opinions, or behaviors are affected by others (Qiu et al. 2018). This means that people are influenced by their social circle and tend to imitate the actions of those in their immediate vicinity. As a result of its wide range of real-world applications, social influence analysis has received considerable attention in the past (Peng et al. 2016) such as domain expert finding (Al-Taie et al. 2018), personal recommendation (Cheng et al. 2016), emotion prediction (Qiyao et al. 2016), and viral marketing (Talukder et al. 2017; Bhattacharya et al. 2019; Menta and Singh 2017) for which it became an important strategy.

2.4 *Viral Marketing*

Viral marketing (VM) is one of the various real-world applications of social influence analysis. According to (Wang and Street 2018), VM is a process of influence diffusion over social networks. It is a relatively recent solution to advertisement within online social networks. It has been applied to business-to-consumer transactions.

Using a social network to spread the word about a product or service is a form of viral marketing. In other words, it employs customers in a market to promote a product (McKay et al. 2019). For example, if a company has a certain number of new products, they could hand them out to a customer, and then the influence model maximization can predict optimal objective to get these products in order to spread the product's influence over a specific network. There is a dearth of new diffusion methods in the literature, particularly for dynamic and massive networks. Additionally, it provides information on the various mining techniques that can be used for viral marketing.

On the other hand, the goal of viral marketing is to minimize marketing cost while maximizing the profit. The main idea of viral marketing is to find a set of customers for giving free samples within the budget B to maximize the expected total sales of the product, in other words use the K-seed set users for influencing other users in the network. Because the majority of the promotional work is done by customers, this type of "word-of-mouth" advertising can be far more cost-effective than more traditional ones. Friends' recommendations are more trustworthy than those made by a company selling the product (Richardson and Domingos 2002). Figure 1 shows the viral marketing process.

2.5 *Influence Maximization Problem*

A social network is depicted as a graph $G\,(V, E)$ with a set of nodes V that represents individuals and a set of edges E that represents the relationship shared among the nodes in the graph. The influence maximization problem takes as input a graph $G\,(V, E)$. The goal of this problem is to identify a subset of users in graph G who have the greatest amount of influence. An initial influence on this problem should yield the maximum number of nodes in the graph G that are influenced by a K-sized seed set, which is the solution to this problem's problem (Du et al. 2019). Figure 2 shows the input and output for influence maximization problem. The important objective of the problem of influence maximization is to find a set of users with that maximum influence in a graph.

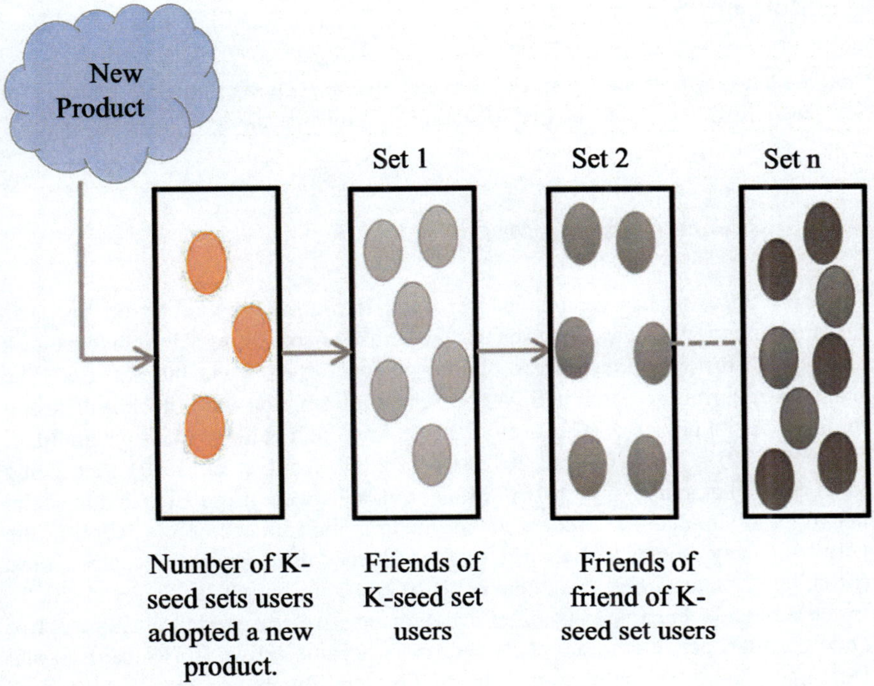

Fig. 1 Viral marketing via social network process

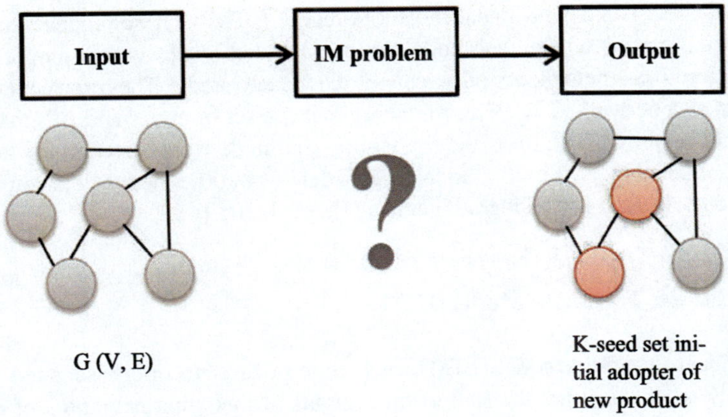

Fig. 2 Input and output of influence maximization problem

3 State of the Art

In this section, we review the most important and most cited existing approaches for influence diffusion, influence maximization, and identification of influential users.

3.1 *Influence Diffusion Model*

Influence diffusion has become an important technique for viral marketing. The Oxford Dictionary defines diffusion as "the spread of something." In social network analysis, diffusion is the process of information diffusion via the network. The majority of current research in SNA focuses on information and influence diffusion in online social networks (Guille et al. 2013; Arnaboldi et al. 2014; Xu et al. 2014; Sun et al. 2019; Dhamal et al. 2016; Gaeta 2018; Kong et al. 2020). According to (AlSuwaidan and Ykhlef 2016), diffusion models were originally used in social networks to simulate the process of information and influence propagation in the network. Many models and algorithms for influence diffusion have been proposed (More and Lingam 2019; Toalombo et al. 2020; Li and Liu 2019; Pan et al. 2020). In these models, each node is either active or inactive over iterations. An inactive node becomes active as more of its neighbors became active. In (Richardson and Domingos 2002), the authors provide the first algorithmic treatment to deal with the influence propagation problem. They built probabilistic models and used these models to choose the best viral marketing plan. Then the authors in (Kempe et al. 2003) studied influence propagation by focusing on the modeling influence by two fundamental stochastic influence cascade graph-based models, named independent cascade model (ICM) and linear threshold model (LTM). These models are based on directed graphs where each node can be activated or not with a monotonicity assumption (i.e., activated nodes cannot be deactivated). They formulated the problem as a network $G = (V, E, p)$, where V is the set of nodes and E is the set of edges between nodes and p is the probability that node v can successfully activate node u, denoted by $p(u,v)$. If a node accepts data from other nodes, it is considered active; otherwise, it is considered inactive (Du et al. 2019).

3.1.1 Linear Threshold Model (LTM)

In the linear threshold model (LTM), each edge or link $e(u,v)$ is associated with a weight $W(u, v)$, such that the sum of the weights of incoming neighbors of node v is less than or equal to 1 and each node v is also associated with a threshold θ_v. The linear threshold model starts with some active nodes with all other nodes being inactive and a random choice of thresholds θ.

The LTM samples the value of v of each user v uniformly at random probability from [0,1]. In step 0, it sets the status of nodes in S as active and others as inactive.

Then, it updates the status of each user iteratively. In step *t*, all nodes that were active in step *t-1* remain active, and any user *v* that was inactive in step *t-1* switches to active. The influence spread of seed set *S* under the LT model (i.e., $\sigma(S)$) is the expected number of activated nodes when *S* is initially activated.

3.1.2 Independent Cascade Model (ICM)

In the independent cascade model (ICM), a probability *p(u,v)* is associated with each edge *e(u,v)*, whereas *u* and *v* are two nodes in the graph. *p(u,v)* is the probability of the ability that *u* succeeds in activating *v*. In this model, a node *v* is activated by each of its incoming neighbors independently by introducing an influence probability *p(u,v)* to each edge *e(u,v)*. This model's diffusion instance unfolds in discrete steps according to influence probabilities and seed sets S at time step 0. Each active node *u* at step *t* will activate each of its outgoing neighbor *v* that is inactive in step *t-1* with probability *p(u,v)*. The activation process can be considered as flipping a coin with head probability *p(u,v)*: if the result is heads, then *v* is activated; otherwise, *v* stays inactive. When there are no more nodes that can be activated, the diffusion instance ends. The expected number of activated nodes when S is used as the initial active node set and the above stochastic activation process is applied is the influence spread of seed set S under the ICM.

3.1.3 Epidemic Model

An epidemic model is a perfect tool for a simplified description of the diffusion strategy that contagious diseases follow in a population. As an epidemic spreads from an infected individual to another healthy one (i.e., non-infected before), the information can also spread from one individual to another through the same network that interconnects them. Epidemic models assume the existence of an implicit network between individuals (i.e., no explicit connections) and assume that exposure to infection (information being diffused) is enough to become infected (informed) and potentially transmit the infection to someone else. The underlying principles of those techniques are the basis of the models used in marketing for the prediction of new product adoption in the communities (Loucif 2016). One of the most popular model SIR or the model of Kermack and McKendrick (Cano 2020) is a mathematical approach created particularly for studying the plague disease that broke out in Bombay. This mathematical model is built upon a set of hypotheses, namely:

(a) In the population, all individuals are sensitive equally to the infection.
(b) The infection leads either to death or to a permanent immunity.
(c) When healthy and infected individuals are living together, there will be a number of healthy individuals who will become infected.

In this model, the individuals can be found in three different states:

Suspected (S) An individual is said susceptible, which means that he is very capable to be infected with the disease. Generally, infections can originate from outside of the population in which the disease spreads (e.g., by genetic mutation, contact with an animal, etc.). Denote *S (t)* the number of individuals who may be infected with the disease at time *t* and *Sf (t)* the current fraction of the population that is susceptible.

$$Sf\ (t) = S\ (t)/N \tag{1}$$

Infected (I) After an individual is affected by the disease, he becomes infectious, i.e., has the ability to infect other susceptible individuals in the population. Let *I (t)* be the number of infected individuals at time *t*. *If (t)* refers accordingly to the infected fraction of the population.

$$If\ (t) = I(t)/N \tag{2}$$

Recovered (R) It refers to the individuals who are either cured of the disease and acquired a full or partial immunity against the infection (can no longer be infected) or removed after being killed by the infection. Let *Rf(t)* be the fraction of the healed (or withdrawn) population where *R(t)* refers to its size.

$$Rf\ (t) = R(t)/N \tag{3}$$

The diffusion of the disease within the population is dynamic: the fractions of susceptible, infectious, and healed individuals evolve over time with respect to the contacts through which the disease passes from infected individuals to healthy ones.

It is worth mentioning that at every moment (Zafarani et al. 2014),

$$1 = Sf\ (t) + If(t) + Rf(t) \tag{4}$$

3.2 *Identification of Influential Spreaders*

A challenging issue in viral marketing is effectively identifying a set of influential users. By sending the advertising messages to this set, one can reach out to the largest area of the network. It is now much easier to identify the network's most influential spreaders (Bhat et al. 2020). In this part, we have selected the most recent work in this field.

The authors in (Okamoto et al. 2008) combine existing methods on calculating exact values and approximate values of closeness centrality and presented a new algorithm to rank the top-k nodes in the network with the highest closeness centrality.

In (Bae and Kim 2014), in an effort to better understand the spread of a node's influence in a network, the researchers developed a new measure called coreness centrality. For their experiment, they used unweighted undirected graph for both real and artificial networks. Their approach is based on the idea that a powerful spreader has more connections to the nodes that reside in the core of network. The K-shell indices of nodes neighbors could be good indicators of its spreading ability. To evaluate their proposed measure, they applied the (SIR) model for investigating an epidemic spreading process. They evaluated the performance of the ranking measures in 12 real networks with different sizes as shown in Table 1.

The study in (Basaras et al. 2013) introduced a new centrality measure; it is a combination of coreness and betweenness centrality. To evaluate their technique's accuracy, they compared it to K-shell decomposition and a baseline measure based solely on the node degree on a large number of complex networks. They used the susceptible-infected-recovered model for an infection originating from both a single spreader and multiple spreaders to investigate the spreading process.

The authors in (Zeng and Zhang 2013) proposed a mixed degree decomposition (MDD) procedure in which both the residual degree and the exhausted degree are considered. By simulating the epidemic spreading process on real networks (Dolphins, Jazz, NetSci, Email, HEP, PGP, TAP, Y2H, Power, Internet, *E. coli*, *C. elegans*, AstroPh), they used the K-shell to generate the influence.

The authors in (Liu et al. 2018) proposed a local h-index centrality (LH-index) method for identifying and ranking the top influential spreaders in networks by calculating the h-indices of the node. The new proposed local h-index (LH-index) method simultaneously considers two factors: the h-index value of the node itself and the h-index values of its neighbors. On the one hand, the h-index of one node indicates the direct influences exerted by its nearest influential neighbors. On the other hand, the h-index values of its neighbors indicate the two-hop indirect influences exerted by further influential neighbors. The performance of their method showed its superiority in both real-world and simulated networks. They adopted the SIR model to evaluate the real spreading ability of the ranking nodes.

The work presented in (Belfin and Bródka 2018) uses multiple measures of centrality to look for overlapping communities and combine them to find a suitable superior seed set. They used degree, eigenvector centrality, and clustering coefficient. The basic idea of the strategy is to find out a fraction of superior nodes of the input network, called superior seed set, around which local communities can be computed.

Finally, the authors of (Bhat et al. 2020) presented the Improved Hybrid Rank algorithm, which combines two centralities, namely, the extended neighborhood coreness centrality and the h-index centrality. For the simulation of their proposed method, they used the SIR (susceptible-infected-recovered) model for both undirected and directed real-world networks. They have tested their algorithm based on various performance matrices like Kendall-Tau's correlation coefficient, spreader's location diversity, and infected scale.

The comparison of identification of influential spreader models is listed in Table 1.

Table 1 The comparison of performance for related identification of influential spreader models

References	Contribution	Used measures	Diffusion model	Graph	Dataset
Okamoto et al. (2008)	New algorithms to rank the top-*k* vertices with the highest closeness centrality for	Combined existing methods on calculating exact values and approximate values of closeness centrality	/	Directed	/
Kitsak et al. (2010)	K-shell centrality	/	SIS/SIR	Directed	LiveJournal.com community The network of email contacts in the Computer Science Department of University College London The contact network of inpatients (CNI) collected from hospitals in Sweden The network of actors who have costarred in movies labelled by imdb.com as adult

Bae and Kim (2014)	A novel measure coreness centrality	K-shell indices	SIR	Unweighted/undirected	Zachary's karate club Lusseau's bottlenose dolphins Jazz musician's network *C. elegans* metabolic network Coauthorship network of scientists (NetScience) Email network of URV Communication network of blogs Western States Power Grid PGP network Collaboration networks of astrophysics (CA-AstroPh) and condensed matter physics (CA-CondMat) Enron email network
Basaras et al. (2013)	A new centrality measure U-power community index (U-PCI)	Coreness and betweenness	SIR	Undirected	CA-CondMat CA-AstroPh From the Stanford Network Analysis Project

(continued)

Table 1 (continued)

References	Contribution	Used measures	Diffusion model	Graph	Dataset
Zeng and Zhang (2013)	Mixed degree decomposition (MDD)	Amelioration of K-shell	SIR	Undirected/directed	Dolphins (friendship) Jazz (musical collaboration) Netsci (collaboration network of network scientists) Email (communication) HEP (collaboration network of high-energy physicists) PGP (an encrypted communication network) AstroPh (collaboration network of astrophysics scientists) CondMat (collaboration network of condensed matter scientists) Nonsocial networks are word (adjacency relation in English text) *E. coli* (metabolic) *C. elegans* (neural) TAP (yeast protein-protein binding network generated by tandem affinity purification experiments) Y2H (yeast protein-protein binding network generated using yeast two hybridization) Power (connections between power stations) Internet (router level)

Ma et al. (2016)	Gravity centrality index	/	SIR	Undirected/unweighted	Facebook (Slavko Žitnik's friendship network in Facebook) NctSci (collaboration network of network scientists) Email (email network of Universitat Rovira i Virgili, URV) TAP (yeast protein-protein binding network generated by tandem affinity purification experiments) Y2H (yeast protein-protein binding network generated using yeast two hybridization) Blogs (the communication relationships between owners of blogs on the MSN (Windows Live) Spaces website) Router (the router-level topology of the Internet) HEP (collaboration network of high-energy physicists) PGP (an encrypted communication network)

(continued)

Table 1 (continued)

References	Contribution	Used measures	Diffusion model	Graph	Dataset
Ahajjam and Badir (2018)	Hybrid rank algorithm	New hybrid centrality measure	SIR	Undirected/unweighted	CondMat (condense matter physics) DBLP bibliography WikiVote The Epinions
Liu et al. (2018)	H-index centrality and extended H-index	H-indices	SIR	Undirected/unweighted	USAir: The network of the US air transportation Blogs: The network of the communication Relationships between owners of blogs on the MSN (Windows Live) Spaces website Email: The network of email interchanges between members of the Universitat Rovira i Virgili Power grid: The network of the power grid of the Western States in the USA
Belfin and Bródka (2018)	Overlapping-based spreader selection	Degree, clustering coefficient, eigenvector centrality	SIR	Undirected	Zachary's karate club Dolphins social network American College Football

Richardson and Domingos (2002)	Influence deep learning (IDL) model	/	/	Neural network learning	Sina Weibo Epinions WikiVote NetHEPT
Bhat et al. (2020)	Improved Hybrid Rank algorithm	Extended neighborhood coreness centrality H-index centrality	/	Directed/undirected	CondMat Brightkite PGP network HyperPhysics P2P Gnutella Football

3.3 Influence Maximization

In this section, we present influence maximization-related research in viral marketing, and we review the important available research progress.

Domingos and Richardson (2001) were among the first to model customers' network value, they used markov random field for modeling the influence between customers such as nodes representing the customers. After that, they extended their previous techniques, achieving a large reduction in computational cost, and applied them to data from a knowledge-sharing site. They founded optimal marketing plan, and they used continuously valued marketing actions and reduce computational cost (Richardson and Domingos 2002). The most cited papers on the matter, maximizing the spread of influence through a social network written by (Kempe et al. 2003), in which the random degree-based and distance centrality algorithms are used as baselines, which led to the development of the greedy algorithm for influence maximization. More generally, they developed an algorithm for selecting the optimal seed set S from nodes in the graph. They proved that the optimization problem is NP-hard under LTM and ICM, and they presented a greedy algorithm that guarantees that the influence spread is within $(1\text{–}1/e - \varepsilon)$ of the optimal influence spread, where e is the base of natural logarithm and ε depends on the accuracy of their Monte Carlo estimate of the influence spread given a seed set. A series of more efficient studies have been done, because the greedy algorithm is infeasible even for medium-sized networks of tens of thousands of nodes and edges (Wang and Street 2018).

The work presented in (Leskovec et al. 2007) exploited submodularity to create an algorithm that can handle large-scale problems, achieve near-optimal placements, and be 700 times faster than a simple greedy algorithm. They proposed the Cost-Effective Lazy Forward (CELF) algorithm based on a "lazy forward" optimization. The obtained solutions are guaranteed to achieve at least a fraction of $1/2(1\text{–}1/e)$ of the optimal solution. They evaluated their algorithm on several large-scale real-world problems, including a model of a water distribution network and real blog data.

Authors in (Goyal et al. 2011) introduced an algorithm called CELF++ that further optimizes CELF by exploiting submodularity property. By avoiding redundant re-calculations of CELF's marginal gains, the algorithm CELF++ has an advantage.

The authors in (Chen et al. 2009) studied the efficient influence maximization in social networks from two complementary directions. One is to improve the original greedy algorithm in (Kempe et al. 2003) and its improvement in (Leskovec et al. 2007) to further reduce running time for the greedy algorithm. After that, they proposed a new degree discount heuristic derived from the independent cascade model that improves influence. They also proposed an algorithm called new greedy algorithm for the influence maximization.

The study in (Chen et al. 2010a) proposed a new heuristic algorithm that is easily scalable to millions of nodes and edges in their experiments. An easy-to-use tunable parameter allows users to balance the running time and the spread of the algorithm's influence in the general ICM. For the experiments, they used four real-world network and a synthetic dataset (NetHEPT, DBLP, Epinions, and Amazon).

The work in (Chen et al. 2010b) shows that to compute the expected influence spread for a given set is P-hard. But it can be expressed as a submodular monotone function of S, which can be used to guarantee the results using a simple greedy algorithm. As an added bonus, we now have the LDAG algorithm, the first-ever scalable heuristic algorithm designed specifically for LT model influence maximization. They firstly disclosed that the computation of influence in directed acyclic graphs (DAGs) can be done in linear time. Based on that, they created a local DAG for every node of the network and restricted the influence of the node in this local area. Gap-filling seed selection was used to update the nodes' incremental influence spread after the DAGs were built, along with an accelerated solution.

The authors in (Barbieri and Bonchi 2014) modeled the viral marketing process of product adoption based on social influence and the feature of the products. They proposed feature-aware propagation model F-TM, and they defined the influence maximization with viral product design (MAXINF-VPD) problem and the study of its properties under F-TM propagation model, using two real-world semantically rich datasets from the domain of social music consumption (Last.fm) and social movie consumption (Flixster).

Squillero and Burelli (2016) explored the problem of influence maximization using a genetic algorithm (GA), which makes use of simple genetic operators commonly found in discrete optimization. They evaluated the genetic algorithm on two large, real-world network datasets from the Stanford Network Analysis Platform (SNAP) repository.

The authors in (Wang and Street 2018) proposed a model in which they quantified influence and tracked its diffusion and aggregation. (MAT) multiple-path asynchronous threshold, for viral marketing on social network. The MAT model captures not only direct influence but also indirect influence passed along messengers and they developed an efficient heuristic IV-greedy to tackle the influence maximization problem. The experiments of the MAT and the IV-greedy were conducted on four real-life networks, and they illustrated an important performance in terms of influence spread and time efficiency.

Saxena and Saxena (2020) proposed an influence maximization model by combining a node connection and its actual past activity pattern. Firstly, they proposed a diffusion model, namely, HAC-Rank algorithm, for the selection of initial adopters. Furthermore, they proposed a new Hurst-based influence maximization for studying the influence spread of seed nodes, wherein the activation of a node depends upon its connections and the self-similarity trend shown by its past activity. The performance of the HAC-Rank has been evaluated under IC, and the HBIM diffusion model achieved an average influence spread of 20.3%. Under the proposed HBIM model, HAC-Rank achieved 49.8% average influence spread in comparison to other state-of-the-art algorithms.

The comparison of maximization influence models is listed in Table 2.

Table 2 The comparison of performance for related influence maximization models

References	Contribution	Diffusion model	Graph	Dataset
Domingos and Richardson (2001)	First algorithm for influence maximization	/	/	Knowledge-sharing site
Kempe et al. (2003)	Approximation algorithm	LTM ICM	Directed/undirected	Large collaboration network
Leskovec et al. (2007)	Efficient approximation algorithm CELE (Cost-Effective Lazy Forward)	ICM	Directed	Water sensor networks (BWSN) Blog data
Chen et al. (2009)	New degree discount heuristic	LTM /ICM	/	Collaboration graphs obtained from the online archival database arXiv.org
Chen et al. (2010b)	First scalable influence maximization problem	LTM	Directed	NetHEPT, DBLP, Epinions, Amazon
Chen et al. (2010a)	New heuristic algorithm that is easily scalable for millions of nodes	ICM	Directed/undirected	NetHEPT, DBLP, Epinions, Amazon
Goyal et al. (2011)	CELF++ algorithm (optimization of CELF algorithm)	ICM	Directed	NetHEPT, Net PYH

Barbieri and Bonchi (2014)	Novel feature-aware propagation model	LTM	Directed	Last.fm, Flixster
Squillero and Burelli (2016)	Influence maximization in social networks with genetic algorithms	ICM	Directed	SNAP repository
Wang and Street (2018)	Developed an effective and efficient heuristic tackle the influence maximization problem	MAT (multiple-path asynchronous threshold)	Directed/undirected	PGP, NetHEPT, WikiVote, *C. elegans*
McKay et al. (2019)	Newer solution for influence maximization using machine learning	Neural network	Neural network	DBLP
Saxena and Saxena (2020)	HAC-rank algorithm	HBIM (Hurst-based influence maximization)	Directed	UC Irvine messages, MathOverflow Linux Kernel mailing list
Ko et al. (2020)	Inductive machine learning method called MONSTOR (Monte Carlo simulation)	ICM	Directed	Extended, WannaCry, Celebrity

4 Deep Learning Approach for Influence Analysis

Recent work in social network analysis and influence analysis has been applied using the deep learning (Najafabadi et al. 2015; Hayat et al. 2019; Gao et al. 2020; Wang et al. 2019; Keikha et al. 2020; Wu et al. 2019; Zhang et al. 2020). A fundamental step for social network analysis using the deep learning is to encode network data into a low-dimensional representation (Tan et al. 2019).

The work presented in (Luceri et al. 2019) studied the impact of social influence on offline dynamics to study human real-life behavior. They used the deep learning technique for modeling social influence and predicting human behavior on real-world activities. They proposed a social influence deep learning framework that combines deep learning with network science for modeling and forecasting social influence on real-life activities, their social influence deep learning (SIDL) framework based on DNNs.

Authors in (McKay et al. 2019) proposed a newer solution for the influence maximization problem using machine learning. Their objective was to create a deep learning model to solve the influence maximization problem for viral marketing in a faster, more efficient, and more current leading algorithm; they are comparing the result of their model named learning algorithm with three main algorithms: the random selector, sum of edge, and greedy algorithm. For their study they built an artificial neural network in order to test their model against the order to other algorithm following they used the real network (DBLP a computer science bibliography website) for obtained the real results on real data. For the comparison of their model and the main other model (algorithm) existed, they compare with two measures the time efficiency and influence spread in the network. Their model reduce the time running and maximize the number of nodes activated. Their model activates 25.46% of the network's threshold, the greedy, sum of edge and the best random algorithm activates more than 4% (2.98%, 2.87%, and 3.78%), these results of smaller network. The result of their model in the large scale network is 48.18%, sum of edge 26.42% and 32.58%. The result experiment proved that their model is more efficient in the total amount of influence spread and in time efficiency.

In (Tian et al. 2020) motivated by the application of viral marketing, first they proposed two topic-aware social influence propagation models based on IC and LT models. Second, they proposed a new graph-embedding network, called Diffusion2Vec, which can extract features for each user in social network automatically. It's also important to note that they came up with a method for calculating the influence of a candidate user based on their embeddings. Finally, they adopted an algorithm of reinforcement learning, called double DQN with prioritized experience replay to train models. For the experiments, they used real-world social network from Twitter.

5 Conclusion

Online social networks are popular services that have been studied heavily in recent years, as more and more people communicate with friends, colleagues, and family through different existing social networks. We found several interesting papers surveying aspects of social networks. In this chapter, we reviewed the major aspects of OSNs, namely, online social network, social network analysis, social influence analysis, and viral marketing, and we defined the problem space in the social influence analysis. We also surveyed the main existing models and algorithms for influence modeling, influence maximization, and influential spreaders for viral marketing. The main objective of this chapter was to summarize the most important algorithms and models of influence analysis for viral marketing for beginners in this area. As we have learned in this chapter, there are many new problems and challenges on social influence analysis in our future work; we aim at proposing a new model of influence spreading or a new method for the identification of influential nodes in online social network for viral marketing. We hope this chapter will be very useful in clarifying this exciting area of research and serve as a solid foundation for readers interested in this field.

References

Abd Al-Azim, N.A.R., Gharib, T.F., Afify, Y., Hamdy, M.: Influence propagation: interest groups and node ranking models. Phys. A: Statist. Mech. Appl. **124247** (2020)

Ahajjam, S., Badir, H.: Identification of influential spreaders in complex networks using hybrid rank algorithm. Sci. Rep. **8**(1), 1–10 (2018)

AlSuwaidan, L., Ykhlef, M.: Toward information diffusion model for viral marketing in business. Int. J. Adv. Comput. Sci. Appl. **7**(2), 637–646 (2016)

Al-Taie, M.Z., Kadry, S., Obasa, A.I.: Understanding expert finding systems: domains and techniques. Soc. Netw. Anal. Min. **8**(1), 57 (2018)

Arnaboldi, V., Conti, M., La Gala, M., Passarella, A., Pezzoni, F.: Information diffusion in OSNs: the impact of nodes' sociality. In: Proceedings of the 29th Annual ACM Symposium on Applied Computing, pp. 616–621 (2014, March)

Bae, J., Kim, S.: Identifying and ranking influential spreaders in complex networks by neighborhood coreness. Phys. A: Statist. Mech. Appl. **395**, 549–559 (2014)

Barbieri, N., Bonchi, F.: Influence maximization with viral product design. In: Proceedings of the 2014 SIAM International Conference on Data Mining, pp. 55–63. Society for Industrial and Applied Mathematics (2014, April)

Basaras, P., Katsaros, D., Tassiulas, L.: Detecting influential spreaders in complex, dynamic networks. Computer. **4**, 24–29 (2013)

Belfin, R.V., Bródka, P.: Overlapping community detection using superior seed set selection in social networks. Comput. Electr. Eng. **70**, 1074–1083 (2018)

Bhat, N., Aggarwal, N., Kumar, S.: Identification of influential spreaders in social networks using improved hybrid rank method. Procedia Computer Science. **171**, 662–671 (2020)

Bhattacharya, S., Gaurav, K., Ghosh, S.: Viral marketing on social networks: an epidemiological perspective. Phys. A: Statist. Mech. Appl. **525**, 478–490 (2019)

Boyd, D.M., Ellison, N.B.: Social network sites: definition, history, and scholarship. J. Comput.-Mediat. Commun. **13**(1), 210–230 (2007)

Cai, Q., Ma, L., Gong, M., Tian, D.: A survey on network community detection based on evolutionary computation. Int. J. Bio-Inspired Comput. **8**(2), 84–98 (2016)
Can, U., Alatas, B.: A new direction in social network analysis: online social network analysis problems and applications. Phys. A: Stat. Mech. Appl. **535**, 122372 (2019)
Cano, C.: The SIR Models, their applications, and Approximations of their Rates (2020)
Chen, W., Wang, Y., Yang, S.: Efficient influence maximization in social networks. In: Proceedings of the 15th ACM SIGKDD International Conference on Knowledge Discovery and Data Mining, pp. 199–208 (2009, June)
Chen, W., Wang, C., Wang, Y.: Scalable influence maximization for prevalent viral marketing in large-scale social networks. In: Proceedings of the 16th ACM SIGKDD International Conference on Knowledge Discovery and Data Mining, pp. 1029–1038 (2010a, July)
Chen, W., Yuan, Y., Zhang, L.: Scalable influence maximization in social networks under the linear threshold model. In: 2010 IEEE International Conference on Data Mining, pp. 88–97. IEEE (2010b, December)
Cheng, Y., Liu, J., Yu, X.: Online social trust reinforced personalized recommendation. Pers. Ubiquit. Comput. **20**(3), 457–467 (2016)
Dabaghi-Zarandi, F., Rafsanjani, M.K.: Community detection in social networks. In: Models and Theories in Social Systems, pp. 273–293. Springer, Cham (2019)
Daud, N.N., Ab Hamid, S.H., Saadoon, M., Sahran, F., Anuar, N.B.: Applications of link prediction in social networks: a review. J. Netw. Comput. Appl. **102716** (2020)
Dhamal, S., Prabuchandran, K.J., Narahari, Y.: Information diffusion in social networks in two phases. IEEE Trans. Netw. Sci. Eng. **3**(4), 197–210 (2016)
Domingos, P., Richardson, M.: Mining the network value of customers. In: Proceedings of the Seventh ACM SIGKDD International Conference on Knowledge Discovery and Data Mining, pp. 57–66 (2001, August)
Du, D.Z., Pardalos, P.M., Zhang, Z.: Nonlinear Combinatorial Optimization, vol. 147. Springer (2019)
Fu, X., Passarella, A., Quercia, D., Sala, A., Strufe, T.: Online social, networks. Comput. Commun. **73**, 163–166 (2016)
Gaeta, R.: A model of information diffusion in interconnected online social networks. ACM Transactions on the Web (TWEB). **12**(2), 1–21 (2018)
Gao, L., Zhou, B., Jia, Y., Tu, H., Wang, Y., Chen, C., Zhuang, H.: Deep learning for social network information Cascade analysis: a survey. In: 2020 IEEE Fifth International Conference on Data Science in Cyberspace (DSC), pp. 89–97. IEEE (2020, July)
Girdhar, N., Minz, S., Bharadwaj, K.K.: Link prediction in signed social networks based on fuzzy computational model of trust and distrust. Soft. Comput. **23**(22), 12123–12138 (2019)
Goyal, A., Lu, W., Lakshmanan, L.V.: Celf++ optimizing the greedy algorithm for influence maximization in social networks. In: Proceedings of the 20th International Conference Companion on World Wide Web, pp. 47–48 (2011, March)
Guille, A., Hacid, H., Favre, C., Zighed, D.A.: Information diffusion in online social networks: a survey. ACM SIGMOD Rec. **42**(2), 17–28 (2013)
Gulati, A., Eirinaki, M.: Influence propagation for social graph-based recommendations. In: 2018 IEEE International Conference on Big Data (Big Data), pp. 2180–2189. IEEE (2018, December)
Hayat, M.K., Daud, A., Alshdadi, A.A., Banjar, A., Abbasi, R.A., Bao, Y., Dawood, H.: Towards deep learning prospects: insights for social media analytics. IEEE Access. **7**, 36958–36979 (2019)
Keikha, M.M., Rahgozar, M., Asadpour, M., Abdollahi, M.F.: Influence maximization across heterogeneous interconnected networks based on deep learning. Expert Syst. Appl. **140**, 112905 (2020)
Kemi, A.O.: Impact of social network on society: a case study of Abuja. Am. Sci. Res. J. Eng. Technol. Sci. (ASRJETS). **21**(1), 1–17 (2016)

Kempe, D., Kleinberg, J., Tardos, É.: Maximizing the spread of influence through a social network. In: Proceedings of the Ninth ACM SIGKDD International Conference on Knowledge Discovery and Data Mining, pp. 137–146 (2003, August)

Kitsak, M., Gallos, L.K., Havlin, S., Liljeros, F., Muchnik, L., Stanley, H.E., Makse, H.A.: Identification of influential spreaders in complex networks. Nat. Phys. **6**(11), 888–893 (2010)

Ko, J., Lee, K., Shin, K., Park, N.: MONSTOR: an inductive approach for estimating and maximizing influence over unseen social networks. arXiv preprint arXiv:2001.08853. (2020)

Kong, X., Gu, Z., Yin, L.: A unified information diffusion model for social networks. In: 2020 IEEE Fifth International Conference on Data Science in Cyberspace (DSC), pp. 38–44. IEEE (2020, July)

Leskovec, J., Krause, A., Guestrin, C., Faloutsos, C., VanBriesen, J., Glance, N.: Cost-effective outbreak detection in networks. In: Proceedings of the 13th ACM SIGKDD International Conference on Knowledge Discovery and Data Mining, pp. 420–429 (2007, August)

Li, D., Liu, J.: Modeling influence diffusion over signed social networks. IEEE Trans. Knowl. Data Eng. (2019)

Liu, Q., Zhu, Y.X., Jia, Y., Deng, L., Zhou, B., Zhu, J.X., Zou, P.: Leveraging local h-index to identify and rank influential spreaders in networks. Phys. A: Statist. Mech. Appl. **512**, 379–391 (2018)

Loucif, H.: *The Analysis of Social Influence in Social Media Networks* (Doctoral dissertation, Université de Bordj Bou Arréridj-Mohamed El Bachir El Ibrahimi) (2016)

Luceri, L., Braun, T., Giordano, S.: Analyzing and inferring human real-life behavior through online social networks with social influence deep learning. Appl. Netw. Sci. **4**(1), 34 (2019)

Ma, L.L., Ma, C., Zhang, H.F., Wang, B.H.: Identifying influential spreaders in complex networks based on gravity formula. Phys. A: Statist. Mech. Appl. **451**, 205–212 (2016)

McKay, D. B., Corse, J. A., & Gonsalves, M. S.: Deep Learning Method for Social Networks (2019)

Menta, V.P.T., Singh, P.K.: Efficient selection of influential nodes for viral marketing in social networks. In: 2017 IEEE, International Conference on Current Trends in Advanced Computing (ICCTAC), pp. 1–6. IEEE (2017, March)

More, J.S., Lingam, C.: A gradient-based methodology for optimizing time for influence diffusion in social networks. Soc. Netw. Anal. Min. **9**(1), 5 (2019)

Najafabadi, M.M., Villanustre, F., Khoshgoftaar, T.M., Seliya, N., Wald, R., Muharemagic, E.: Deep learning applications and challenges in big data analytics. J. Big Data. **2**(1), 1 (2015)

Okamoto, K., Chen, W., Li, X.Y.: Ranking of closeness centrality for large-scale social networks. In: International Workshop on Frontiers in Algorithmics, pp. 186–195. Springer, Berlin, Heidelberg (2008, June)

Ortiz-Gaona, R.M., Postigo-Boix, M., Melús-Moreno, J.L.: Extent prediction of the information and influence propagation in online social networks. Comput. Math. Organ. Theory. (2020)

Otte, E., Rousseau, R.: Social network analysis: a powerful strategy, also for the information sciences. J. Inf. Sci. **28**(6), 441–453 (2002)

Pan, T., Li, X., Kuhnle, A., Thai, M.T.: Influence diffusion in online social networks with propagation rate changes. IEEE Trans. Netw. Sci. Eng. (2020)

Peng, S., Wang, G., Xie, D.: Social influence analysis in social networking big data: opportunities and challenges. IEEE Netw. **31**(1), 11–17 (2016)

Pourhojjati-Sabet, M., Rabiee, A.: A soft recommender system for social networks. arXiv preprint arXiv:2001.02520. (2020)

Qiu, J., Tang, J., Ma, H., Dong, Y., Wang, K., & Tang, J.: Deepinf: social influence prediction with deep learning. In Proceedings of the 24th ACM SIGKDD International Conference on Knowledge Discovery & Data Mining (pp. 2110–2119) (2018, July)

Qiyao, W., Zhengmin, L., Yuehui, J., Shiduan, C., Tan, Y.: Ulm: a user-level model for emotion prediction in social networks. China Univ. Posts Telecommun. (2016)

Richardson, M., Domingos, P.: Mining knowledge-sharing sites for viral marketing. In: Proceedings of the Eighth ACM SIGKDD International Conference on Knowledge Discovery and Data Mining, pp. 61–70 (2002, July)

Saxena, B., Saxena, V.: Influence maximization in social networks using Hurst exponent-based diffusion model. In: 2020 10th International Conference on Cloud Computing, Data Science & Engineering (Confluence), pp. 167–171. IEEE (2020, January)
Singh, N., Malik, A., Maini, O., Rajput, G.: Identification of influence propagation metrics in social networks. In: 2019 International Conference on Automation, Computational and Technology Management (ICACTM), pp. 224–227. IEEE (2019, April)
Squillero, G., Burelli, P.: Applications of Evolutionary Computation: 19th European Conference, Evo Applications 2016, Porto, Portugal, March 30–April 1, 2016, Proceedings, Part I, vol. 9597. Springer (2016)
Sun, J., Tang, J.: A survey of models and algorithms for social influence analysis. In: Social Network Data Analytics, pp. 177–214. Springer, Boston, MA (2011)
Sun, Q., Li, Y., Hu, H., Cheng, S.: A model for competing information diffusion in social networks. IEEE Access. **7**, 67916–67922 (2019)
Talukder, A., Layek, M. A., & Hong, C. S.: A Novel Approach of Viral Marketing in Social Networks, 1265–1267 (2017)
Tan, Q., Liu, N., Hu, X.: Deep representation learning for social network analysis. Front. Big Data. **2**, 2 (2019)
Tian, S., Mo, S., Wang, L., Peng, Z.: Deep reinforcement learning-based approach to tackle topic-aware influence maximization. Data Sci. Eng. **111**, 1–11 (2020)
Toalombo, M., Wang, B., Xu, H., Xu, M.: A novel greedy fluid spread algorithm with equilibrium temperature for influence diffusion in social networks. IEEE Syst. J. (2020)
Towhidi, G., Sinha, A.P., Srite, M., Zhao, H.: Trust decision-making in online social communities: a network-based model. J. Comput. Inf. Syst., 1–11 (2020)
Wang, W., Street, W.N.: Modeling and maximizing influence diffusion in social networks for viral marketing. Appl. Netw. Sci. **3**(1), 6 (2018)
Wang, F., She, J., Ohyama, Y., Wu, M.: Deep-learning-based identification of influential spreaders in online social networks. In: IECON 2019-45th Annual Conference of the IEEE Industrial Electronics Society, vol. 1, pp. 6854–6858. IEEE (2019, October)
Wu, J., Sha, Y., Jiang, B., Tan, J.: DSINE: deep structural influence learning via network embedding. In: Proceedings of the AAAI Conference on Artificial Intelligence, vol. 33, pp. 10065–10066 (2019, July)
Xu, W., Wu, W., Fan, L., Lu, Z., Du, D.Z.: Influence diffusion in social networks. In: Optimization in Science and Engineering, pp. 567–581. Springer, New York, NY (2014)
Yuan, S., Zhang, Y., Tang, J., Hall, W., Cabotà, J.B.: Expert finding in community question answering: a review. Artif. Intell. Rev. **53**(2), 843–874 (2020)
Zafarani, R., Abbasi, M.A., Liu, H.: Social Media Mining: An Introduction. Cambridge University Press (2014)
Zeng, A., Zhang, C.J.: Ranking spreaders by decomposing complex networks. Phys. Lett. A. **377**(14), 1031–1035 (2013)
Zhang, Y., Li, S., Yu, Z., Zhang, F., Lu, H.: A 2020 perspective on "predicting the influence of viral messages for VM campaigns on Weibo". Electron. Commer. Res. Appl. **40**, 100949 (2020)

The Role of E-Learning in the Algerian Open University to Achieve the Development of Human Capital

Oussama Nabil Bessaid and Chahrazed Benyahia

Abstract Educational institutions play a key role in the development of human resources and the improvement of their capacities and knowledge, which reflects positively on economic and social development. Therefore, we must give importance to these institutions and activate the pioneering role of universities, especially in the context of globalization and the knowledge revolution. The most important of which is e-learning that works to develop the capacities of human resources and their skills as a knowledge capital and the main element for generating and disseminating information, and the source of creativity and innovation. All this at a distance and without the costs of mobility to obtain knowledge, in addition to being done quickly and accurately. Therefore, the focus must be on an open university that uses e-learning in all aspects to develop Human Capital and achieve the goals of sustainable development.

Keywords E-learning · Algeria · Open university · The development · Human capital

1 Introduction

In order to develop the economies of countries and their societies, it is necessary to invest effectively in human capital to achieve sustainable development.

Education is the basis and the axis of the development process. Countries, including Algeria, have recently worked on the development of a strategy for the development of education and the dissemination of knowledge, particularly in the light of the dissemination of information technology communication and the

O. N. Bessaid (✉)
University of Belhadj Bouchaib, Ain temouchent, Algeria
e-mail: bessaidoussamagrh@gmail.com

C. Benyahia
University Aboubekr Belkaid, Tlemcen, Algeria
e-mail: chahrazed.benyahia@univ-tlemcen.dz

S. Sedkaoui et al. (eds.), *International Conference on Managing Business Through Web Analytics*, https://doi.org/10.1007/978-3-031-06971-0_12

scientific, cognitive, and technological revolution and the knowledge society, which has led to the emergence of modern teaching methods to keep pace with these developments.

The e-learning process is one of the main subjects that countries seek to achieve in order to develop their economy, cope with global economic progress, and improve the conditions of their society, leading them to invest in human capital, which is the engine of development.

Today's world is undergoing a change in all economic, social, and technological fields, which has led to a new vision of the training process in order to keep pace with these developments; modern trends resting mainly on human competencies because they are the source of innovation and continuous improvement. It is therefore necessary to train people to achieve a higher level of human competencies, which play a key role in the process of economic and social development of countries. Therefore, special attention should be paid to this by improving the performance of training institutions, in order to invest effectively and develop human capacities and competencies, which are the key factor in socio-economic development.

In this way, we pose the following problem: how can training develop human competencies and contribute to socio-economic development? and what are the mechanisms to activate it?

Problematic

Through it, we ask the following problematic question: What is the role of e-learning in the development of intellectual capital and the achievement of sustainable development?

The Importance of Studying

The important role that e-learning and its institutions play in investing in intellectual capital and achieving sustainable development.

Study Objectives

- Learn more about online learning methods as modern methods
- Highlighting the modern university as an essential pillar of investment in human capital
- Highlight the role of e-learning in the development of intellectual capital and the achievement of sustainable development

Study Approach

This study used the descriptive method, given the nature of the subject, to clarify the concepts related to e-learning and investment in human capital and its relation to sustainable development. The analytical method was used to interpret and analyze the results.

1.1 *Introduction to E-Learning*

1.1.1 Detailed Regulations Concerning Conducting Classes in the Form of E-Learning[1]

As a modern supplement or even an alternative to traditional education, e-learning is becoming increasingly popular. It makes high-quality education readily available, no matter where you are or what time of day it is. An original teaching material, such as syllabi and lectures and interactive exercises and instructional videos and many other multimedia contents that make distance learning more effective and also meet the expectations of students is used. The fact that a new type of education allows for almost instantaneous verification of knowledge is noteworthy.

Due to the method's characteristics, it is possible to perform tasks in two different ways. Education that does not require the presence of a teacher is known as "distance learning" or "distance education." The source of the material to study and examiner is in fact a computer connected to the Internet. The second way is "blended learning." To support the teaching process, e-learning is used here. Classes are still held in the old-fashioned way, with teachers and students physically present, alongside the electronic mode of communication.

E-learning allows a manual selection of the preferred format for the delivery of knowledge and its transmission rate. Educational platforms capable of facilitating e-learning classes are the most advanced and easy-to-use technology. Distance learning platforms are the professional Internet service with a teaching focus. You can access training materials on a separate website that the platform creates for you. The site is password-protected, so you will need both your login ID and the one you've been given.

1.1.2 University Twinning and Networking[2]

UNITWIN stands for "University Twinning and Networking," and it was envisioned that the UNITWIN/UNESCO Chairs Program would help advance research, education, and program development in all of UNESCO's areas of expertise by fostering inter-university collaboration and university networks. The UNITWIN program aspires to be timely, forward-looking, and effective in influencing social and economic development. These projects have proven useful in creating new teaching programs, inventing new ideas through research, and in enriching existing university programs while respecting cultural diversity up to this point in their

[1]Dorota Górska, E-learning in Higher Education, *The Person and the Challenges Volume 6 (2016) Number 2, p. 35–43,* DOI: https://doi.org/10.15633/pch.1868, p. 36.

[2]BOUKELIF Aoued, The role of e-learning in Algerian universities in the development of a knowledge society?, https://www.academia.edu/26627862/The_role_of_e_learning_in_Algerian_universities_in_the_development_of_a_knowledge_society

history. UNITWIN networks and UNESCO Chairs as have a dual role as "think tanks" and "bridge builders" between the academic world and civil society.

- Realignment of UNITWIN's priorities with UNESCO's medium-term strategy (2008–2013) in April 2007;
- Readjust geographic imbalance which is now in favor of the North;
- Stimulate triangular North-South-South cooperation;
- Creation of regional or sub-regional poles of innovation and excellence;
- Cooperation with the United Nations Educational, Scientific and Cultural Organization (UNESCO) (UNU). It is the goal of the International University Cooperation to foster intellectual exchange and cooperation among universities and academics around the world by establishing formal links between them and facilitating the exchange of information within and across borders.

By helping Member States establish centers of excellence, it aims to bridge the knowledge gap and significantly reduce the brain drain. The International Union of Catholic Universities (IUCN) is attempting to meet the challenges of globalization by advancing new information technologies to build capacity and increase knowledge for the advancement of education, science & technology, humanities & culture and communication.

1.1.3 National System of Distance Learning Network

New educational approaches are being implemented in the training process to compensate for the lack of supervision, while also ensuring high-quality training in accordance with the standards set forth by quality assurance. Our national system of distance integration phase has begun, and the key to its success is the creation of a chain of knowledge that extends beyond the academic world and reaches a wider audience that includes people with special needs, senior citizens, patients in the hospital, and those undergoing rehabilitation, among others. As of now, the National System of Distance Learning Network is made up of video conferencing and electronic learning platforms that are scattered throughout the majority of training institutions. The National Research Network provides access to this network.

1.2 Human Development and Economic Integration

1.2.1 Curriculum and Vocational Education and Training[3]

Enhanced vocational education and training should not be seen as a one-size-fits-all solution to the challenges posed by globalization. To determine if countries will benefit from globalization, there are only two things to consider.

[3]George S. Mouzakitis, The role of vocational education and training curricula in economic development, Procedia Social and Behavioral Sciences, Available online at www.sciencedirect.com WCES-201, 2 (2010) 3914–3920.

A more open trading system will be shaped by two factors: first, national responses to globalization's demands; and second, international rules and processes that are shaped in response to those responses. UN Public Sector Global Sector Report (2001). Individuals and nations alike can benefit from the advantages of globalization through market expansion, interdependence of global economies, global operations, workforce mobility, and a global marketplace. Moreover, there are strategies that will support these endeavors (for example, IBM's Globalization Team). As it is, developing and managing a global organization necessitates the development and management of people who can think, lead, and act from a global perspective, and who must possess a global mind and global competencies. The end result of all of this is a well-designed vocational course curriculum that forms the basis of a well-structured vocational program. VET systems are critical components of countries' economic development strategies. Economic competitiveness can only be achieved or maintained if the workforce's skills and knowledge are constantly improved. As a result, the importance of vocational education and training is widely recognized.

As a result, both organizations and their employees must adapt their training practices. In addition, TVET needs to be reformed in order to enhance the supply of competencies and better match the demand for them. It is only through the creation of TVET programs based on an appropriate vocational curriculum that such reforms will be successful. Curriculum, in general, is the road that leads to the development of professional knowledge and skills, making it easier to move from the classroom to the workplace. The obvious benefits of a vocational education include: (a) It provides instruction for many different fields that require technical competencies rather than academic knowledge, (b) it allows students to focus solely on training for a career and (c) a major advantage is that it provides flexible programs available from a variety of sources. To put it another way, UNESCO defines curriculum as the arrangement of learning sequences to produce specific, intended outcomes, while curriculum development is a set of practices aiming to introduce planned changes in search of better outcomes....[4] It is a standard practice to design two separate curricula in order to achieve a more effective outcome. Transience education is recommended for the employed workforce. In the past few years, the rate of change has increased at an unprecedented rate, and many areas of the curriculum are undergoing fundamental changes.

There are three main goals for those who are already employed: (a) to bridge the gap between required professional knowledge and competencies and what is already available; (b) to apply learning to new situations; and (c) to cultivate an entrepreneurial mindset. However, a group of 36 educators and policymakers were consulted before moving forward with the final design of the transience curriculum to decide on the fundamental aspects of the curriculum that should be taken into account.

[4] Ibd.

The curriculum should ensure that students can use technology, think creatively and independently, develop and communicate their own worldviews and beliefs, succeed in a variety of activities, gain knowledge and understanding, make well-informed decisions, communicate in a variety of settings, and work cooperatively at the end of the instructional period. "Possibility to transfer learning from one problem to another within the course, from one year in school to another, between school units and home, and from school to workplace."[5]

1.2.2 Knowledge Economies and Educational Reform[6]

Fast changing knowledge economies call for new core competencies among all learners in the society. At the heart of these are longstanding "soft competencies" such as communication, collaboration, and teamwork. To these can be added others such as the ability to create, apply, share, and distribute knowledge; to convert tacit knowledge into explicit and formally codified knowledge that is easily transferable to others; to employ effortless use of advanced information technology; to work in teams that may be socially and psychologically heterogeneous; to have the capacity to reskill and retrain as circumstances demand; to be able to participate in networks and develop the social capital that creates the learning and develops the resilience to cope with change; to cultivate a positive, opportunistic and entrepreneurial orientation to change; and to become committed to continuous and lifelong learning far beyond the years of formal education. For the World Bank, these sorts of skill sets constitute the "New Basics" of innovation and education in the knowledge economy that are built on the old, eternal and still inalienable basics of literacy and numeracy.

For those whose views of educational and social reform extend beyond the knowledge economy to the knowledge society, and to a productive integration of the two, the desired core competencies that define the heart of educational reform are wider still.

Thus, the "New Basics" in Queensland, Australia include the development of life pathways and social futures; multiliteracies and communication media; active citizenship; environment and technologies. To these might be added yet other new basics such as emotional literacy and conflict resolution, especially in post-conflict societies, or education for sustainable development almost everywhere – for there is no work without a world in which to do it. The implications of a shifting emphasis in knowledge economies toward these previously underutilized "soft competencies" and the emergence of innovation-oriented new basics alongside them are that educational policy and reform strategies will move increasingly *beyond*

[5]ibd.

[6]Andy Hargreaves, Paul Shaw, Knowledge and Skill Development in Developing and Transitional Economies An analysis of World Bank/DfID Knowledge and Skills for the Modern Economy Project, http://siteresources.worldbank.org/EDUCATION/Resources/278200-1126210664195/1636971-1126210694253/DFID_WB_KS_FinalReport_7-31-06.pdf

improving *access*, or changing governance structures and financing, *toward* making policy interventions that significantly affect and improve the nature and *quality of teaching* and learning.Most government efforts to intervene directly in the details of learning and instruction have been neither widespread nor especially effective – particularly where more creative knowledge economy capacities and competencies are concerned. The most interventionist programs in literacy and numeracy have tended to concentrate on low level literacy competencies that yield the strongest results in the more simple areas of learning with predominantly younger children; and their purported successes may well have been illusory – for example, reflecting test items that have been redesigned to make them progressively easier over time. Intervention in more high skill knowledge economy areas of teaching and learning needs to be just as intentional and insistent yet it also needs to be more strategically indirect – creating the *conditions, incentives, and coaching for high capacity teaching and high quality learning* of a transformational nature to emerge. In this respect, greater pedagogical quality requires improved teacher quality, and changed institutional conditions and capacity to develop it. These issues apply not just to the period of schooling through childhood and adolescence, but to learning, lifelong through university, other kinds of post-compulsory education, and all other formal and informal education far beyond these domains.

1.2.3 Socio-economic Development and Human Resource Development[7]

As a method of human resources management, policies promoting social and economic development are becoming increasingly popular. In order to improve the well-being of their workforce, companies are turning to corporate social responsibility (CSR). The majority of CSR is devoted to education, training, and the improvement of educational infrastructure. The development of infrastructure and technological advancement is an essential part of social progress. It boosts productivity and employee satisfaction for businesses. A globalized company's HR managers are responsible for ensuring that their employees are properly compensated, that they have access to value-added benefits, and that their environmental practices are sound.

As of 1990, it has been observed that the vast majority of businesses in developed countries such as the United Kingdom and the United States rely on UNDP and "Human Development Reports". Developing human resources necessitates addressing socioeconomic issues, which are common in these countries. For future strategic planning, the results of these trials are consulted. As a result of these measures, the number of companies' employees or other stakeholders joining together will rise.

[7]Nada Krypa (Tapija), Social Economic Development and the Human Resources Managemen, *Academic Journal of Interdisciplinary Studies Vol 6 No 1 March 2017, p 75.*

Firms' human resource management policies can have an impact on the socioeconomic development of different countries, according to Abdeljalil and Cohen. There has been the observation in the past that economic development strategies do not include social well-being. Human resources management strategies aim to improve the well-being of employees and stakeholders, as well as the economy as a whole. The political and social unrest in various countries is taken into account by international human resource managers. Because of improved employee satisfaction as a result of CSR or socio-economic welfare strategies, the quality of human capital will go up in the future. It leads to a high level of productivity and the retention of employees who are committed to the company.

Productivity is the key to success in today's globalized business environment. In addition, educational level, management policy, and training modules are factors that affect employee productivity. As Kramar pointed out, socioeconomic development has the unintended consequence of raising productivity at the expense of wages. Appraisal and motivation are two value-added proportions that make this possible. There is evidence that wages will rise in the global labor market in order to achieve high productivity, according to Gilmore and Williams. To help new graduates get their feet on the career ladder, HR professionals are implementing tactics like increasing on-the-job training. As a result of such policies, local businesses and workers benefit. From now on, the country's economy will grow and develop.

1.3 *A Case Study of the University of Continuing Education – Algeria*

1.3.1 Presentation of the University[8]

The University of Continuing Education is a national university whose head office is located in Dali Ibrahim in Algiers.

The university is distributed and located in all the states of Algeria through what are called continuing education centers with their headquarters located in the capital of each state and on the campus of the traditional university.

Al-Takween University was established in 1991 and was called the Second Chance University because its objective is education and training, for people who do not have the chance to enroll in traditional university.

This is due to several reasons, since its justifications, decisions and composition are addressed to all segments of society, in particular workers hired, where the

[8]Ounis Abdelmadjid, Virtual Universities And Their Role In Promoting Higher Education And Community Service (Algeria Case Study), Economic dimensions, Volume 9, Numéro 12,019, P128/129.

permanent stewardship of university professors and other contractual supervisors and collaborators continued.

To borrow university, Phd and Master degrees as for the majors in which the university supervises teaching and training, these are:

A-1 The majors taught in the person and after official working hours, that is to say after five evenings and Saturday, such as the branches of the arts, natural sciences, and economic studies, which are known as the branches of the year and are preparatory school for students who have not obtained a baccalaureate, i.e., they will accept to take an exam entry for the first academic year, to the University of Training and Continuing Education, and join one of the following majors:

1 legal branches; 2 economic and travel sciences branches; 3 branches in foreign languages; 4 branches in science.

A-2 Specializes in distance education: for example, Business law; international economic relations law.

Students are required to attend on Saturdays, to attend gatherings led by teachers to explain the extent of the lesson and to respond to questions and requests from students. In addition to this, the University undertakes training to continue the education and training task after official working hours, i.e., after five in the evening and during the second day of the weekend, which is Saturday.

B- How to submit courses to the University of Continuing Education:

In terms of education, the university depends on:

The first type is represented by the presentation of attendance courses at the steps and halls, as is the case in conventional universities, but after hours of departments, so workers can join the steps and the rooms.

The second type is represented by distance education and training, where educational support is provided in the form of publications, CDs, which students can see before meeting teachers, after which students and teachers meet on the second day of the weekend, i.e., Saturday, to enrich the lesson and to explain and discuss it.

Students and teachers are assigned to study units.

The third type is known as on-demand training, and dojo for workers and employees of companies and institutions of economic and administrative, public or private, and this is done by submitting a training request to the Virtual Universities Unit and a support role for higher education and community service (case study Algeria).

Among the units responsible for establishing the constitution and controlling administrative and procurement techniques, as well as image-based techniques in general, such as budget, accounting, personnel management, and automated media.

C- Distance learning at university:

Distance education is one of the main functions of the University for Training and Continuing Education, as it supervises staff a Bert Management administrator of the director of distance education in the central administration or centrality to direct the training university communication, in coordination with the heads of the distance education departments at the training center level.

Administrative staff, in coordination with the teachers, undertake the task of fixing lessons, in the form of publications, CDs, to be distributed to students and teachers of continuing education centers by distance education departments, visualize and revise them, then meet at weekends with teachers and students to discuss and enrich the lessons.

1.3.2 E-Learning and Virtual University in Algeria[9]

After gaining independence, Algeria faced challenges on several levels: economic, political, and according to that, education had to be given the importance it deserved, and Algeria worked to build educational institutions and the free adoption of educational democracy, but the objectives are great and the possibilities are limited. Therefore, this is where the idea came from to create a center that works to generalize education by correspondence and that is aimed at all those who wish it. The National Center for Education was created and completed by correspondence in 1969. Among the most important concepts related to online learning we find:

- The concept of distance learning.
- Direct learning, open learning.

Learning via the Internet involves using simultaneous or recorded lessons, video techniques (image and sound), electronic display techniques (video conferences broadcast via the Internet).

Through the results of some research and studies in this context, it has been found that the teachers are sufficiently familiar with the concept of e-learning. They also have a good knowledge of the most important concepts, making it possible to say that the theoretical knowledge of this modern technology is not a problem for teachers, because they are either familiar with what is published about them in scientific journals and books, or because they use these techniques in additional research.

The application of online learning at university differs from one university to another and from one department to another. In addition, it is increasingly used in scientific and technical disciplines, specifically in scientific and applied disciplines more than in theoretical scientific disciplines, and less so in literary disciplines.

Moreover, its use in general is average, because the presentation techniques such as the technique of "data of data," the presentation of information, and the

[9] Saidani Salami, Noureddine Dahmar, Sawsen Seki, Algerian experience in the field of e-learning and virtual universities – a critical study, http://virtuelcampus.univ-msila.dz/facshs/wp,p11/12

preparation of conferences in the form of PowerPoint, have evolved in a remarkable way, but the availability of courses and their availability on the Internet is still somewhat modest. For example, some teachers resort to the presenting their courses on their own blog instead of the location of the university because of its weakness and its lack of modernization.

In higher education, the national distance education project has been launched, with the aim of overcoming the shortcomings of supervision on the one hand, but also to improve the quality of the training, in line with the quality guarantee requirements, according to sources revealed by the University of Oran, which launched this project as part of the integration of new training and education methods. It aims to achieve objectives divided into three phases, namely:

The first step:[10]

It is advanced by the use of technologies, such as video conferencing, to absorb the growing number of learners, while improving the level of education and training, and this will be in the short term.

The second step:

In this step, modern educational technologies are adopted, in particular WAP, and this means online or e-learning, in order to achieve quality assurance in the medium term.

The third step:

This is the integration stage, during which the distance education system is approved and published through "distance" education by the Knowledge Channel, the scope and benefits of which are extend far beyond the academic field, targeting a large audience of learners who wish to deepen their knowledge and others who need specialized information, even hospitalized and recovered patients, and other segments of society who want more knowledge.

Distance education is currently based on a network platform for video conferences and e-learning distributed in the majority of higher education establishments, and access to this network is possible via the National Research Network, ARN. Thirteen higher education establishments will be a transmission and reception site at the same time, while 64 other establishments will be a host site, which will cover the distance learning projects of the 77 higher education establishments spread over the national territory, including universities, university centers and colleges, while the scientific and technical research center will be the focal point of the project.

In addition, the video lectures will be broadcasted from the universities of Ben Youssef Ibn Khedda and Houari Boumediene in Algiers, Saad Dahlab and Baji Mokhtar in Annaba, and Kasdi Merbah Ouargla, and Abdel Rahman Mira in Bejaia and Hadj Lakhdar from Batna and Mentouri in Constantine and Farhat Abbas Setif, as well as the universities of Essenya, Oran and Abu Bakr Belkaid from Tlemcen,

[10]Ibd.

next to the Center for the Development of Advanced Technologies and the Research Center for Scientific and Technical Media.

1.3.3 The Case of the University of Oran1 Ahmed Ben Bella in the Development of E-Learning (Table 1)

Table 1 E-learning at the University of Oran1 Ahmed Ben Bella

Number of teachers who participated in the training sessions: approximately 430
Number of platform users via the Moodle Mobile application: 3216
Number of PhD students' email accounts: 723
Number of teacher/researcher email accounts; university staff; laboratories; structures and services; researchers (Domaine univ-oran1.Dz): 1361

Source: https://elearn.univ-oran1.dz/mod/
Modified on: Saturday August 7, 2021, 20:50

2 Conclusion

Our study has shown that e-learning is an effective investment of human capital and works to increase its capacities and skills and increase its productivity, and to improve knowledge and the social domain, through educational establishments. The virtual university is a basic pillar in this, providing lessons and knowledge and conferences that aim to open up the socio-economic and cultural environment, to increase awareness to keep abreast of developments, changes and events going on in the world, which leads to an increase in employment opportunities and income and the improvement of people's quality of life and health conditions, which is one of the most important aspects of achieving sustainable development.

Recommendations and Suggestions

- Build a solid technological base.
- Communication network development.
- Effective orientation of open universities.
- Rely on real talent to build an effective knowledge base.

References

Boukelif, A.: The role of e-learning in Algerian universities in the development of a knowledge society? https://www.academia.edu/26627862/The_role_of_e_learning_in_Algerian_universities_in_the_development_of_a_knowledge_society

Salami, S., Dahmar, N., Seki, S.: Algerian experience in the field of e-learning and virtual universities – A critical study. http://virtuelcampus.univ-msila.dz/facshs/wp,p11/12

Website: https://elearn.univ-oran1.dz/mod/page/view.php?id=10213

Web Analytics and Social Media Monitoring

Soraya Sedkaoui, Rafika Benaichouba, and Khalida Mohammed Belkebir

Abstract The widespread use of social networks has resulted in an exponential increase in data and content that can be used to generate insights and valuable information. However, dealing with such data is considered difficult and necessitates appropriate tools. This chapter describes current web monitoring techniques, emphasizing how these tools handle social networks' data, intending to develop a culture for web analytic tools in Algeria. The study looks at different specialized tools for analyzing social networks activity that have a lot of potential for public relations monitoring and basic and applied research.

Keywords Web analytics · Social networks · Text analysis · Data

1 Introduction

The growth of information and communications technologies (ICT), mainly the development and spread of social networks and mobile devices such as smartphones and tablets, contributes to an increase in the amount of data and structured and unstructured multimedia content of people and organizations (Sedkaoui 2018). Social networks connect users to create specific interest groups, and others need to know more information about them and their profiles.

The internet directly modified access to information, reconsidered the notions of space and time, and made possible new business models, which generated ongoing debates on associated problems. The generic term "social networks" consists of a new public space that enables new ways of relating. In particular, social networks have changed the way organizations and people interact. The introduction of social networks resulted in significant changes and transformations in business activities,

S. Sedkaoui (✉) · R. Benaichouba · K. M. Belkebir
University of Khemis Miliana, Khemis Miliana, Algeria
e-mail: s.sedkaoui@univ-dbkm.dz; r.benaichouba@univ-dbkm.dz; r.benaichouba@univ-dbkm.dz

S. Sedkaoui et al. (eds.), *International Conference on Managing Business Through Web Analytics*, https://doi.org/10.1007/978-3-031-06971-0_13

particularly in the production and exchange of information and content by users, which has gained popularity among activities.

In the digital world, users simultaneously participate in networks, share information recommendations, and seek advice from peers and strangers. Users create, initiate, exchange, and use facts, opinions, evaluations, impressions, feelings, rumors, and experiences concerning products, brands, services, and problems. The profiles of organizations in different social networks proliferated, especially in recent years, seeking to follow the path of users' attention.

There are currently many tools available for monitoring user activity on social networks. Some are simple tools, while others are more complex. They both can be free and involve a significant investment. Some are extremely simple to use, while others necessitate extensive installation and implementation assistance. Several studies recognized web analytics as a valuable tool for monitoring social networks.

Web monitoring techniques are reflected in most of the bibliography surveyed as the promise that will revolutionize the way business is done on the web (Kaushik and Sukhwal 2010), and therefore they are mostly related to web business companies. Globally, these tools combine a qualitative and quantitative approach to developing insights and meaningful information for strategic decision-making.

The effort of service providers or tool providers in training their clients and producing self-learning material about what to expect on social networks demonstrates the embryonic state of companies in social networks and, more importantly, the evaluation of results. Web monitoring techniques seek to answer fundamental questions about the website visitors to which they have access: What exactly did they do? How did they do it? What motivated them to act in this manner? These questions are, to varying degrees, measurable using web analytics tools.

The overarching goal of this research is to investigate the expectations of a select universe of organizations regarding social networks, what tools they use to record and evaluate their activity in this new space, and how they incorporate that information into their businesses. What goals do businesses set for their social media presence, how do they intend to achieve them, and how much do the tools they use assist them in obtaining valuable information about the results obtained? What features do the most popular monitoring tools on the market provide?

The purpose also consists in exploring the situation in Algeria and inquiries about the goals and forms that these take when translated into communication and marketing operations on social networks, the criteria for selecting a monitoring tool, and the evaluation of the results of the activities carried out, as well as the effectiveness of the instrument used.

The rest of this chapter is organized as follows: Sect. 1 describes the chronology and the evolution of social networks. In Sect. 2, the necessary foundations for analyzing social network data will be mentioned. Section 3 explains the importance of analyzing data generated by social networks. Section 4 details and compares the main online tools usually used. Section 5 presents aspects that distinguish the perspective of social networks by identifying some challenges related to users' data security and privacy. By indicating conclusions and considering the growth in the development of social network analysis, thanks to the advancement of new technologies, future lines of research will be described in Sect. 6.

2 The Evolution of Social Network Analysis

Social networks are digital tools that support communication and interaction between users. These digital tools, such as wikis, blogs, and microblogging, among others, facilitate the creation and exchange of content and allow communication from many to many. Currently, social networks are used by many people around the world. Among the most popular social networks, we can mention Facebook, Twitter, and LinkedIn, which encourage their users to create a list of friends, followers, and contacts to generate indirect connections.

Mentioning the history of social networks is difficult because their origins are diffuse, and their evolution has occurred rapidly. Table 1 will list the most significant events in social networks chronologically.

A social network is defined as an organization or structure formed by the interactions of various actors (people, institutions, organizations, societies, etc.) that must possess or be linked to certain particularities or common features in order to interact with each other. Throughout history, the structure of social networks has evolved, acquiring new paradigms and typologies. Currently, the networks that have emerged due to new technologies, such as the internet and social interaction, and which are the focus of research stand out.

With the generation of Web 2.0, approximately in 2004, there is an explosion in the field of social networks where the themes and content cover all areas and

Table 1 Social networks' chronology

Year	Description
1971	Sent of the first email between two computers
1978	Ward Christensen and Randy Suess create a BBS (Bulletin Board Systems) to notify, publish, and share information with their friends
1994	GeoCities is launched, allowing people to develop their own websites
1997	AOL instant messenger is launched which allows users to chat; in addition, blogging began, and Google was launched
1998	Friends reunited was born; it is a social network in the United Kingdom very similar to Classmates
2000	It reaches the figure of 70 million connected computers
2003	LinkedIn, MySpace, and Facebook were born
2006	The microblogging social network Twitter was launched
2010	Facebook has 550 million users. Twitter registers 65 million tweets daily
2012	Facebook has more than 800 million users, and Twitter has 200 million

Source : Authors' elaboration

where new ways of establishing and maintaining social relationships are generated (Sedkaoui 2018).

When people talk about social networks, they refer mainly to platforms that facilitate fluid and instantaneous communication spaces between peers. In this sense, Facebook, Twitter, Instagram, and LinkedIn, among others, made possible the creation and development of online communities where people can share as much personal information as they wish, opinions, concerns, and suggestions (Sedkaoui and Khelfaoui 2020).

One of the phenomena paid for by these platforms is the enormous amount of data that can be easily created, edited, and shared. What was once hierarchical and restrictive is now symmetrical and collaborative. The reproduction conditions of content have changed and are now multidirectional, multipolar, and synchronous.

For this work, we can distinguish three stages in the evolution of network theory:

- ***1930–1970:*** During this period, the search for new methods and techniques to measure and quantify social relations in various social groups gradually led to a convergence between the followers of the anthropological structural-functionalist school and the mathematicians who developed the graph theory, thus laying the foundations for the theoretical, methodological, and technical bases of social network analysis.
- ***1970–1980:*** During this decade, there was a break with the aforementioned schools due to two fundamental innovations: (i) the introduction of the concept of structural equivalence and (ii) the introduction of multidimensional scaling and block modeling techniques. This gave rise to the concept of structural equivalence, which allowed incorporating isolated nodes (individuals), until then not considered in the analysis, and simultaneously analyzing the nodes and their connections. With this accumulation of principles, theoretical concepts, methodologies, and techniques of its own, network analysis seemed to have come of age.
- ***From 1980 to nowadays:*** From 1980, the analysis of social networks has progressed enormously, in both its theoretical-conceptual and methodological and technical aspects, with the support of computational tools (Sedkaoui and Gottinger 2017). The analysis of social networks offers vast possibilities by combining traditional data collection techniques (questionnaires, in-depth interviews, observation, and document records) with more modern techniques, whose use has been greatly facilitated with the development of computer programs.

Technological tools have strengthened the development by merging the ideas and opinions offered on social networks, taking advantage of their great potential to obtain data and carry out surveillance tasks.

Forming market-oriented capabilities of emerging technologies such as artificial intelligence, data analytics, and machine learning is necessary for the search for improvement, productivity, and competitiveness in different economic sectors and to provide an opportunity that allows market knowledge to improve and encourage companies to generate new and outstanding experiences with their clients.

With the appearance of the social networks Facebook and Twitter, the users' dissatisfaction with the services began to be visible, and the value attributed to a message in the companies' accounts began to have relevance in society. This constitutes pressure for companies since they must have qualified personnel and a strategy to interact with customers on this social network.

The introduction of new technologies and the rapid emergence of different interdisciplinary fields of activity present new challenges for organizations. Globalization has led companies to search for tools that allow them to stay ahead of the events that may arise, which means staying in a world market where the strongest competitors survive.

3 The Role of Analytical Tools

Social media analytics is collecting, analyzing, and tracking social data. The analysis of social networks is the mapping and measurement of the relationships and flows between people, groups, organizations, or teams being necessary for the decision-making processes (Wasserman and Faust 1994; Sedkaoui 2018). Conducting a social media analysis allows focusing time, efforts, and budgeting more effectively.

There is no doubt that to know what is happening in the social fabric, it is necessary to have advanced tracking instruments to carry out diagnoses and monitor actions in social networks. The approach from an observational methodology is fundamental since what happens in the interactive social environment is collected.

Monitoring social networks and the metrics of marketing actions on these platforms is receiving increasing attention from academic researchers and professionals who must position a linked product or service.

Monitoring social networks data refers to the discovery of knowledge in text documents to find important tracking patterns about the customer and, in this way, satisfy the requirements associated with the acquisition of a good or service (Kamruzzaman et al. 2010). Therefore, using the information and knowledge generated through the analysis of large amounts of data is an opportunity to obtain benefits and generate value. In the literature, it consists of the following steps (Table 2).

To analyze text, it is essential to use techniques (see Fig. 1) such as natural language processing (NLP), machine learning, and knowledge management (Andrews and Fox 2007), which are executed through different steps in processes of extraction, grouping, association, summaries, or visualization (Sukanya and Biruntha 2012; Sedkaoui 2018).

The type of learning classifies advanced techniques, supervised and unsupervised (Halibas et al. 2018; Sedkaoui and Khelfaoui 2020), such as decision trees, Bayesian algorithms, genetic algorithms, neural networks, or support vector machines (SVM) to handle documents whose data structure is complex (Shukri et al. 2015). These techniques can be presented in Fig. 1.

The main methods used to analyze social networks data are:

Table 2 Steps to discover knowledge in text

Author(s)	Step 1	Step 2	Step 3	Step 4
Kim et al. (2017)	Data collection	Natural language processing	Text analysis	Visualization
Halibas et al. (2018)	Data extraction	Information preprocessing	Application of algorithms	
Sukanya and Biruntha (2012)	Text preprocessing	Text mining	Text analysis	
Shukri et al. (2015)	Data collection	Data preprocessing	Application of machine learning models	

Source: Authors' elaboration

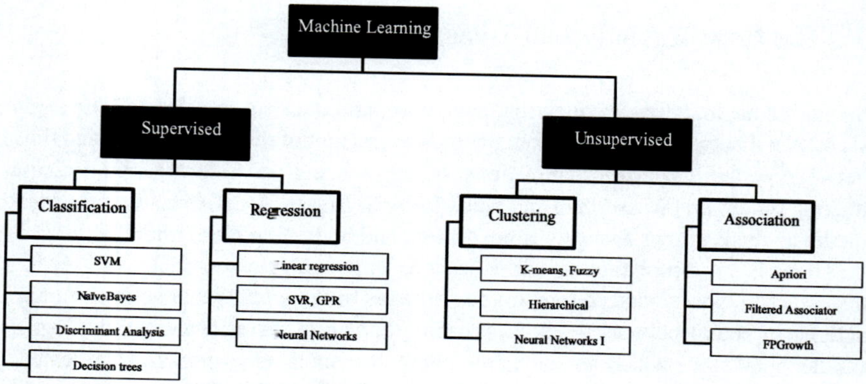

Fig. 1 Supervised and unsupervised learning techniques. (Source: Authors' elaboration)

3.1 *Text Mining*

The importance of text mining stems from the need to extract information from unstructured documents automatically (Sedkaoui 2018; Halibas et al. 2018). The concept of text mining emerged at the end of the 1990s when the interest in investigating this topic began to be highlighted and considered as one of the main methods to extract and discover relevant information from unstructured text documents. As time went by, digital documentary information grew (Kim et al. 2017), and studying social networks became a transcendental factor for the work of companies.

The basic concepts and the mathematical formulation of the techniques used in text mining are:

Matrix of terms and documents: It is a matrix that stores a set of terms in its rows and documents in its columns. The number of times of co-occurrence of each term is counted with these matrices. The relationship is expressed using a document-term matrix called N of dimension $[mxn]$, where m is the number of terms in the collection of documents (Bagga and Baldwin 1998), n is the total number of documents in the matrix, and an element a_{ij} of matrix A is a representation between the term i and the document d.

The column vectors of these matrices indicate whether the texts are similar or not.

Term frequency (tf)*:* It is a measure to quantify the importance of a word in a document. Theoretically, it is the appearance of a term in a document, where tf_{ij} is the same number of repetitions of the term t_i in the document d_j.

$$\mathrm{tf} = \frac{t_i}{\mathrm{longitudocument}_j} \tag{1}$$

Inverse document frequency (idf)*:* Presents the importance of a term i of document j in a set of documents N, where n_i is the number of documents that contain the term i, as shown in f.

$$\mathrm{idf} = \log\frac{N}{n_i} \quad \forall n_i > 0 \tag{2}$$

tf idf: The statistic tf – idf measures the weight of a term in a document taking into account the frequency in other documents.

$$\mathrm{tf} - \mathrm{idf} = \mathrm{tf} * \mathrm{idf} \tag{3}$$

3.2 Sentiment Analysis

Sentiment analysis or opinion mining is the computational study of people's opinions, evaluations, attitudes, and emotions about a product, brand, organization, company, event, or person (Liu and Zhang 2012). It is one of the research areas with the greatest participation in natural language processing and is widely studied in web and text mining. Sentiment analysis provides companies with a means to estimate the degree of acceptance of a product or service and determine strategies that allow them to improve the perception of their quality.

The sentiment found in the comments or criticisms provides information that is valuable for the development of activities such as monitoring social networks, brands, the voice of the customer, customer service, trend detection, and research of markets that belong to the daily operation of an organization (Miner et al. 2012). The importance of sentiment analysis coincides with the growth of social networks on the web; individuals and organizations use public opinions as to the main means of decision-making (Liu and Zhang 2012). Sentiment analysis systems are being applied in different sectors and social domains because opinions are central to human activities and are key influencers of behaviors at the social level.

Poeczea et al. (2018) developed a sentiment analysis study on the comments made on social networks by the followers of a group of gamers, who shared through Facebook the videos published on their YouTube accounts about the theme of videogames and their experience in them. The classification of feelings was carried out using the supervised learning method K-NN (K nearest neighbor). The study determined consistency between the results obtained through sentiment analysis and the metrics used by Facebook.

Struweg (2020), based on user perception, has modeled the South African national health office based on Twitter users and their interactions in Twitter. Broek-Altenburg and Broek-Altenburg and Atherly (2019) have identified consumers' feelings about health insurance by analyzing Twitter, using sentiment analysis in visualization form. Duwairi et al. (2014) analyzed comments or tweets as positive, negative, or neutral feelings in Jordan.

These techniques have been used to examine, for example, public policy opinions or international news items. They have also been used in marketing to assess users' perceptions and opinions of products and services (Lin and Geertman 2019; Sedkaoui 2018).

3.3 *Visual Analytics*

The global vision of conducting visual analytics is to turn information overload into an opportunity for organizations to process data and information transparently in an analytical process. Visual analytics is a consequence of scientific and informational visualization fields that include technologies from other fields, including knowledge management, statistical analysis, cognitive science, and decision science, among others (Wong and Thomas 2004; Sedkaoui 2018).

4 Online Social Network Analysis Tools

Due to space limitations, various monitoring tools will be analyzed in order to adjust the qualities of each one. Some tools can monitor conversations across the network on a specific topic, brand, or person, and others focus on analyzing certain applications simultaneously.

For this, a specific search was carried out as an example, such as the "Covid-19" due to its spread since 2020. Faced with this situation, some tools will be analyzed, mainly:

4.1 Google Trends and Google Insights

This application provides information on the relevance of search terms on the internet without specifically indicating the social content, allowing viewing graphically the evolution of searches in a certain period and analyzing social behaviors or possible seasonality, among other information. In the upper part of the graph shown, the search variables appear over time, and in the lower graph, the news published about Covid-19. The vertical axis represents the frequency with which the term has been searched globally (Fig. 2).

The second tool, Google Insights, has a great similarity in idea and functionality to Google Trends, but it offers more data on keywords and specifies the geolocation of crawls. It is possible to compare search volume patterns in some geographical regions, time intervals, categories, and properties, analyzing which areas are most interested in the term examined. Graphically, the result limited to 2020 is accessed and represents the frequency with which the term has been searched in different Algerian regions (Fig. 3).

Google interfaces are straightforward, manageable, and reliable as they belong to the Google laboratory, providing beneficial information.

Countless tools work on Twitter to manage the account, manage followers, know what is being talked about, follow hashtags, calculate influence, manage events, etc. Among them are Klout, Twendz, Twitter Sentiment, TwitterCounter, etc.

The tool's accuracy and limitation currently depend on the Twitter search API. One of its restrictions is that it will mostly go back to the last 1500 tweets or 30 days passed for each keyword.

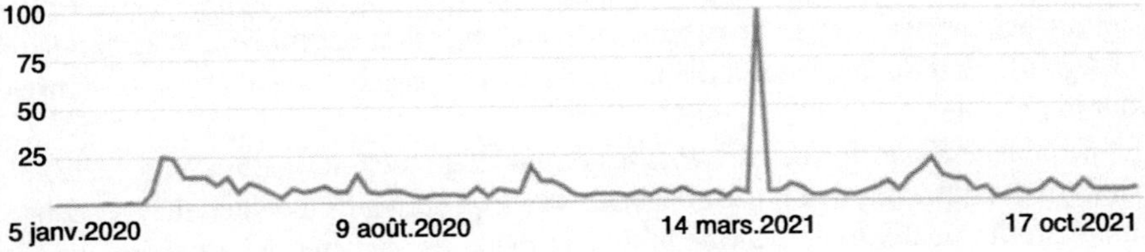

Fig. 2 The evolution of the term "Covid-19" in Algeria since 2020. (Source: Google Trends)

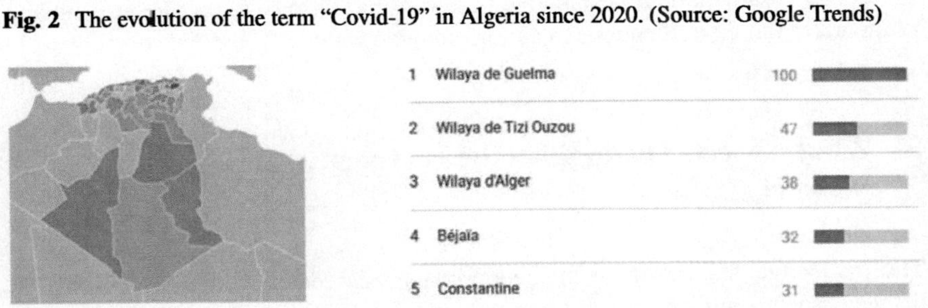

Fig. 3 Distribution by region of the volume of searches for the term "Covid-19". (Source: Google Insights)

The search area allows to enter the desired parameters and contrast up to four keywords or hashtags simultaneously, which is very useful for benchmarking. It is also possible to control the number of searches made and represented, the language, and the location, and it is possible to access different graphics through the random function.

4.2 *Twitter Analytics and Facebook Insights*

On the one hand, Twitter Analytics consists of analyzing the behavior of followers, the statistics of publications, influence within the social network, etc. This tool gives statistics that tweets have such as followers, the profile visits, etc. Being internal to the social network, this tool only statistically measures the data of its own social network. On the other hand, the main function that Facebook Insights give is to collect all possible statistical information about what happens around pages on the social network Facebook. For this, a page through the user profile must be created.

Although the objectives that each one has are somewhat different in some of the cases, the most complete is Facebook Insights. This tool provides the most information according to the objective set as a measurement tool. It is true that Twitter Analytics has a similar objective and like Facebook it displays a lot of data. However, it does not offer as many possibilities. For example, a business that Facebook has, so they are multiplied by turn the data that has a social network to the detriment of the other. It is right that both show the data related to social networks with the greatest accuracy and clairvoyance possible.

We can also find Social Mention, which is a totally free search engine, with content generated by the user in blogs, microblogs, forums, images, videos, news, comments, events, and the possibility of searching all at once. Social Mention is in charge of the influence, not of the users, but of the search terms that can be carried out in its tool.

Similarly, *Klout*, for its part, focuses on making a definition through its calculations of the influence that the user has on the social networks to which they associate their profile in the tool. This definition is made through the Klout index, and it would show how influential someone is toward his followers, profile, etc. Table 3 summarizes these clarifications in a more suitable way.

5 What Aspects Distinguish the Perspective of Social Networks?

The distinctive feature of social network analysis focuses on the relationships between social entities and the patterns and implications of these relationships (Wasserman and Faust 1994). Instead of analyzing the behaviors of individuals,

Table 3 Comparison between Twitter Analytics, Facebook Insights, Klout, and Social Mention

Functions	Facebook Insights	Twitter Analytics	Socialmention	Klout
Scope of publications	+	+	+	+
Interactions (likes, comments)	+	+	+	+
Export account data	+	+	+	−
Nbr of mentions	+	+	+	−
Nbr of publications	+	+	−	−
Nbr of followers	+	+	−	−
Visits to profile	+	+	−	−
Videos	+	+	−	−
Trends	+	−	+	−
Paid version	+	+	−	−
Types of followers	−	−	−	+
Schedule posts	+	−	−	+
Creation of events	+	−	−	−
Private message statistics	+	−	−	−
Recommendations to other	+	−	−	−
Impact on social media	−	−	−	+
Nbr of hashtags in publications	−	−	+	−

Source: Authors' elaboration
(+), allowed; (−), not allowed

attitudes, and beliefs, the analysis of social networks must focus its attention on social entities or actors' interactions and how these interactions affect the structure and content of the companies.

In the analysis of social networks, the most important is not so much the descriptive or attributive data of the social actors, but above all the relational data, that is, the relationships that are established between the different actors that are part of a social structure.

According to Kadushin (2013), there are two fundamental principles of social networks:

Propinquity: consists of being in the same place at the same time.

Hemophilia: if two people share the same characteristics, there is a high probability that they will connect. The opposite is also true; if two people are connected, there is a greater probability that they share common characteristics or attributes.

These principles allow understanding how the relationships between actors (individual and collective) within a given network are structured, depending on the position they occupy and the role they play.

Both confidentiality and privacy have great importance in social networks since disclosure and the misuse of personal information can cause damage to the lives of the people involved. However, handling social network analysis's potential means dealing with some challenges. Privacy in the context of social networks has several categories (Sedkaoui and Gottinger 2017; Zhang et al. 2010):

- *User identity anonymity:* the protection of a user's identity and changes of identity in different types of social networks can vary. There is no user identity anonymity as many Facebook applications depend on connecting to their profile and their public identities.
- *Privacy of the user's personal space:* the visibility of the user profile can vary from one social network to another; MySpace allows users to choose whether their profile is public or only for friends; Facebook instead allows users of a subnet to see the profile unless the owner has restricted this access.
- *Communication privacy:* in addition to the personal data that a user can disclose in the digital space, the user can also disclose personal information. In this data, there may be connection time, longitude, latitude, and IP address.
- *Authentication and data integrity:* it is mentioned that most social networks support their pre-existing social relationships with real life (Boyd 2007). This data stored in the social network can be modeled with a social graph.

The tragedy of the Covid-19 pandemic has reshaped how we move forward in our daily lives. Our dependence on the internet grew exponentially in 2020 with the home office, virtual classes, and online concerts, meetings, and parties. The use of technology has been deeply integrated into consumers' daily lives, from communication to daily tasks. The growing connectivity of people and the internet of things have brought a new perspective on the protection of personal data in the world.

6 Conclusion

The most valuable feature of a tool for analyzing and monitoring social networks is that it is free and virtual, allowing it to work in the most challenging areas with problems of a geographical or ideological sector, combining the communication system with social motivation. Social networks are more than a meeting point, requiring the use of different virtual applications capable of coordinating and manifesting the behavior of people and brands, transforming the results into an intelligent component.

In Algeria, many businesses are increasingly engaging in social networks, but their goals are not always clear, and neither the indicators nor the monitoring devices are relevant. For Algerian businesses, the analysis of social networks presents theoretical, methodological, and technical alternatives, which can open up new social research possibilities.

Furthermore, it should be noted that clients respond actively to the events that Algerian companies publish regularly. It is observed that customers come to this social network to comment on their experiences and level of satisfaction with the services or products offered by these companies, demanding improvements in the technical and functional care processes.

The possibility of graphing social structures and measuring their properties from sophisticated software programs of many of these tools allows understandably and succinctly to access the results of communication, collaboration, transaction, valuation, or any other relationship developed through face-to-face or virtual means. Great activity and participation in the networks have been observed, highlighting some opinion leaders and that is where the analysis of social networks selects and filters the information to distinguish knowledge.

The development of social analysis networks allowed acquiring knowledge and tools that can be applied in any industry and context of interest. Obtaining these skills will allow the industrial engineer to improve company processes using traditional tools, and by using data science techniques, it will allow positive innovation in the work of the Algerian organizations and, thus, improve their competitiveness.

To use the new tools available in the technology market, it is necessary to use cloud service platforms to manage and store the data obtained. Building a visual analytics application would allow automating the interaction between the generated data and the business, thus facilitating decision-making.

References

Andrews, N.O., Fox, E.A.: Recent Developments in Document Clustering. Department of Computer Science, Virginia Polytechnic Institute State (2007)

Bagga, A., Baldwin, B.: Entity-based cross-document coreferencing using the vector space model. In: Proceedings of the 36th Annual Meeting of the Association for Computational Linguistics and 17th International Conference on Computational Linguistics. Association for Computational Linguistics (1998)

Boyd, N.B.E.: Social network sites: definition, history, and scholarship. Int. Rev. Res. Open Distance Learn. **12**(3), 210–230 (2007)

Broek-Altenburg, E.M., Atherly, A.J.: Using social media to identify consumers' sentiments towards attributes of health insurance during enrollment season. Appl. Sci. **9**(10), 2035 (2019)

Duwairi, R., Marji, R., Shaaban, N., Rushaidat, S.: Sentiment analysis in Arabic tweets. In: Proceedings of the 5th International Conference on Information and Communication Systems (2014)., Irbid, Jordan, 1–3 April

Halibas, A.S., Shaffi, A.S., Mohamed, M.A.K.V.: Application of text classification and clustering of Twitter data for business analytics. In: 2018 Majan International Conference (MIC). IEEE (2018)

Kadushin, C.: Understanding Social Networks: Theories, Concepts and Findings. Oxford University Press, New York (2013)

Kamruzzaman, S., Haider, F., Hasan, A.R.: Text classification using data mining. arXiv preprint, arXiv1009 4987 (2010)

Kaushik, V., Sukhwal, P.: Efficacy of self developed environmental information empowerment package (IEP) in transferring knowledge to farm families. Int. J. Hum. Ecol. **30**(1), 71–73 (2010)

Kim, Y., Ju, Y., Hong, S., Jeong, S.R.: Practical text Mining for Trend Analysis: ontology to visualization in aerospace technology. KSII Trans. Internet Inf. Syst. **11**(8), 4133–4145 (2017)

Lin, Y., Geertman, S.: Can social media play a role in urban planning? A literature review. In: Computational Urban Planning and Management for Smart Cities, pp. 69–84. Springer (2019)

Liu, B., Zhang, L.: A survey of opinion mining and sentiment analysis. In: Mining Text Data, pp. 415–463. Springer (2012)
Miner, G., Elder IV, J., Fast, A., Hill, T., Nisbet, R., Delen, D.: Practical Text Mining and Statistical Analysis for Non-structured Text Data Applications. Academic (2012)
Poeczea, F., Ebsterb, C., Strauss, C.: Social media metrics and sentiment analysis to evaluate the effectiveness of social media posts. Proc. Comput. Sci. **130**, 660–666 (2018)
Sedkaoui, S.: How data analytics is changing entrepreneurial opportunities? Int. J. Innov. Sci. **10**(2), 274–294 (2018)
Sedkaoui, S., Gottinger, H.W.: The internet, data analytics and big data. In: Gottinger, H.W. (ed.) Internet Economics: Models, Mechanisms and Management, pp. 144–166. eBook Bentham Science Publishers, Charjah (2017)
Sedkaoui, S., Khelfaoui, M.: Sharing Economy and Big Data Analytics. ISTE-Wiley, London (2020)
Shukri, S.E., Yaghi, R.I., Aljarah, I., Alsawalqah, H.: Twitter sentiment analysis A case study in the automotive industry. In: 2015 IEEE Jordan Conference on Applied Electrical Engineering and Computing Technologies (AEECT). IEEE (2015)
Struweg, I.: A twitter social network analysis: the south African health insurance bill case. In: Conference on e-Business, e-Services and e-Society. Skukuza, South Africa (2020)
Sukanya, M., Biruntha, S.: Techniques on text mining. In: 2012 IEEE International Conference on Advanced Communication Control and Computing Technologies (ICACCCT). IEEE (2012)
Wasserman, S., Faust, K.: Social Network Analysis. Methods and Applications. Cambridge University Press, Cambridge (1994)
Wong, P.C., Thomas, J.: Visual analytics. IEEE Comput. Graph. Appl. **5**, 20–21 (2004)
Zhang, C., Sun, J., Zhu, X., Fang, Y.: Privacy and security for online social networks: challenges and opportunities. IEEE Netw. **24**(4), 13–18 (2010)

COVID-19-Related Information Classification: A Case Study Based on Algerian Online Discussion

Benfredj Rima, Bouziane Abderraouf, and Nouioua Farid

Abstract During a crisis, the public trends of information consumption and sharing on online social networks (OSN) have changed. Users used various OSN platforms to understand the situation, acquire needed information, and exchange their behaviors and opinions about emergencies. Research demonstrated that filling the need of information is useful, but it is more useful identifying the key features that can predict the propagation of this information. Some studies analyze the circulation of information during a health emergency like COVID-19. Only a few of them target to show COVID-19 information types. Thus, this study aims to shed new light on the information that circulated on OSN on the COVID-19 pandemic; first, this chapter discusses the spread of information in general on social networks, and then it theoretically studies the comments on social networks on the COVID-19 in order to classify them into three different types or categories. For this, we have chosen the Algerian society, on the basis of their online discussions concerning the COVID-19 pandemic. Identifying OSN information types and understanding how it is being propagated are very beneficial to improve appropriate information publishing strategies.

Keywords COVID-19 · Online social networks · Information diffusion · Information types

1 Introduction

In our time, the exchange of information has become one of the most important social activities. The invention of the internet and online social networks in our world improved this exchange and allowed information to be easily spread from one to another through interactions between users of online social networks (Zhang

B. Rima (✉) · B. Abderraouf · N. Farid
Departement mathématique et informatique, Université de Bordj Bou Arreridj, Bordj Bou Arreridj, Algeria
e-mail: Rima.benfradj@univ-bba.dz

S. Sedkaoui et al. (eds.), *International Conference on Managing Business Through Web Analytics*, https://doi.org/10.1007/978-3-031-06971-0_14

and Yu 2020). Guille et al. (2013) define an online social network (OSN) as a consequence of using a web service called a social networking site (SNS), which allows the user to (i) create his or her profile in the form of a page and post messages and (ii) thus create social relationships by explicitly connecting to other users. On the other hand, information propagation between various users in the course of time in a social network is called information diffusion (Singh 2019). In fact, OSNs play an important role in the diffusion of information due to the ease of access to a very large source that allows unlimited spread.

OSNs have been used to disseminate informations in various situations, and have proved to be very powerful like the 2008 US presidential elections whose Twitter had a major role (Guille et al. 2013). Another instance, the 2010 Arab Spring (Guille et al. 2013), Collective Action and Social Movement, taking the example of 22nd February 2019 "El Hirak" in Algeria whose social networks were the only means used to make people's voices heard, they allowed demonstrators to share and disseminate their information and opinions on a large scale.

Currently, the world is facing a situation never seen before that is the emergence of a new coronavirus (SARS-CoV-2) (Rock 2020). Coronaviruses are "a large family of viruses which may cause illness in animals or humans. In humans, several coronaviruses are known to cause respiratory infections. The most recently discovered coronavirus 2 (SARS-CoV-2) causes COVID-19" (Sakib 2020). From December 2019, China has become the country of origin of this new coronavirus, COVID-19 has spread rapidly in China, and since then, it has infected other countries around the world. Many models link the influence of the media on the spread of epidemics, and there is evidence that there is a clear correlation between the two. These studies assume that media effects slow down the rate of transmission, thus allowing the correct identification of preventive measures, and with a causal relationship, this helps to reduce the spread of diseases (Shanlang et al. 2020). Other researchers explore that the media impacts increase with the number of people infected (Sun et al. 2011; Liu et al. 2007; Cui et al. 2008). When the number of cases increases and with a higher probability that the disease will be transmitted quickly, the dissemination of information about this disease can reduce their spread and create an interesting dynamic of disease transmission (Liu et al. 2007; Cui et al. 2008).

In this time of crisis, we have been able to show that social networks can be used as an awareness and prevention tool during a pandemic where the whole world has adopted social network platforms as reliable sources of information about COVID-19. We notice that in the first 4 months of 2020, since the outbreak of the COVID-19 pandemic, OSNs like Facebook, Instagram, Snapchat, Twitter, WhatsApp, and YouTube posted and shared a lot of information and news reports about it and have played a positive role in promoting effective strategies to urge and help individuals cope with social and physical distancing (Hussain 2020). Likewise, the establishment of online social groups has also proven to help people stay connected during this pandemic.

Beyond being a research tool, online social networks alongside other media platforms can be used robustly and as a reliable source for COVID-19, of which

Facebook has been the most used social network (Olum and Bongomin 2020). Twitter has also proven its worth to positively impact COVID-19 in terms of information sharing, encouraging social distancing, and helping people to face emotional changes (Thelwall and Thelwall 2020).

Abd-Alrazaq et al. (2020) aimed to identify the main topics posted by Twitter users related to the COVID-19 pandemic. They collected coronavirus-related tweets between February 2, 2020, and March 15, 2020, using a set of predefined search terms ("corona," "2019-nCov," and "COVID-19"), which are the most widely used scientific and news media terms relating to the novel coronavirus. They obtained 167,073 English tweets with coronavirus-related terms, after analyzing these tweets; they identified 12 topics and grouped them into 4 themes: the origin of COVID-19, the source of a novel coronavirus, the impact of COVID-19 on people and countries, and the methods for decreasing the spread of COVID-19.

During a health emergency like the COVID-19 pandemic, social media users accessed different platforms to gain needed information and exchange their opinions at unprecedented scale (Li et al. 2020). This could highlight that user trends in information consumption and sharing on social media will be able changed (Pulido et al. 2020).

There are relatively different studies on COVID-19-related information, and only a few of them classified COVID-19-related information into different categories. In order to do these studies, researchers used various social network data, such as Twitter, and based on different discussions around the world, no one of them takes into consideration the Algerian context, in other words, the Algerian online discussion. In addition, we notice that they didn't classify the COVID-19-related information according to the meaning of the information and that's what made the terms used in their classifications so closer to each other; also, they didn't make a classification per subject, i.e., according to the purpose and the objective of this information.

To fill these research gaps, we focused on Algerian online publication and followed the posts and comments posted by different Algerian accounts in order to deal with this pandemic. In this work, we aim to categorize COVID-19-related information into three categories; Firstly, if the information that circulates in an online social network during a pandemic and its content intended to provide news in the medical and health sense, it is classified into a category called "Medical Information"; besides, this category collects all information that have a medical relationship and provide a self-protection, for example, news of the evolution of the pandemic, awareness and prevention measures, and also all medical services provided during this epidemic. Secondly, we put the "Notification and Donations" category. This category includes preventive measures taken by the state and the different kinds of offers of help proposed since the outbreak of the pandemic. Lastly, we have "Rumors and Counter-Rumors" category; during the spread of an epidemic, there is a high probability that many rumors will invade social media, and, in some cases, we find that the concerned authorities refute these rumors, so-called counter-rumors.

Filling information needed during sudden epidemics such as COVID-19 is critical. But the most important and useful of that is identifying the key features that can predict the propagation scale of this information (Abd-Alrazaq et al. 2020). Identifying information types and predicting its propagation and adoption are useful at the same time for the public and for the authorities. It would benefit the concerned authorities to sense the mood of the public to ensure the publish of different types of information since the crisis outbreak, based on the needs of the public; it would then help them come up with proper emergency response strategies. As the result, it allows them to fill information gaps between the authority and public to solve the problem of information need (Yan and Pedraza-Martinez 2019).

2 Related Works

When a disease spreads among people, online social network is shaken by a lot of information about the disease, which spreads immediately. The information helps in understanding the situation during emergencies, and it is useful for both the public and authorities to guide their responses (Li et al. 2020). Research has proven that information improves the public's awareness of the virus (Shanlang et al. 2020).

However, different researchers studied COVID-19-related information, and only a few of them have categorized it into different categories. L. Shanlang et al. (2020), in order to find the correlation between information diffusion and the spread of COVID-19, divided the information into two categories. One is information on COVID-19 that the public wishes to know. The other is the self-protection information during the epidemic search by the public in order to protect themselves from the virus. Moreover, L. Li et al. (2020) categorized seven types of COVID-19-related information: (1) caution and advice; (2) notifications and measures taken; (3) donations of money, goods, or services; (4) emotional support; (5) help seeking; (6) doubt casting and criticizing; and (7) counter-rumor.

We have noticed through these categorizations that the meaning of these concepts is closer to each other, and from here, we have found an idea to make a categorization according to subject, i.e., classify the information propped and published on the social media since this emergence is related to the COVID-19 pandemic according to their senses and specifically according to their objectives.

3 COVID-19-Related Information Classification

On February 17, 2020, an Italian man arrived in Algeria; and a week later, Algeria confirmed its first case of COVID-19 (Boukhatem 2020). This news came down like a shock to the hearts of many Algerians and caused panic among the Algerian society. This pandemic began to spread and penetrate among Algerian people. As no cure has been found for this deadly pandemic at that time, the relevant authorities need to provide a number of measures and precautions to reduce the spread of the pandemic, and OSN was their first destination to disclose their instructions.

During a crisis, various types of information are posted at an enormous volume and rapid rates. These different types of information have a different benefit which helps the concerned authorities or individuals acquire a high-level understanding of the emergency situation (Abd-Alrazaq et al. 2020).

In this work, we rely on Algerian online discussions in order to classify COVID-19-related information in general in three different categories. Firstly, we have *the Medical Information category*. As we mentioned earlier, health organizations have no other choice other than educating people about the need to be careful. For this, and to ensure the dissemination of a gigantic mass of precautions, medical teams have used social networks as a means of communication and sharing a list of advices such as staying at home, social distancing, frequent handwashing, wearing masks, etc. in order to raise awareness and protect their healthcare providers. Many Algerian medical teams set up online applications, pages, and also groups which have various names such as "Coronavirus COVID-19 Informations and Sensibilisation DZ" with 589 K members, "Algerian Doctors," and "COVID-19 (SARS CoV2) Algérie DZ" only for information purposes to help smooth the care of people by the emergency services during the pandemic of COVID-19. In addition, there are worldwide tools which are provided free of charge, taking the example of MaladieCoronavirus.fr, a website set up by the Pasteur Institute to self-test with coronavirus; this site is also made available to all citizens to inform, guide, and advise them how to protect themselves and others on a daily basis.

Sharing this type of information allows the authorities to explain the face of the virus and ensure a wide diffusion of information. On the other hand, sharing this type of information helps people to gain a lot of medical information to protect themselves from the virus. The hashtag option has a significant effect on the diffusion of medical information; it makes its propagation faster, deeper, and more widely spread. The best proof of that the Algerian hashtags for example '#أقعد في داركم' it means stay at home, which make the buzz and invaded Facebook pages, also '#ريح في داركم' hashtags which lead the Algerian trend from the virtual platform Twitter, in order to convince the citizen to not be reckless and cynical, and protecting himself is a saving for the lives of others.

Secondly, we have **Notification and Donations type**. This category has two sections: one is the **Notification section**, which includes all posts aimed to inform the public about the details of the epidemic since its emergency such as the update of the daily infected and death and recovery cases and rates. Furthermore, this section includes a number of decisions and instructions announced by the president himself and other relevant authorities to reduce the spread of coronavirus, including the decision to close all educational institutions on March 12, 2020, as well as decisions on the containment time and the wilayas concerned. These news and information have been widely disseminated on OSN compared to other means of communication.

Sharing this type of information is useful for both the public and authorities to understand the situation, to plan relief efforts accordingly, and to guide their responses and behaviors.

The COVID-19 pandemic is like any crisis that causes human and material loss, but what made this pandemic lighter is that it has seen many offers of help since it begun; those aids are donations of money, goods, and services (Abd-Alrazaq et al. 2020). The second category is divider in two section, Notification section and Donation type, the last collect all the information which aids the public and people who need help to know what kinds of donations are available. At this point, we can refer to the voluntary campaigns initiated by some young Algerians to fight the epidemic, including Khansa Khalil hashtag, a young man from Laghouat, "أقعد في داركـ أنا نصر فلكـ.#" His goal is to buy the needs of citizens, especially those with chronic diseases, instead of going out. Sharing the donations informations had a significant effect, it helps to get collective support and empathy, taking the example of the hashtag "nous somme tous blida" 'posts to show all their compassion and solidarity with their brothers in Blida.

Each event has consequences, whether physical, moral, or material. From there, we liked to combine notifications and donations information on the same category because we observe that every decision made during this pandemic causes many consequences and it requires the authorities and the public to take the necessary measures so-called offers of help.

Finally, we have the last category **Rumors and Counter-Rumors**. The term fake news found and distributed in online settings has been increasingly used over the last couple of years, especially linked to the US presidential election of 2016 (Li et al. 2020). Following Y. Wang et al. (2019), false information or fake news describe any form of falsehood, including rumors, hoaxes, myths, conspiracy theories, and other misleading or inaccurate shared or published content. Rumor is an information that both is false and means damage, which prospers in social media. Most often, it is used for political targets (Alam et al. 2020). In addition, research shows that in the OSN, rumors diffused significantly farther, faster, deeper, and more broadly than the truth in all information categories (Vosoughi et al. 2018) and, on average, were 70% more retweeted than veracious information (Li et al. 2020).

In this context, the spread of false information about the COVID-19 pandemic is rapidly developing (Li et al. 2020). The industry of rumor and misinformation is at all times a crime against societies and their harmony, but in times of crisis, as is the case today with coronavirus, it could cause a lot of problems, because it goes beyond the objectives of incitement, maneuver, and spread of panic and fear, and destroy people's immune capacity in the case of the deadly virus.

Algeria is one of the few countries that have coexisted with the popular movement "El Hirak" for more than a year, which in turn has witnessed a huge amount of spreading false information. With the advent of the COVID-19 virus, political rumor promoters have changed their direction to pandemic promoters by making a large number of rumors, including the rumor of the death of Wali from Mascara province, while he was infected and did not die. Moreover, a French channel belonging to the French Ministry of Foreign Affairs launched a serious statement that in Algeria, the Chinese medical aids have directed to the military sector instead of directing it to citizens. This information was unfortunately frequently picked up by opponents of the government, while the latter considers it fake news, and there's no evidence to prove it and the Algerian Foreign Ministry was forced to summon the French ambassador to Algeria to protest. Another no less serious information of the previous is the video disseminated on a Facebook page "Algeria live-الجزائر لايف," which includes a woman claiming that quarantined people leave the tourist complex "the Andalusians" in Oran before finishing 15 days. The government considers it false and misleading information.

The health sector, like other sectors, has not been spared from rumors during this health emergency, which includes rumors circulating on social media about the COVID-19 pandemic, which sparked a wave of outrage such as "Coronavirus is transmitted through the air" while it is transmitted by the infected through sneezing or coughing, as well as the information that "All cases infected with this virus are dangerous and deadly." This information is false and misleading, as international studies have shown that 80% of the infected are recovering spontaneously and the mortality rate is between 2% and 3%. Among other things, there are references to promoting fake cures, like drinking bleach and vinegar to cleanse the body of the virus and also eating garlic and oregano to protect yourself from infection. Sharing this type of false remedies, rumors, and conspiracy theories is intended to spread panic, influence people's opinions, and mislead society with false allegations.

In the case of the rumor launched by users who have a higher number of followers or come from developed cities, inevitably, it will see a wide spread, and this requires the concerned authority refuting this false information through counter-rumors. **The counter-rumors** and rapid answers to these rumors help the public learn the truth and reduce the confusion caused by rumors (Abd-Alrazaq et al. 2020). Further, in the case of dangerous situations like COVID-19, publishing these types of rumors encourages the spread of panic and fear among people, and that requires the necessary procedures. WHO and the relevant authorities, in order to refute many COVID-19-related rumors, issued several statements which are called "counter-rumors" that mislead all these rumors, considering them false information.

Besides, counter-rumor is crucial to help authorities direct their efforts and limit the spread of false information that may cause panic, mistrust, and other problems in the society. In fact, it also aids health authorities to be up-to-date on how social media users share information and publish more posts from their official accounts (Li et al. 2020).

4 Conclusion

With the outbreak of the COVID-19 pandemic, there are many trials on the use of social networks to address this pandemic. Thus, it is important to recognize the critical role of online social networks to disseminate information to a large number of people in the same time of health emergency. Online users during this situation tend to understand the situation and to fill their needs of information. For that, we noticed a wide and unlimited sharing of a huge amount of information. The findings of this chapter indicate the necessity of classifying these various information into different types. We have used the Algerian context to assert that the classification that is based on the Algerian online discussion is different. In addition, we aim to clarify that identifying these types of information and understanding how it is being propagated are critical and help to improve information publishing strategies especially during sudden emergency situations; Classifying information into different type are critical and helps to improve information publishing and understanding how it is being propogated, in the covid-19 situation, it helps learn how to organize covid-19 related posts to ensure the publications of information and based on the needs of the public. Also, it benefits the authorities to come up with proper emergency response strategies. In general, it is useful for both authorities and individuals. Moreover, this theoretical work will be beneficial for researchers or practitioners who aim to build emergency response programs and crisis information systems (Zhang and Yu 2020).

This chapter has some limitations that we aim to overcome in future work. In the present work, we suggest a theoretical classification of information that needs a data analysis, and results demonstrate this categorization. For that, firstly, we plan to provide a framework for a deeper understanding. We will also collect information extracted from real social networks and use other classification methods to identify other keys that may influence the diffusion and adoption of this information. Furthermore, we are working on information adoption and diffusion process modeling that can help the Algerian public during such health-related emergencies.

5 Disclaimers

This work is not intended to make any form of political or social commentary whatsoever on the current situation of COVID-19 spread in Algerian or on the strategies applied to face it. The analysis done is purely based on deductions made of the data set at hand.

References

Abd-Alrazaq, A., Alhuwail, D., Househ, M., Hai, M., Shah, Z.: Top concerns of tweeters during the COVID-19 pandemic: a surveillance study. J. Med. Internet Res. **22**(4) (2020). https://doi.org/10.2196/19016

Alam, F., Shaar, S., Nikov, A., Mubarak, H., Martino, G.D.S., Abdelali, A., Dalvi, F., Durrani, N., Sajjad, H., Darwish, K., Nakovi, P.: Fighting the COVID-19 Infodemic: modeling the perspective of journalists, fact-checkers, social media platforms, policy makers, and the society. arXiv. (2020)., Available: http://arxiv.org/abs/2005.00033

Boukhatem, M.N.: Novel coronavirus disease 2019 (COVID 2019). Outbreak in Algeria: a new challenge for prevention. J. Commun. Med. Health Care. **41**(2), 2–7 (2020). [Online]. Available: https://www.ncbi.nlm.nih.gov/pubmed/32064853

Cui, J., Sun, Y., Zhu, H.: The impact of media on the control of infectious diseases. J. Dyn. Diff. Equat. **20**(1), 31–53 (2008). https://doi.org/10.1007/s10884-007-9075-0

Guille, A., Hacid, H., Favre, C., Zighed, D.: Information diffusion in online social networks: a survey. ACM SIGMOD Rec. **956**(2) (2013). https://doi.org/10.1145/2503792.2503797

Hussain, W.: Role of social media in COVID-19 pandemic. Int. J. Front. Sci. **4**(2) (2020). https://doi.org/10.37978/tijfs.v4i2.144

Li, L., Zhang, Q., Wang, X., Zhang, J., Wang, T., Gao, T., Duan, W., Tsoi, K.K., Wang, F.: Characterizing the propagation of situational information in social media during COVID-19 epidemic: a case study on Weibo. IEEE Trans. Comput. Soc. Syst. **7**(2), 556–562 (2020). https://doi.org/10.1109/TCSS.2020.2980007

Liu, R., Wu, J., Zhu, H.: Media/psychological impact on multiple outbreaks of emerging infectious diseases. Comput. Math. Methods Med. **8**(3), 153–164 (2007). https://doi.org/10.1080/17486700701425870

Olum, R., Bongomin, F.: Social media platforms for health communication and research in the face of COVID-19 pandemic: a cross sectional survey in Uganda. medRxiv, 2020.04.30.20086553 (2020). https://doi.org/10.1101/2020.04.30.20086553

Pulido, C.M., Villarejo-Carballido, B., Redondo-Sama, G., Gómez, A.: COVID-19 infodemic: more retweets for science-based information on coronavirus than for false information. Int. Sociol. **35**(4), 377–392 (2020). https://doi.org/10.1177/0268580920914755

Rock, G.: Theme: information on the Covid-19 pandemic. Transfus. Apher. Sci. **59**(3), 102791 (2020). https://doi.org/10.1016/j.transci.2020.102791

Sakib, F.S.: COVID-19. Poster. (2020). https://doi.org/10.13140/RG.2.2.22188.00645

Shanlang, L., Chao, M., Ruofei, L., Junpei, H., Ruohan, X., Aini, Y.: Research on the influence of information diffusion on the transmission of the novel coronavirus (COVID-19). medRxiv, 2020.03.31.20048439 (2020). https://doi.org/10.1101/2020.03.31.20048439

Singh, S.S.: A survey on information diffusion models in social networks. Commun. Comput. Inf. Sci. **956**, 426–439 (2019. , ICAICR 2018, Springer Singapore, 2018). https://doi.org/10.1007/978-981-13-3143-5_35

Sun, C., Yang, W., Arino, J., Khan, K.: Effect of media-induced social distancing on disease transmission in a two patch setting. Math. Biosci. **230**(2), 87–95 (2011). https://doi.org/10.1016/j.mbs.2011.01.005

Thelwall, M., Thelwall, S.: Retweeting for COVID-19: consensus building, information sharing, dissent, and lockdown life. arXiv, 0–1 (2020). [Online]. Available: http://arxiv.org/abs/2004.02793

Vosoughi, S., Roy, D., Aral, S.: News on-line. Science (80). **1151**, 1146–1151 (2018)

Wang, Y., McKee, M., Torbica, A., Stuckler, D.: Systematic literature review on the spread of health-related misinformation on social media. Soc. Sci. Med. **240**, 112552 (2019). https://doi.org/10.1016/j.socscimed.2019.112552

Yan, L., Pedraza-Martinez, A.J.: Social media for disaster management: operational value of the social conversation. Prod. Oper. Manag. **28**(10), 2514–2532 (2019)

Zhang, J., Yu, P.S.: Information Diffusion Broad Learning through Fusions, pp. 315–349. 9 9.1, no. June. Springer Nature Switzerland AG 2019 (2020). https://doi.org/10.1007/978-3-030-12528-8_9. Journalists, Fact-Checkers, Social Media Platforms, Policy Makers, and the Society, [Online]. Available: http://arxiv.org/abs/2005.00033

User Similarity and Trust in Online Social Networks: An Overview

Aya Zouaoui, Meriem Laifa, and Samir Akrouf

Abstract In the last 10 years, the rapid development of online social networks has displaced traditional interpersonal relationships in favor of online communications. When it comes to enhancing interactions and ensuring the safety of such online communications, having a foundation of trust is critical. A user's trustworthiness is influenced by a variety of factors, including how similar they are to each other. There are numerous uses for the concept of user similarity. In the context of online trust, this chapter provides an overview of the user similarity concept. We provide a literature classification based on well-established studies that includes characteristics, data, and applications that researchers in this field use to measure similarity. Finally, we look at other studies that examine the importance of trust and similarity in social networks.

Keywords User similarity · Trust · Online social network · Literature classification · Metrics

1 Introduction

Online social networks (OSNs) may be defined as applications or online platforms that people can incorporate into their daily practices and express themselves, such as Facebook, Twitter, LinkedIn, Snapchat, and Instagram. A social network is generally represented by a set of nodes that are individuals or organizations, and the relationships between these nodes are represented by edges. Among the advantages of online social networks is allowing users to create profile pages, share information, and connect with others on a variety of topics (personal, professional, etc). It also

A. Zouaoui (✉) · M. Laifa
Department of Computer Science, Bordj Bou Arreridj University, Bordj Bou Arreridj, Algeria
e-mail: aya.zouaoui@univ-bba.dz; meriem.laifa@univ-bba.dz

S. Akrouf
Department of Computer Science, Mohammed Boudiaf University, M'sila, Algeria
e-mail: samir.akhrouf@univ-msila.dz

S. Sedkaoui et al. (eds.), *International Conference on Managing Business Through Web Analytics*, https://doi.org/10.1007/978-3-031-06971-0_15

allows to build and develop social relationships with individuals of similar interests around the world. Hence, people are not constrained anymore by the traditional geographic boundaries which became an enduring part of everyday life (Dabeeru 2014).

Resulting from the persistent increase in the number of OSN users and to the variety of their activities, a massive amount of information is generated, creating what is known as "information overload" phenomenon (Toffler 1970). This phenomenon makes it hard for users to make decisions and for decision-makers to estimate users' preferences that are essential for different applications (Tsakalakis and Koutsaki 2018; Malhotra 1982; Jin et al. 2019). This massive information is coming from online interactions between users. The online interactions are connected to the important element which is trust. Trust is an important link between individuals, members, or societies. Any relationship is built on trust, and people need to feel physically and emotionally safe. In an online context, users find it hard to evaluate trust in people, services, or products before making any decision. Trusting someone does not necessarily mean sharing the same preferences or interests with them (Ziegler and Golbeck 2007). However, trust may include certain information about personal identity and depends on the similarity of users' profiles for evaluation (Kalai et al. 2016). In this context, it is hard to estimate users' preferences and detect the similarities or the differences between users, that is why researchers rely on users' profiles on social networks and their behavioral data (Bhattacharyya et al. 2011; Mohammad et al. 2012; Garmsiri and Hamzeh 2015; Chakravarty et al. 2016).

In the following sections, we talk about some of the most works that studied the correlation between trust concept and user similarity – according to users' personal information or behaviors – in online social networks. Though the description of similarity by its definitions in different applications is adopted by many researchers in their literature studies, they focus on metrics which are used to measure user similarity and overview its features. In our overview, we aim to complete the research gap in the literature of user similarity. We categorize user similarity based on three criteria: (i) characteristics-based similarity, (ii) data-based similarity, and (iii) applications-based similarity. Each in turn can be further classified:

Based on characteristics into three categories: social attribute, social structure, and social activity.
Based on data: profile similarity, behavior similarity, and network similarity.
Based on application: P2P system, location prediction, link prediction, and recommendation system.

In the remaining sections of this chapter, we describe the fundamentals of similarity where we provide a definition of the user similarity concept and we explain briefly its metrics. In Sect. 3, we present our user similarity literature classification. We address trust and user similarity-related work in Sect. 4, and we conclude in Sect. 5.

2 Similarity Fundamentals

2.1 *Defining User Similarity*

User similarity is the resemblance between two users who have the same characteristics, interests, or behavior. The relationship among users in online social networks can be established in similarity indications (Maurya and Singh 2016). This last one may be defined as when users share common attributes between each other like common preferences or connections with the same people.

User similarity on online social networks can be perceived from two angles: a network similarity and a profile similarity (Akcora et al. 2013; Akcora et al. 2011). From the first angle, similarity refers to the resemblance in the graph structure between two nodes according to their connections. It has stemmed from early work on document classification (Akcora et al. 2011; Cover and Thomas 1991; Deshapande and Karypis 2004). From the second angle, profile similarity is the set of similar personal data stored in profile items of two social network users. The different attributes which represent personal information, such as age, education, or job, show the nature of the relationship between users.

2.2 *User Similarity Metrics*

In the literature, there are many user similarity metrics used to extract or compute the similarity between users. However, our intention in this chapter is to define the metrics that can be used to estimate how users are similar by such numeric attribute vectors as contacts, age, etc. and such semantic information as posts, share, or comments that can be found within the OSNs.

Cosine Similarity It is defined as the angle between both users when they are represented in the item space. It uses two vectors that contain multiple attributes to calculate the cosine angle. Accordingly, the cosine similarity can take on values between -1 and $+\,1$. If the vectors point to the exact same direction, the cosine similarity is $+1$. If the vectors point to opposite directions, the cosine similarity is -1 (Maurya and Singh 2016; Xiao 2015). Cosine similarity is used in various applications (Lee and Tukhvatov 2018). used cosine similarity to find like-minded people with the same interests in social networks. In neighbor-based recommender systems, the most similar neighbors are calculated by using cosine similarity between the two users (Bellogín and De Vries 2013).

Jaccard Similarity The Jaccard index and distance are two metrics used in statistics to compare similarity and diversity between samples (Niwattanakul et al. 2013; Hossny and Mitchell 2019). They are named after the Swiss botanist Paul Jaccard (Choi et al. 2009). The Jaccard similarity is the relationship between two sample data. It returns the result of the division between numbers of attributes that are common between them, divided by the total number union of attributes selected for similarity measure (Maurya and Singh 2016; Xiao 2015). In a recent work (Verma and Aggarwal 2020), the authors showed that the Jaccard index is the measure of the asymmetric binary similarity between two data objects with binary attributes in data mining jargon and in a recommender system.

Pearson Similarity It is the correlation between the rating patterns between two users. It takes into account the variety of individual user reviews. It is represented in an item space like the cosine, but it is invariant to displacement, for example, if the x object was moved to x + 1, the Pearson similarity would not change. The Pearson correlation coefficient subtracts the average rating score from each rating, thereby eliminating the individual subjective differences (Xiao 2015). It gives a value ranging from −1 to 1; thus, it can represent a negative correlation when below 0 (Maurya and Singh 2016; Akcora et al. 2011; Bellogín and De Vries 2013).

Levenshtein Similarity The distance of Levenshtein measures the difference between two input string attributes. It is equal to the minimum number of characters that must be removed, inserted, or replaced to move from one string to another (Kang 2015; Singh et al. 2015; Haldar and Mukhopadhyay 2011). Levenshtein is suitable to use for determining the similarity between information to estimate the similarity between users. In (Po 2020), authors showed word order similarity between sentences using the Levenshtein distance. The paperwork suggests that methods based on distance measures are the best in similarity-based information.

Jaro-Winkler Similarity Jaro-Winkler is a metric measuring a distance between two sequences and calculates the similarity between characters. It is proposed in 1990 by William E. Winkler of the distance metric Jaro. The result of this metric is a measure between 0 and 1, where 0 represents the absence of similarity and 1 the equality of the chains compared. This is particularly suited for dealing with short strings such as names or passwords (Maurya and Singh 2016). In the identity resolution problem in the online social network, users tend to create a profile with multiple attributes (Yadav et al. 2019). focused on developing a novel framework for matching and merging redundant user profiles with the Jaro-Winkler similarity techniques on Twitter social media.

Mutual Information Mutual information is a measure of the degree of relatedness between two variables. In online social networks, these variables represent comments, tags, or conversations between users. In (Markines et al. 2009), authors proved that mutual information can estimate social tag similarity with higher accuracy compared to overlap, Jaccard, Dice, and cosine similarity. In (Zheng 2019) to measure co-citation similarity, the mutual information is calculated based on researchers' publications and their co-citation frequencies.

Table 1 represents the formulas of each described metric.

Table 1 Similarity metrics' formula

Similarity metrics	Formula
Cosine similarity	Scos (u, v) = cos(Θ) = Au. Av ‖Au‖ ‖ Av ‖
Jaccard similarity	Sj (u, v) = | u ∩ v | | u U v |
Pearson similarity	$Sp(u, v) = \frac{\sum (u_i - u^-)(v_i - v^-)}{\sqrt{\sum (u_i - u^-)^2} \sqrt{\sum (v_i - v^-)^2}}$
Levenshtein similarity	Sl (u, v) = (|u|, |v|) length of u and v.
Jaro-Winkler similarity	Sj (u, v) = 0 if m=0 $\frac{1}{3}\frac{m}{\|u\|} + \frac{m}{\|v\|} + \frac{m-t}{m}$
Mutual similarity	Smi (u, v) = ∑au∈Au ∑av∈Av p (au, av) log p (au; av) p(au) p(av)

3 Similarity Literature Classification

In online social networks and different applications such as recommendations and link prediction, researchers introduced user similarity in their searches that addressed several similarity perspectives and research areas (Liben-Nowell and Kleinberg 2007; Cho et al. 2011; Liu and Lee 2010; Raad et al. 2011). In our overview, we classify the reviewed works into three main categories as shown in Fig. 1: based on characteristics of similarity, based on the type of user data, and based on similarity application.

3.1 *Characteristics-Based Similarity*

Social network platforms use varied features or characteristics of similarity to represent a user (Xiao 2015), and they can be categorized into three classes as follows.

3.1.1 Social Attributes

Social attributes are all the features derived from users' basic information. They include age, job, gender, location, education, interest, etc. All the profile information can be explored to investigate users' characteristics in OSNs, such as homophily (Maurya and Singh 2016; Xiao 2015). These social attributes provide a clear comprehension of users' characters. For example, (Raad et al. 2011) used interests and available profiles information to recommend new items such as home page skin,

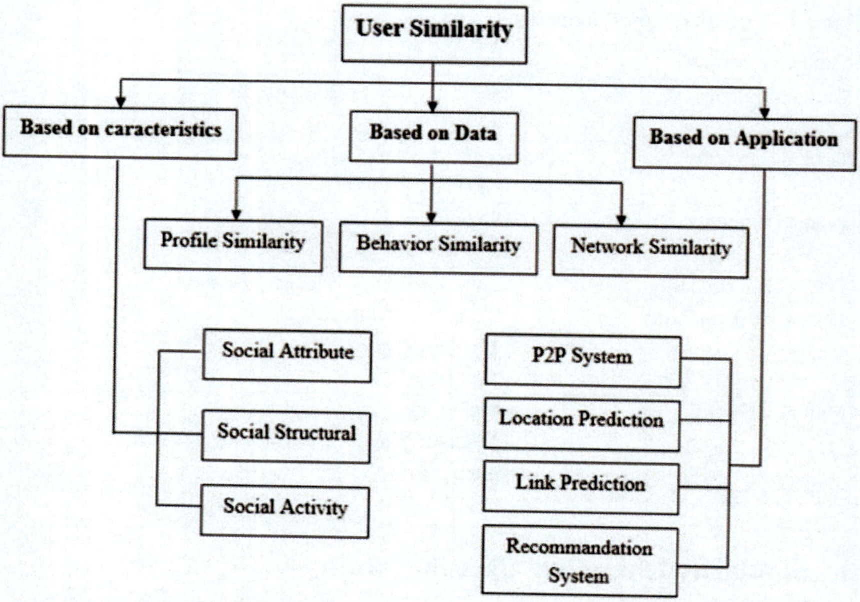

Fig. 1 User similarity literature classification.

background music, and virtual appliances for home page users (Backstrom et al. 2010). used a preference attribute in user profile matching to discover the biggest possible number of profiles that refer to the same physical user. We cannot rely only on these attributes because some OSN platforms (e.g., Facebook) limit access to users' profile information, while in other networks, access is possible.

3.1.2 Social Structural

In this category, features are defined on the basis of the links present in the users' social graph. Researchers have studied these links which reflect the relationship between users in a social network, for example, in a geographic relation when the user can interact with others who have the same location as him (Liang and Zhang 2012; Abrol and Khan 2010), in a mutual friend relationship when users have the same age or similar interests (Li et al. 2019), and in a follower-following relation which directs the information flow and reflects users' influence (Xiao 2015; Kraus et al. 2015).

3.1.3 Social Activity

Users' behavior on social networks represents what users do online and covers various social activities. These activities are varied depending on the objectives and areas of application (Backstrom et al. 2010; Nguyen et al. 2018). For example, a user can chat with friends, publish posts and comment on others, share information, make new relationships, etc. These activities offer valuable information when studying the properties of information flows in OSNs (Xiao 2015) and identifying the interests extracted from a user based on his/her messages, comments, page visits, and scoring (Backstrom et al. 2010).

3.2 *Data-Based Similarity*

Because of the variety of data used to calculate or measure user similarity, we classify the similarity into three categories: profile similarity, behavior similarity, and network similarity.

3.2.1 Profile Similarity

A user profile is a representation of information about an individual user such as gender, age, contacts, and others. Profile similarity is to determine how much social network users are similar by comparing personal information stored in the profile items of users. Most of the research worked on profile similarity for various aims. For example, (Dabeeru 2014) calculated the similarity score and compared between the profiles to predict whether one person is associated with another. Authors take as attributes hometown, friend list, gender, and city, and used the decision-making algorithm for deciding the match, degree of closeness, and interaction level in a social network among user profiles. The analysis-based profile similarity can help to evaluate trust between users in OSNs (Maurya and Singh 2016). The analysis based profile similarity can help to evaluate trust between users in OSNs, (Maurya and Singh 2016) proposed model evaluates trust used preferred attributes among profile attributes which are location, education, interest, and friends. They applied all similarity metrics. Measuring profile similarity for computing the user similarity among users in OSNs allows the use of classical metrics such as cosine similarity and Pearson correlation coefficient. Concerning features of user similarity, social attributes adopt more because they refer to all basic information of users.

3.2.2 Behavior Similarity

Determining the similarity between users also refers to other resources' information. These resources represent the activities of users within online social networks.

Among different information, there are comments, posts, shares, likes, forwarding, tagging, communicating, etc. (Malekmohammad and Hadi Khosravi-Farsani 2016) estimated the similarity between users based on their behavior on the social network Twitter. The considered behaviors are activities including posting, liking, commenting, sharing, and joining a group. Authors in (Mohammad et al. 2012) proposed a method that employs user similarities to extract trust values. This method used behavior information to calculate user similarity which are posts and comments from the social network Facebook. They have shown that they can use the behaviors of users to estimate or predict trust in OSNs. Social activity characteristics were used in behavior similarity.

3.2.3 Network Similarity

In this category, identifying users who are similar in online social networks is based on the structure of social networks. The network similarity refers to the user's position in the network graph with the features as vertex nodes, vertex degree, clustering coefficient, eigenvector centrality, and average shortest path length (Patel et al. 2018). The network similarity uses structural metrics such as Jaccard and Dice, SimRank, and Katz metric to compute the user similarity. Estimating network similarity between a target user and stranger user, (Akcora et al. 2013) counted the number of mutual friends between the two users, taking into account friendship edges among the mutual friends. In (Garmsiri and Hamzeh 2015), authors calculated user similarity by using the connection between two users on the graph and the k-nearest neighbor method using indirect links to evaluate trust between users. They used vertices of graph features to calculate the maximum link between trustor and trustee, and the similarity between users is computed by weight of paths in the graph.

3.3 Applications-Based Similarity

Various applications employ the similarity between users to achieve specific goals. One of these applications is peer-to-peer (P2P) networks. Identifying users who have the same interests is an important task to discover the content in a P2P social network. There are many studies benefiting from a variety of techniques for establishing peer-to-peer relationships based on their similarity (Lu et al. 2008; Liben-Nowell and Kleinberg 2007; Nguyen et al. 2018). Recommendation systems are another type of application of user similarity. We can recommend services or products to a user based on his or her personal interests that are similar to others' interests (Systems 2015; Qiao et al. 2017). The user similarity is also found in the location prediction system where the location of users becomes a kind of information essential to understanding the spatial structure of online social networks. They also recommend new users that are similar geographically with other users and establish a relationship with them. For example, the social recommender system can suggest new friends share a GPS data point with others. As the calculation of the similarity between users of a social network, it allows to estimate their next location and trajectory (Sarwar et al. 2013; Khan et al. 2016). In the link prediction application, authors employed membership in interest groups as an information to predict links for new users. They applied user similarity features such as number of common groups and size of common groups (Abrol and Khan 2010; Phukseng and Sodsee 2017).

4 Related Work of User Similarity and Trust

Many studies on trust evaluation have been achieved by researchers by developing a wide variety of trust and user similarity models, especially in OSN context. In order to evaluate trust between users, (Xiao 2015) proposed a trust computing system for simulating trust among two directly connected users on an OSN. This system contains three trust computing components based on profile similarity, information reliability, and social opinions. In the first one, to compute the similarity, 40 friends' profiles were collected via OSNs. They treated seven categories which are calculated by two methods: characteristics-based profile similarity, using an algorithm based on ontology structure and cosine similarity, and concept-based profile similarity. In (Mohammad et al. 2012), researchers proposed a new method that employs user similarities to extract trust values. They calculated user similarity from some information resources which are user personal information, the posts, and the comments shared by the user and shared text via text mining techniques. They have shown that there is a correlation between trust and similarity from the experimental results obtained by this method, which are sufficiently acceptable. Authors (Maurya and Singh 2016) proposed a model which evaluated and computed trust between users based on profile similarity analysis and preferred attribute among profiles. They used different similarity metrics, cosine, Levenshtein, Jaro-Winkler, Jaccard, Dice, Monge-Elkan, and letter, and different social and structural attributes such as identity data and social graph information. They found that cosine metrics are better than others. On the other hand, other researchers calculated trust for recommendation systems by considering user similarity and social trust, focusing on the relationship trust between users in the network (Davoudi and Chatterjee 2018; Ayub et al. 2020; Golbeck 2009). It was based on a trust propagation and concerning an item-rating similarity among users within the social network.

Another work in this area has confirmed the existence of a correlation between trust and user similarity. Ziegler and Golbeck (Ziegler and Golbeck 2007) presented two frameworks for analyzing the correlation between interpersonal trust and interest similarity based on evidence of socio-psychological research. Moreover, (Pooria Taghizadeh Naderi and Fattaneh Taghiyareh 2020) investigated some attributes of profile similarity and its effect on trust between users. The results showed a strong and positive correlation. Predicting the trust between two users with similarity measures, (Zhan and Fang 2011) introduced the LookLike algorithm. The results of the LookLike algorithm are used as new features for supervised classifiers to predict the trust/distrust label. They chose a list of similarity measures as Jaccard coefficient, Adamic/Adar, and preferential attachment to examine the study on four real-world trust network datasets. The results demonstrated that there is a strong correlation between users' similarity and their opinion on trust networks. In [54], authors presented a trust computing system for simulating trust between two directly connected users on social networks. They are based on three components, profile similarity, information reliability, and social opinions, in order to evaluate the volume of trust between two users. Measuring profile similarity is dependent on semantic similarity which can be obtained by exploring the similarities between concepts that are obtained from the characteristics of profiles.

5 Conclusion

Online social networks have an important role in people's life. But the lack of credibility and anonymization makes it difficult for the users to build trust relations among them on the network. In this context, user similarity can play a role as a factor to facilitate making the decision in different contexts such as link prediction within social networks and trust between users in OSNs. In this chapter, we gave a comprehensive overview of user similarity. We first presented what is user similarity in online social networks. We also identified similarity metrics used in social context. Then, we set out the user similarity literature classification based on previous works on three categories. Characteristics-based similarity as social attributes, structural and activity, Data based similarity: profile similarity, behavior similarity, and network similarity, and application-based similarity like in P2P systems, recommendation systems, link prediction, and location prediction. We next reviewed some studies on user similarity and trust in OSNs where researchers proved that there is a correlation between these two concepts in online communities. For our future work, we aim to propose a trust evaluation model based on a hybrid model of user similarity, using users' profile data and their behavioral data as well.

References

Abrol, S., Khan, L.: TweetHood: Agglomerative Clustering on Fuzzy k – Closest Friends with Variable Depth for Location Mining, pp. 153–160 (2010)

Akcora, C.G., Carminati, B., Ferrari, E.: Network and Profile Based Measures for User Similarities on Social Networks, pp. 292–298. IEEE (2011)

Akcora, C.G., Carminati, B., Ferrari, E.: User similarities on social networks. Soc. Netw. Anal. Min. **3**(3), 475–495 (2013)

Ayub, M., Ghazanfar, M.A., Mehmood, Z., Alyoubi, K.H., Alfakeeh, A.S.: Unifying user similarity and social trust to generate powerful recommendations for smart cities using collaborating filtering-based recommender systems. Soft. Comput. **24**(15), 11071–11094 (2020)

Backstrom, L., Sun, E., Marlow, C.: Find Me If You Can: Improving Geographical Prediction with Social and Spatial Proximity (2010)

Bellogín, A., De Vries, A.P.: Understanding Similarity Metrics in Neighbor-Based Recommender Systems ACM International Conference Proceedings Series, pp. 48–55 (2013)

Bhattacharyya, P., Garg, A., Shyhtsun Felix, W.: Analysis of user keyword similarity in online social networks. Soc. Netw. Anal. Min. **1**(3), 143–158 (2011)

Chakravarty, S., Yadavl, A., Sibel, R.: On evaluating the effectiveness of rating similarity-based trust. Soc. Netw. Anal. Min. **6**(1), 1–13 (2016)

Cho, E., Myers, S.A., Leskovec, J.: Friendship and mobility: user movement in location-based social networks. In: The 17th ACM SIGKDD International Conference on Knowledge Discovery and Data Mining, pp. 1082–1090 (2011)

Choi, S.-S., Cha, S.-H., Tappert, C.C.: A survey of binary similarity and distance measures. In: WMSCI 2009 – 13th World Multi-Conference on Systemics, Cybernetics. Informatics, Jointly with 15th International Conference on Information Systems Analysis and Synthesis ISAS 2009 – Proceeding, vol. 3, no. 1, pp. 80–85 (2009)

Cover, T.M., Thomas, J.A.: Elements of Information Theory. Wiley-Interscience, New York (1991)

Dabeeru, V.A.: User profile relationships using string similarity metrics in social networks. arXiv preprint arXiv, 1408.3154 (2014)

Davoudi, A., Chatterjee, M.: Social trust model for rating prediction in recommender systems: effects of similarity, centrality, and social ties. Online Soc. Netw. Media. **7**, 1–11 (2018)

Deshapande, M., Karypis, G.: Item-based top-n recommendation algorithms. ACM Trans. Inf. Syst. **22**, 143–177 (2004)

Garmsiri, S., Hamzeh, A.: New graph based trust similarity measure. Department of Computer Science and Engineering, Shiraz University, shiraz, Iran. Ciancia e Nat. **37**(December), 339 (2015)

Golbeck, J.: Trust and nuanced profile similarity in online social networks. ACM Trans. Web. **3**(4), 33 pp (2009)

Haldar, R., Mukhopadhyay, D.: Levenshtein Distance Technique in Dictionary Lookup Methods: An Improved Approach (2011)

Ahmad Hany Hossny and Lewis Mitchell. (2019) Event detection in Twitter: A keyword volume approach. IEEE Int. Conf. Data Min. Work. ICDMW, vol. 2018(January), pp. 1200–1208

Jin, Z., Zhangwen, W., Naichen, N.: Helping consumers to overcome information overload with a diversified online review subset. Front. Bus. Res. China. **13**(1), 1–25 (2019)

Kalai, A., Abdelghani Wafa, C., Zayani, C.A., Amous, I.: LoTrust: a social trust level model based on time-aware social interactions and interest's similarity. In: 14th Annual Conference on Privacy, Security and Trust (PST), vol. 2016, pp. 428–436 (2016)

Kang, S.-S.: Word similarity calculation by using the edit distance metrics with consonant normalization. J. Inf. Process. Syst. **11**(4), 573–582 (2015)

Khan, F., Fatima, M., Alvi, U.T., Jilani, T.: Comparative Study of Similarity Measures in Link Prediction Using Facebook Data. **14**(2), 132–143 (2016)

Kraus, N., Carmel, D., Keidar, I., Orenbach, M.: NearBucket-LSH: Efficient Similarity Search in P2P Networks. Springer, Cham (2015)

Lee, J.Y., Tukhvatov, R.: Evaluations of similarity measures on VK for link prediction. Data Sci. Eng. **3**(3), 277–289 (2018)

Li, R., Wang, S., Chang, K.C.C.: Multiple location profiling for users and relationships from social network and content. Proc. VLDB Endowment. **5**, 1603–1614 (2019)

Liang, G., Zhang, A.: Pseudo cold start link prediction with multiple sources in social networks. In: SIAM International Conference on Data Mining, pp. 768–779 (2012)

Liben-Nowell, D., Kleinberg, J.: The link-prediction problem for social. Networks. **58**(7), 1019–1031 (2007)

Liu, F., Lee, H.J.: Expert Systems with applications use of social network information to enhance collaborative filtering performance. Expert Syst. Appl. **37**(7), 4772–4778 (2010)

Lu, L., Antonopoulos, N., Mackin, S.: Managing Peer-to-Peer Networks with Human Tactics in Social Interactions, pp. 217–236. Springer (2008)

Malekmohammad, A., Hadi Khosravi-Farsani, H.: Structural and non- structural similarity combination of users in social networks. J. Comput. Secur. Struct. **3**(1), 43–52 (2016)
Malhotra, N.K.: Information load and consumer decision making. J. Consum. Res. **8**, 419 (1982)
Markines, B., Cattuto, C., Menczer, F., Benz, D., Hotho, A., Stumme, G.: Evaluating Similarity Measures for Emergent Semantics of Social Tagging, pp. 641–650. ACM (2009)
Maurya, A., Singh, M.P.: Trust evaluation on social media based on different similarity metrics. Int. J. Database Theory Appl. **9**(12), 101–110 (2016)
Mohammad, H., Zadeh, H., Shahriari, H.R.: Using user similarity to infer trust values in social networks regardless of direct ratings. In: 2012 9th International ISC Conference on Information Security and Cryptology, vol. 2012, pp. 66–72 (2012)
Nguyen, T.H., Tran, D.Q., Dam, G.M., Nguyen, M.H.: Estimating the similarity of social network users based on behaviors. Vietnam J. Comput. Sci. **5**(2), 165–175 (2018)
Niwattanakul, S., Thongchai, J.S., Naenudorn, E., Wanapu, S.: Using of Jaccard coefficient for keywords similarity. Lect. Notes Eng. Comput. Sci. **2202**, 380–384 (2013)
Patel, A., Paradkar, S., Parmar, T.: User-Based News Recommendation System Using Twitter, pp. 2175–2178 (2018)
Phukseng, T., Sodsee, S.: Calculating trust by considering user similarity and social trust for recommendation systems. In: Proceedings of the 2017 12th International Conference on Intelligent Systems and Knowledge Engineering. ISKE 2017, vol. 2018, pp. 1–6 (2017)
Po, D.K.: Similarity based information retrieval using Levenshtein distance algorithm. Int. J. Adv. Sci. Res. Eng. **06**(04), 06–10 (2020)
Pooria Taghizadeh Naderi and Fattaneh Taghiyareh: LookLike: similarity- based trust prediction in weighted sign networks. In: 2020 6th International Conference on Web Research ICWR, vol. 2020, pp. 294–298 (2020)
Qiao, J., Li, S., Lin, S.: Location prediction based on user Mobile behavior similarity. In: 2017 IEEE 23rd International Conference on Parallel and Distributed Systems (ICPADS), vol. 3, pp. 7–10 (2017)
Raad, E., Chbeir, R., Dipanda, A.: User profile matching in social networks to cite this version. In: User Profile Matching in Social Networks, pp. 297–304 (2011)
Sarwar, G.., Ullah, F., Lee, S. Temporal-aware Location Prediction Model Using Similarity Approach. no. 3ca, pp. 239–243 (2013)
Singh, S.P., Kumar, A., Darbari, H., Chauhan, S., Srivastava, N., Singh, P.: Evaluation of similarity metrics for translation retrieval in the Hindi-English translation memory. Int. J. Adv. Res. Comput. Commun. Eng. **4**(8) (2015)
Systems, R.: Council for Innovative Research. **14**(9), 6118–6128 (2015)
Toffler, A.: Future Shock. Bantam (1970)
Tsakalakis, G., Koutsaki, P.: Improved user similarity computation for finding friends in your location. Hum. Centric Comput. Inf. Sci. **8**(1), 1–17 (2018)
Verma, V., Aggarwal, R.K.: A comparative analysis of similarity measures akin to the Jaccard index in collaborative recommendations: empirical and theoretical perspective. Soc. Netw. Anal. Min. **10**(1) (2020)
Xiao, H.A.N.: Mining User Similarity in Online Social Networks: Analysis, Modeling and Applications. (2015)
Yadav, S., Sinha, A., Kumar, P.: Multi-attribute identity resolution for online social network. SN Appl. Sci. **1**(12), 1–15 (2019)
Zhan, J., Fang, X.: A novel trust computing system for social network. In: IEEE International Conference on Privacy, Security, Risk, and Trust, and IEEE International Conference on Social Computing (2011)
Zheng, L.: Using mutual information as a cocitation similarity measure. Scientometrics. **119**(3), 1695–1713 (2019)
Ziegler, C.-N., Golbeck, J.: Investigating interactions of trust and interest similarity. Decis. Support. Syst. **43**(2), 460–475 (2007)

Part III
Web Analytics, Big Data and the Internet of Things

Security Issues in the Internet of Things

Abderrezzak Sebbah and Benamar Kadri

Abstract The IoT is considered as the future of the Internet has made everyone's life comfortable and convenient that connects billions of objects all together. This network is based on different kinds of sensors, actuators, and integrated devices communicate with each other. They are used in industry, agriculture, computer security, smart home, transportation or health, as well as outside for environment control. These benefits may be associated with enormous risks of security loss making the network subject for several attacks. The present work set in this manuscript analyzes and discusses the most recent state of the art on authentication, security, and attacks in IoT.

Keywords IoT · Security · AMQP · Insider attack · Replay attack

1 Introduction

The Internet has evolved over the years, the Internet of Things is a wireless sensor network where devices are connected to each other wirelessly, and they are able to converse with each other without any human interaction (Gupta et al. 2020). The Internet of Things is a complex notion between RFID chips and the middleware. The IoT is composed of many complementary elements each with their own specificities; these elements (objects) allow capturing, store, process, and transfer data in physical environments but also between physical contexts and virtual universes. These benefits may be associated with enormous risks of security loss making the network subject of several attacks. In this paper, we analyze:

A. Sebbah (✉)
Department of Computer Science, and Laboratories STIC, University of Abou Bekr Belkaid, Tlemcen, Algeria

B. Kadri
Department of Computer Science, University of Abou Bekr Belkaid, Tlemcen, Algeria

S. Sedkaoui et al. (eds.), *International Conference on Managing Business Through Web Analytics*, https://doi.org/10.1007/978-3-031-06971-0_16

1. The most recent state of the art on authentication, security, and attacks in IoT.
2. Providing a comparison table of related works of both: the authentication protocols and the classification of attacks in the IoT.

This paper is organized as follows: Sect. 1 gives an introduction; then, in Sect. 2 we exhibit security of Internet of Things, we present security services and IoT layers and IoT protocols; after that, in Sect. 3 we present security issues in Internet of Things environments; and then, in Sect. 4 we analyze the related works of authentication protocols for the IoT. Finally, we conclude this work with a general conclusion in Sect. 5.

2 Security of IOT

2.1 Security Services

In this section we will present the objectives and requirements of the security:

- Confidentiality: this aspect guaranties that only the sender and the receiver of a given message can read it. This aspect is generally ensured using encryption techniques.
- Integrity: is defined as the property that gives to the receiver and the sender the possibility to detect the any alteration on the exchanged message or the way to ensures that the data is modified by the authorized person only.
- Authentication: This objective allows verifying the identity claimed by an entity, or the origin of a message, or data. This aspect is generally ensured using biometry or asymmetric encryption, etc.
- Freshness: this aspect deal with the possibility that attackers can use old messages to break the session with the communicating parties.
- Availability: Refers to the ability of a user to access data in any kind of situation in the correct format within a specified amount of time.

The IoT ecosystems can be customized to each community of users; therefore developers, professionals, and operators all must participate in the guaranty of security:

Firstly, the confidentiality and integrity and the stored information must be ensured by various encryption techniques, on the other hand, user confidentiality, understood as the ability to guarantee and ensure data protection and user anonymity. Beyond data security and quality is a condition for widespread adoption (Sicari et al. 2016).

Secondly, the authentication and authorization mechanisms must be provided to prevent unauthorized users or devices from accessing the system.

2.2 *IoT Layers*

The security on the Internet of Things comes in four key levels as shown in Fig. 1, namely, the perception, network, support, and application layers.

The first level is the physical layer, made up of physical elements of the IoT system which collects all kinds of information. This level makes it possible to improve security with the guarantee of confidentiality and authentication of the information exchanged between the nodes (Suo et al. 2012; Mendez Mena et al. 2018).

The second level is the network layer which is responsible for the transmission of the inputs obtained by the perception layer, at this level, existing security mechanisms such as authentication and confidentiality and completeness of data.

The third level is the support layer, which requires high application security, such as cloud computing and secure multi-stakeholder computing, encryption algorithms and protocols, improved system security technologies and antivirus, etc. (Suo et al. 2012).

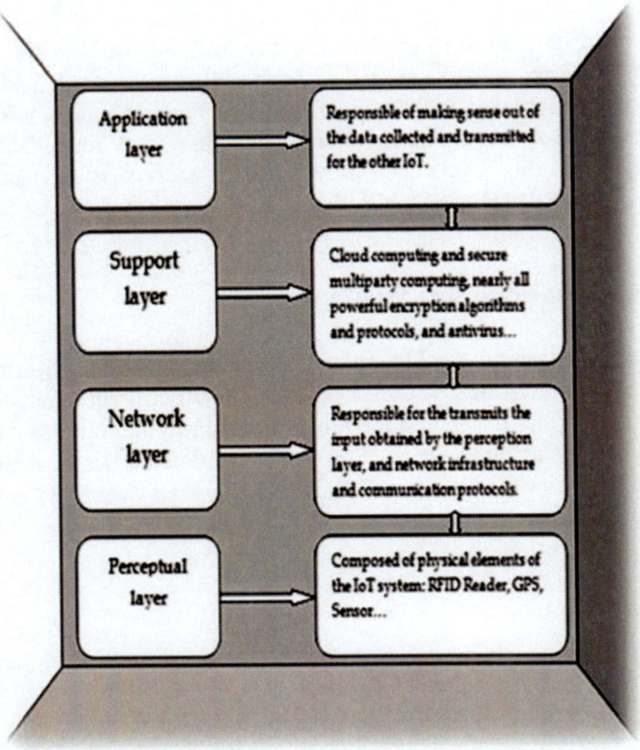

Fig. 1 IoT layers

Lastly the fourth level, also called the application layer, is responsible for making sense out of the data collected and transmitted for the other IoT layers; the needs for security solutions at this level are the authentication and establishment of a key in heterogeneous networks (Mendez Mena et al. 2018).

2.3 *IoT Protocols*

In this section we present the IoT protocols divided into two categories IoT-to-cloud and IoT-to-fog.

2.3.1 IoT-to-Cloud

The interconnected IoT to cloud include AMQP and DDS protocols:

2.3.1.1 Advanced Message Queuing Protocol (AMQP)

AMQP is an open standard protocol; this protocol is based on the use of the publication/subscription message model which routes and records and exchanges messages within a broker with a set of policies for component wiring. It can also be considered as a software layer protocol; it uses TCP for reliable transport and provides a security mechanism using TLS for encryption and authentication (Dizdarević et al. 2019; DATAFLAIR Team 2019).

2.3.1.2 Data Distribution Service (DDS)

DDS is a decentralized protocol based on peer-to-peer communication and reliable and real-time data exchange. It uses a publication-subscription interaction model. This protocol does not depend on the broker component and multicast to transmit high quality QoS to applications. DDS uses UDP by default, but it can also support TCP for security. For the DDS mechanism, TLS is used over TCP, and DTLS is used over UDP (Dizdarević et al. 2019).

2.3.2 IoT-to-Fog

The interconnected IoT to fog include MQTT and CoAP protocols:

2.3.2.1 Message Queue Telemetry Transport Protocol (MQTT)

This protocol is based on architecture: publication/subscription instead of the HTTP request/response paradigm is one of the lightest messaging protocols. The communication in the network between the sender and the recipient is done via the broker. MQTT uses TCP for reliable transport (Bashir and Mir 2017).

2.3.2.2 Constrained Application Protocol (CoAP)

CoAP is a lightweight protocol that supports the request/response paradigm designed for the IoT system based on HTTP protocols which uses UDP for reliable transport. Also it provides a security mechanism based on DTLS; the exchange of messages between the client and the server occurs only by encryption keys. This is divided into two logically different layers, the first layer is called the request/response layer which implements the RESTful paradigm. And the second layer is called structural layer intended to retransmit lost packets (Dizdarević et al. 2019).

3 Security Issues in Internet of Things Environments

3.1 IoT Issues

Regarding their size, applications, and the environment of deployment, the IoT confront and study several shortcomings such as heterogeneity, impact of the physical world, security and privacy of individuals' standard definition, and the battery capacity and computing power, complexity (Yang et al. 2017).

3.1.1 Battery Life Extension

In general, IoT devices are powered and are characterized by limited battery power, which makes the use of advanced and complex security protocol very costly. Therefore, any developed security mechanism must take into consideration the aspect of battery power. In some situation another problematic appears in which it is impossible to recharge the battery; therefore it is suitable to find a way to generate energy from the environment (light, heat, vibration, wind).

3.1.2 The Standards

The Internet of Things is the center of all modern technologies and developments, although the major challenge is how to manage the heterogeneity of object and standards coupled with multitudes of application.

3.1.3 Heterogeneity

The Internet of Things is composed by the diversity of hardware and software components used to build objects. They do not use the same operating systems and do not have the same communication interfaces, leading to significant technical heterogeneity.

3.1.4 Security/Privacy

The modern security technologies (encryption, authentication, key exchange, signature, etc.) are inadequate to the IoT environment which makes objects vulnerable to several attacks making the property and people's life in danger; recent works have proposed various standards and protocols to guarantee the security and user privacy, but these works still insufficient due the variety of attacks against sensors.

3.2 Security Threats Models

IoT can be subject of multiple attacks; in this section we are going to present a non-exhaustive list of attacks.

Any two communicating parties over an insecure channel such as user or smart devices are not considered as trusted entities. An attacker, say A, can eavesdrop the exchanged messages and also can modify or delete the message contents during transmission (Shuai et al. 2019); hence it becomes vital to analyze all possible attacks against IoT (Dhillon and Kalra 2017):

The replay attack: A replay attack also known as playback attack. In this attack an adversary intercept exchanged messages during a long period and retransmits them. These messages are injected into the transmission channel during the authentication in order to impersonate the server or the gateway. The use of timestamps can limit the effect of this attack against nodes (Dojen et al. 2008).

Eavesdropping attack: An adversary listens the exchanged communication in order to capture your passwords, credit card details, and information during the communication. The captured data can also be used to learn about the used security protocols, and a symmetric encryption with a periodic key update can stop this kind of attack.

Impersonation attack: In this attack, an adversary spoofs successfully the identity of one of the legitimate parties in a communication channel (Adams n.d.). For example, the attacker node diffuses false routing information to access the confidential data of the authentic nodes to become a legitimate node on the network.

Man-in-the-middle attack: This type of attack is considered as a real-time attack, in which an adversary takes place between the communicating parties and intercepts all the exchanged data and eventually modifies or delete it depending on the intention of the attacker. This attack can be executed during authentication phase which may allow the attacker to get access to all the future communication, in conventional network the use of digital certificate can stop this attack.

Denial-of-service attack: This attack aims to consume the resources of a node or a server by making it treat useless information or requests; in other cases the attacker can also target the network bandwidth in order to stop or limit the access to a server or gateway.

Stolen smart device attack: It is an attack for the purpose of extracting the secret credentials by means of power analysis attack. With these secret parameters, it is possible to get to know the user's password and even the session key.

Parallel session attack: In this attack, the attacker attempts to use previously hacked messages to establish parallel sessions of the protocol. Or the attacker can succeed by using two or more protocol executions running simultaneously to circumvent the security objectives of the protocol. This attack occurs in the absence of an authentication mechanism (Jurcut et al. 2014).

Password-change attack: For this attack, an adversary might try changing the password of the user during the password changing phase; the attacker gets access to the password and becomes a legal user making a legal user unable to access the service (Dhillon and Kalra 2017).

Gateway node bypassing attack: In the absence of any security protocol, the attacker might try to bypass the gateway node and connect with an IoT node directly without getting authenticated by the gateway node and access the services or sensitive data.

Offline guessing attack: In this attack, an adversary tries to guess the password of a legal entity using offline dictionary attack or brute force attack to take control of the system (Dhillon and Kalra 2017).

The dictionary attack: This attack has the purpose of trying to reach a targeted list of weak passwords. In other words, the attacker tries a limited number of key combinations that have a high chance of success to reach the real password.

Insider attack: An attacker gets access to a network or a computer and with legitimate privilege in order to gain access to a user's information like IDs and passwords.

Node compromise attack: Is one of the most common and damaging attacks of WSN and IoT, where the attacker might try to dig into nodes with the intention of extracting useful data (extracting private keys from sensor nodes). These attacks can obtain the hash of password used as a complement for guessing attack.

4 Related Works of Authentication Protocols for the IoT

In this section we compare, examine, and analyze the most known and recent works between 2014 and 2019 on IoT security Table 1.

Muhamed Turkanović et al. (2014). Propose a user authentication protocol for WSNs tailored in IoT; the scheme ensures mutual authentication between the user, sensor node, and the gateway node (GWN). The auteur chose hash and XOR computations for the scheme designed to be lightweight; the scheme protects against of various popular attacks like replay attacks, privileged insider attacks, stolen verifier attacks, stolen smart card and smart card breach attacks, impersonation attacks, many logged-in users with the same login id attacks, GWN bypassing attacks, password-change attacks, and DoS attacks. They chose the fundamental security benefits in the IoT key agreement, mutual authentication, and user anonymity.

Based on the weaknesses in the scheme (Turkanović et al. 2014), Mohammad Sabzinejad Farash et al. (Farash et al. 2016) found that (Turkanović et al. 2014) has some vulnerability, and negative point in the security is susceptible to some cryptographic attacks. Furthermore, Farash et al. proposed an efficient user authentication and key agreement scheme (UAKAS) for HWSN. Has affecting the functionality of the registration or authentication process of both the user and sensor node. In this scheme it is guaranteed the confidentiality and authentication only with a simple and low-cost computation the XOR operation and the hash. But their scheme was still insecure.

(Dhillon and Kalra 2017) present multifactor user authentication protocol is efficient in terms of computational cost compared to previous schemes like (Turkanović et al. 2014). Based on the above observation, it is worth noticing that most existing user authentication schemes proposed for WSNs and IoT have several drawbacks and they are secure against various known attacks.

Mohammad Wazid et al. (2018) presented a lightweight three factor smart card, password and personal biometrics, remote user authentication scheme for the hierarchical IoT network called the (UAKMP), and their analysis for various known attacks and examined also the sensing nodes capture attack exception.

In 2018, (Chen et al. 2018) proposed scheme uses the XOR operations, hash operations, and only four elliptic multiplications and proposed a lightweight privacy protection user authentication and key agreement scheme tailored for the IoT environment scenario based on low-capability devices. They provide the security features like user anonymity, sensor anonymity, perfect forward secrecy, and excellent resistance to the loss of synchronization problem. Most importantly, the password-change phase has been modified to prevent against an offline password-guessing attack and user being tracked. The scheme ensures authentication and key establishment scheme between users and sensors. The proposed scheme gains the computation cost and more efficient the communication cost.

But Abderrezzak Sebbah et al. (2020) found that (Chen et al. 2018) has some vulnerability against various attacks. Furthermore, the work of Abderrezzak Sebbah

Table 1 Comparison of authentication protocols and classification attacks for the IoT

Paper	Network model	Goals	Keys steps	Resilience against and limitations	Communication and computation costs
Turkanović et al. (2014)	Trust entity (GWN) communication (between user and node and gateway node)	Key agreement Mutual authentication Password protection	Pre-deployment phase Registration phase Login phase Authentication phase Password-change phase The dynamic node addition phase	Replay attack Privileged insider and stolen verifier attack Stolen smart card and smart card breach attack Impersonation attack GWN bypassing attack Many logged-in users with the same login id attack Password-change attack Denial-of-service attack User anonymity	Computation: User = 7TH Sensor = 5TH GWN = 7TH Communication:2720 bits
Farash et al. (2016)	Communication between user and sensor node and gateway node (trust entity)	Mutual authentication Session key agreement Change dynamic node	Pre-deployment phase Registration phase Login and authentication phase Password-change phase	Replay attack Privileged insider attack Man-in-the-middle attack insider and stolen verifier attack smart card attack User impersonation attack sensor node impersonation attack GWN bypassing attack many logged-in users with the same login id attack password-change attack DoS attack Offline password attack Traceability protection password User anonymity Sensor node anonymity	Computation: User = 11TH Sensor = 7TH GWN = 14TH

(continued)

Table 1 (continued)

Paper	Network model	Goals	Keys steps	Resilience against and limitations	Communication and computation costs
Wazid et al. (2018)	Communication is the type USD If a user wants to access a sensing node corresponding to an application in IoT, he needs to first send his login request to the GWN. The GWN then contacts the accessed sensing node via CH cluster	Mutual authentication Session key agreement Fast wrong input detection	Offline sensing node registration Registration of each user User login Authentication and key agreement Password and biometric update New sensing node deployment	User impersonation attack GWN impersonation attack Sensing device impersonation attack Privileged insider attack Forward secrecy Replay attack Man-in-the-middle attack Stolen verifier attack and smart card attack Session specific temporary information attack GWN bypassing attack Resilient against sensing device capture attack User anonymity Sensor anonymity	Computation: User = 13TH + Tfe+2TE/TD Sensor = 4TH + 2TE/TD GWN = 5TH + 4TE/TD communication: 2592bits
Chen et al. (2018)	Communication between user and sensor node and gateway node (trust entity)	User being tracked Propose an authentication Session key agreement between users and sensors	Registration phase of the user Registration phase of the sensor Login and authentication phase Password-change phase	Offline dictionary attack Excellent resistance to the loss of synchronization problem Perfect forward secrecy User anonymity Sensor anonymity User anonymity to sensor Loss of synchronization Protect the privacy of the data Users being untraceable Sensors being untraceable	Computation: User = 5TH + TMUL sensor = 4TH + 2TMUL GWN = 8TH Communication: 396byte

Gupta et al. (2018)	Network model consist a three-entity the server, gateway/mobile terminal, and the wearable devices	Mutual authentication Session key agreement	System setup Registration phase Authentication and key-establishment Password-change	Perfect forward secrecy Replay attack User impersonation attack Sensing device impersonation attack Y Gateway impersonation attack Node capture attack Offline guessing attack Privileged insider attack Man-in-the-middle attack User anonymity Sensor anonymity	Computation: User = 7TH + 4TXOR sensor = 4TH + 4TXOR GWN = 5TH + 3TXOR Communication: 3808bits
Zhou et al. (2019)	Network model consist a user and sensor and trust entity that all smart devices (sensors), are connected through the internet Trust third party	Mutual authentication Unforgeability of message Session-key	Initialization phase User registration phase User login and request phase User authentication and key agreement phase	Unlink ability Forward security Resistance against impersonation attack Resistance against replay attack Resistance against stolen verifier attack Resistance against man-in-the-middle attack	Communication: 1344bits

TH: time for a hash operation. TD/E, time for symmetric key decryption/encryption. TMAC, the time for performing MAC operation. THMAC, the time for performing HMAC operation. TMUL, ECC multiplications; TXOR, time for XOR operation et al. proposed an authentication scheme with three factors using ECC and fuzzy extractor to guarantee a various security against various attacks.

Recently (Gupta et al. 2018) introduced a lightweight authentication and key establishment for wearable device using simple Xor and hash function, which guarantee the security against various knowing attacks; the proposed scheme gains the communication and provide security but more efficient in term computation cost.

In 2019, (Zhou et al. 2019) provide unlink ability for IoT environment to protect users privacy and achieves anonymity and others knowing attacks based on bilinear pairings.

5 Conclusion

In this paper, we have concisely reviewed security in the IoT, and we presented at the beginning a general view on the IoTs followed by a state of the art and analyzed the security services in IoT. Additionally, we have analyzed also the security layer the perceptual layer, network layer, support layer, and application layer. Then we discussed the IoT protocols, and we give comparison in form of a tabular of the authentication protocols proposed for IoT at large.

In sum, our work analyzed related works of authentication and privacy protocols that were made for the IoT.

References

Adams, C.: Impersonation attack. In: Encyclopedia of Cryptography and Security, pp. 286–286 (n.d.)

Bashir, A., Mir, A.H.: Securing publish-subscribe services with dynamic security protocol in MQTT enabled internet of things. Int. J. Secur. Appl. **11**(11), 53–66 (2017)

Chen, Y., López, L., Martínez, J.-F., Castillejo, P.: A lightweight privacy protection user authentication and key agreement scheme tailored for the internet of things environment: LightPriAuth. J. Sensors. **2018**, 1–16 (2018)

DATAFLAIR Team.: 4 Key IoT Protocols – Learn In Great Detail. Data-Flair (2019)

Dhillon, P.K., Kalra, S.: Secure multi-factor remote user authentication scheme for internet of things environments. Int. J. Commun. Syst. **30**(16), e3323 (2017)

Dizdarević, J., Carpio, F., Jukan, A., Masip-Bruin, X.: A survey of communication protocols for internet of things and related challenges of fog and cloud computing integration. ACM Comput. Surv. **51**(6), 1–29 (2019)

Dojen, R., Jurcut, A., Coffey, T., Gyorodi, C.: On establishing and fixing a parallel session attack in a security protocol. In: Badica, C., Mangioni, G., Carchiolo, V., Burdescu, D.D. (eds.) Intelligent Distributed Computing, Systems and Applications. Studies in Computational Intelligence, vol. 162, pp. 239–244. Springer, Berlin/Heidelberg (2008)

Farash, M.S., Turkanović, M., Kumari, S., Hölbl, M.: An efficient user authentication and key agreement scheme for heterogeneous wireless sensor network tailored for the Internet of Things environment. Ad Hoc Netw. **36**, 152–176 (2016)

Gupta, A., Tripathi, M., Shaikh, T.J., Sharma, A.: A lightweight anonymous user authentication and key establishment scheme for wearable devices. Comput. Netw. **149**, 29–42 (2018)

Gupta, S., Vyas, S., Sharma, K.P.: A survey on security for IoT via machine learning. In: 2020 International Conference on Computer Science, Engineering and Applications (ICCSEA) (2020)

Jurcut, A.D., Coffey, T., Dojen, R.: Design requirements to counter parallel session attacks in security protocols. In: 2014 Twelfth Annual International Conference on Privacy, Security and Trust (2014)

Mendez Mena, D., Papapanagiotou, I., Yang, B.: Internet of things: survey on security. Inf. Secur. J. Glob. Perspect. **27**(3), 162–182 (2018)

Sebbah, A., Kadri, B.: A privacy and authentication scheme for IoT environments using ECC and fuzzy extractor. In: 2020 International Conference on Intelligent Systems and Computer Vision (ISCV), pp. 1–5 (2020)

Shuai, M., Yu, N., Wang, H., Xiong, L.: Anonymous authentication scheme for smart home environment with provable security. Comput. Secur. **86**, 132–146 (2019)

Sicari, S., Rizzardi, A., Miorandi, D., Cappiello, C., Coen-Porisini, A.: A secure and quality-aware prototypical architecture for the internet of things. Inf. Syst. **58**, 43–55 (2016)

Suo, H., Wan, J., Zou, C., Liu, J.: Security in the internet of things: a review. In: 2012 International Conference on Computer Science and Electronics Engineering (2012)

Turkanović, M., Brumen, B., Hölbl, M.: A novel user authentication and key agreement scheme for heterogeneous ad hoc wireless sensor networks, based on the Internet of Things notion. Ad Hoc Netw. **20**, 96–112 (2014)

Wazid, M., Das, A.K., Odelu, V., Kumar, N., Conti, M., Jo, M.: Design of Secure User Authenticated key Management Protocol for generic IoT networks. IEEE Internet Things J. **5**(1), 269–282 (2018)

Yang, Y., Wu, L., Yin, G., Li, L., Zhao, H.: A survey on security and privacy issues in internet-of-things. IEEE Internet Things J. **4**(5), 1250–1258 (2017)

Zhou, Y., Liu, T., Tang, F., Tinashe, M.: An unlinkable authentication scheme for distributed IoT application. IEEE Access. **7**, 14757–14766 (2019)

A Requirement Elicitation Method for Big Data Projects

Chabane Djeddi, Nacer Eddine Zarour, and Pierre-Jean Charrel

Abstract The late correction of requirements costs up to 200 times as much as correction during requirement engineering. Requirements for systems do not arise naturally; they need to be engineered and done with great care and precision. Many software projects fall due to the elicitation of the requirements; for big data, it become more complicated due to its own specific characteristics; eliciting the requirements for big data must undertake the specific characteristics of big data such as (volume, variety, etc.); in fact, there are many research recommendation and perspectives in the literature to undertake big data characteristics, but unfortunately the studies to fill this gap are so rare. We analyzed the literature to identify the big data concepts that existing requirements engineering (RE) methods does not support. After, we elaborate an adopted KAOS method to make sure the elicitation of the requirements. We apply this new method (BKAOS) to an illustrative scenario to show its uses. We see that BKAOS is more suitable to elicit the requirements for big data projects and catch the right requirements; therefore, it reduce the effort and facilitate data analysis.

Keywords Big data · Requirements engineering · KAOS · KAOS extension · Formal checking

1 Introduction

Industries become more and more interested in the high potential of big data, government agencies announced major plan to accelerate big data research and application (Chen et al. 2014), Google processes data of hundreds of PB, Baidu

C. Djeddi (✉) · N. E. Zarour
LIRE Laboratory, Constantine 2- A. Mehri University, El Khroub, Algeria
e-mail: chabane.djeddi@univ-constantine2.dz; nasro.zarour@univ-constantine2.dz

P.-J. Charrel
IRIT Laboratory, Toulouse 2 Jean Jaurès University, Toulouse, France
e-mail: charrel@univ-tlse2.fr

S. Sedkaoui et al. (eds.), *International Conference on Managing Business Through Web Analytics*, https://doi.org/10.1007/978-3-031-06971-0_17

processes data of ten of PB, and Facebook generates log data to over 10 PB per month. Therefore, big data projects their one requirement.

The requirements can be specific when the software has to manage data that is very large in size and complex (Eridaputra et al. 2014). We must preform requirements engineering (RE) with great care and precision (Dardenne et al. 1993); authors (Anderson 2015; Arruda 2018; Arruda and Madhavji 2018; Eridaputra et al. 2014; Madhavji et al. 2015; Noorwali et al. 2016; Otero and Peter 2015; Sangeeta and Kapil 2016) emphasize on the necessity to undertake big data characteristics by the RE methods. Specifying the precision details of requirements is important for the successful construction of software (Dardenne et al. 1993; Sangeeta and Kapil 2016); one of the main challenges encountered when attempting to engineer big data requirements is dealing with its own characteristics (Anderson 2015; Eridaputra et al. 2014; Noorwali et al. 2016; Otero and Peter 2015), and the need to address those characteristics in the specification of big data requirements is clear and evident (Anderson 2015; Arruda 2018; Noorwali et al. 2016). Taking the specific characteristics such as volume, variety, etc. while dealing with requirements become necessary (Arruda 2018; Arruda and Madhavji 2018; Eridaputra et al. 2014; Sangeeta and Kapil 2016), and those characteristics must be representable and modelled in requirements notations, so that solution design can be created to meet the specification (Arruda and Madhavji 2018; Madhavji et al. 2015). Unfortunately, few researches were made to fill this gap, (Eridaputra et al. 2014) who propose to model requirements for big data application using goal-oriented approach and (Djeddi et al. 2018) who propose an extension of iStar for big data projects.

To represent and model the requirements notations for big data, we develop an extension of the KAOS method. KAOS (van Lamsweerde 2000a) (Knowledge Acquisition Automated System or Keep All Object Satisfied) is a goal-oriented requirements engineering (GORE) method and commonly used for requirements elicitation. KAOS is classified as general-purpose modelling language (GPML); in ours case we need a domain-specific modelling language (DSML) (Brambilla et al. 2017). The need to extend existing RE method for each domain is clear; that explains why there is such amount of extensions for RE methods to fit every specific domain (Ali et al. 2014; Guzman et al. 2016; Lockerbie et al. 2012; Mazón et al. 2007; Morandini et al. 2017).

We start the research by giving first question; after other questions arise, we answer the questions in the rest of the paper: (i) does the existing RE method support big data project elicitation? (ii) what are the specific characteristics of big data to be undertaken by RE methods? and (ii) can we develop an extension of an existing RE method to undertake big data characteristics? In this work we contribute at the development of big data field and requirements engineering (RE) by the construction of a new adopted RE method; this one allows a better elicitation of the right requirements.

2 Literature Review

Requirements engineering (RE) and big data are briefly discussed in this section.

2.1 *Requirements Engineering*

One of the most important criterion for the success of any software is the degree of satisfaction of the goals set by the stakeholders (Horkoff et al. 2016). The objective of RE is to have precise specifications of software behavior, to evolve over time (IEEE Computer Society et al. 1997), and to know the requirements of the stakeholders and to verify them in order to reach an agreement on the requirements; moreover, RE also helps to know the limits of our system (Attarha and Modiri n.d.). To fulfill this, we perform four steps (Ramingwong 2012): (i) there are typically five sub-steps in the requirements elicitation process (Zowghi and Coulin 2005). Learning about the application domain is essential: locating the origins of the demand for goods and services; identifying and evaluating the various stakeholders choosing the methods, techniques, and tools to be employed; and consulting with various stakeholders and other potential sources to gather requirements. (ii) We focus on the review, understanding the elicited requirements and verifying their quality in terms of accuracy, completeness, clarity, and consistency during requirements analysis and negotiation. (iii) As a foundation for evaluating and controlling future products and processes, requirements documentation can be considered (system design, system test cases, and validation) (Attarha and Modiri n.d.). (vi) Requirements validation is done for controlling the quality; it means confirming that requirements are complete and well-written and supply needs of customer. It is possible that this phase will be repeated in the future because of deficiencies, gaps in requirements, additional information, and other issues. That is why, in the software life cycle test phase, the implemented software product is tested on the basis of its requirements. We are primarily concerned with the elicitation of requirements in this paper.

We find in the literature (Zowghi and Coulin 2005) that there are three basic approaches of RE: (i) goal-based approaches; the fundamental premise of goal-based approaches (GORE) is high-level goals. The system's objectives are represented by these goals, which are broken down into sub-goals and then refined to the point where even the most basic requirements can be elicited (e.g., by using *and-or* relationships). (Zowghi and Coulin 2005); several methods can be considered as belonging to GORE: iStar Framework (Yu n.d.), NFR (Chung and do Prado Leite 2009), and KAOS (van Lamsweerde 2000a); among all GORE methods, KAOS and iStar have been the most cited (Werneck n.d.). In our work, we choose to extend KAOS because it is very used in academic research. (ii) Scenario-based approaches story and specific descriptions of current and future processes are used in scenario-based approaches. Scenarios are similar to use cases in that they don't typically consider the internal structure of the system and must be developed incrementally and interactively. Obviously, when using scenarios, it is critical to gather all of the possible exceptions for each step (Zowghi and Coulin 2005). (iii) Viewpoint-based approaches which can be modified or mixed to create new ones; the viewpoint-based approaches aim to model the domain from different perspectives in order to develop a complete and consistent description of the target system (Zowghi and Coulin 2005). Initially, the requirements are opaque, informal, and only expressed through personal views. These views reflect the skills, objectives, and roles of each participant. The elicitation activity is, therefore, a collective activity. The expression of multiple views allows for better elicitation of requirements.

2.2 Big Data

There has been a frenzied, haphazard, and unorganized growth of the big data concept in academic and professional literature. (De Mauro et al. 2019). There is no single definition of big data, despite the appearance of a slew of definitions. The term "big data" refers to a dataset that is too large to be processed using conventional software (Baig et al. 2019; Chen et al. 2014; Lin et al. 2019). One way to look at big data is from the perspective of the infrastructure. There is a significant amount of data that is characterized by (volume, velocity, variety, veracity, and value) when viewed from the perspective of the analysis of the data. (iii) From a business perspective, big data is viewed as events. Big data can be viewed as a byproduct that is directly applicable to improving the quality of one's work (Otero and Peter 2015; Sangeeta and Kapil 2016). While data storage is an issue, the most pressing issue is the need for rapid analysis of heterogeneous data sets (Madden n.d.).

The most important properties of big data are (i) we have structured, semi-structured, and even unstructured data that can't be manipulated using traditional systems because of the variety of data that can be manipulated (Chen et al. 2014; Katal et al. 2013). (ii) Volume is an important consideration when developing the big data concept, as current data handling systems are unable to handle the zettabytes of data routinely handled by the world's largest corporations (Chen et al. 2014; Katal et al. 2013). (iii) The velocity of incoming data from various sources is so critical; consequently, traditional systems have difficulty dealing with the situation (Katal et al. 2013). (vi) The value: the stored data is important. Incorrect data can be returned to decision makers if a user performs queries on stored data or misuses already available data (Katal et al. 2013). (v) Because big data is compiled from a variety of different sources, it is crucial to ensure the data's integrity and avoid it ending up in unmanageable situations that the correlation and links between the data are maintained (Katal et al. 2013).

3 KAOS and Its Application

In this session, we explain the KAOS method as well as their models, we give an illustrative scenario, and we perform KAOS on the illustrative scenario to show its application.

3.1 KAOS

Getting high-quality requirements is difficult and critical. Surveys have shown that RE is becoming increasingly recognized as a critical area of study and practice in the field of software engineering (van Lamsweerde 2000b, p. 00). GORE is an approach to requirements engineering (RE) dealing with intentionality in accordance with the relations among different actors. KAOS (van Lamsweerde 2000a) (Knowledge Acquisition Automated System or Keep All Object Satisfied) have been receiving many references as being one of the important GORE proposals (Werneck n.d.). It is a goal-oriented requirements engineering (GORE) method and commonly used for requirements elicitation. KAOS consists of four models: goal model, object model, responsibility model, and operational model.

The goal model is a set of goals organized in a top-down hierarchy; we perform a top-down and/or decomposition for each goal until we find sub-goals that can be assigned to an agent. The object model defines objects of interest. The responsibility model shows for each agent its responsibilities describing the requirements under his responsibility and expectation assigned to the agent. The operation model consists of describing the agents' behaviors that are necessary to reach the requirements; behaviors are expressed in terms of operations and tasks performed by the agents.

3.2 The Illustrative Scenario

We take an example of the presidential elections in Algeria. In order to increase the chances of a candidate's success, a camp's community has decided to create a big data project to study the opinions of people, which will allow them to identify the key points on which to focus in order to launch targeted advertisements. They collect data from social networks and analyze them to know the essential points in the opinion of the different categories of people. On this basis, they draft a presidential strategy and present it to the public; following that, they solicit feedback from the public in order to make revisions to the strategy and conduct targeted advertising. Large amounts of data (structured, semi-structured, and even unstructured) are manipulated in a short period of time, making this a big data project that cannot be processed using traditional systems. The application of KAOS on the illustrative scenario.

The goal-oriented process of KAOS is developed through activities: (i) identification of goals, (ii) formalization of goals, (iii) modelling of objects and identification of state variables, (iv) detection and resolution of goal conflict levels, (v) refining of goals and identification of agent responsibilities, (vi) generation of obstacles and resolution to goal fulfillment, and (vii) derivation of operation requirements from system goals (Werneck n.d.).

Figure 1 shows the application of KAOS's goal model on the example.

Figure 2 shows the application of KAOS's responsibility model on the example.

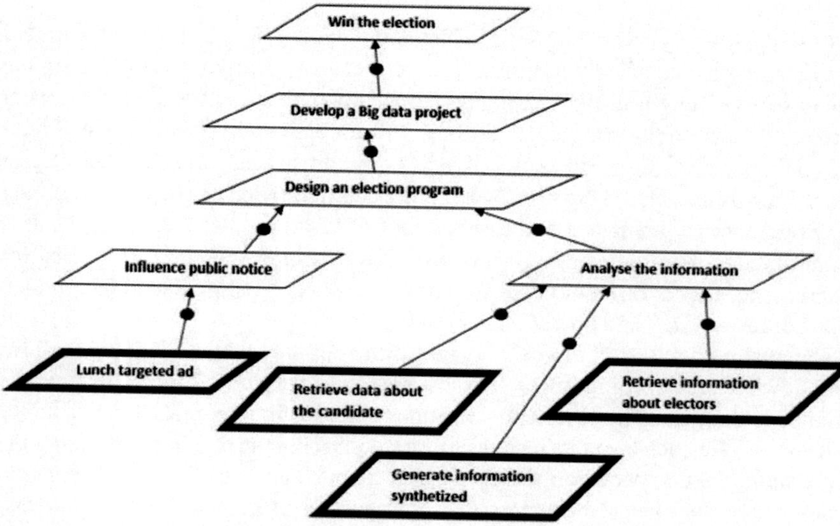

Fig. 1 The application of KAOS's goal model on the example

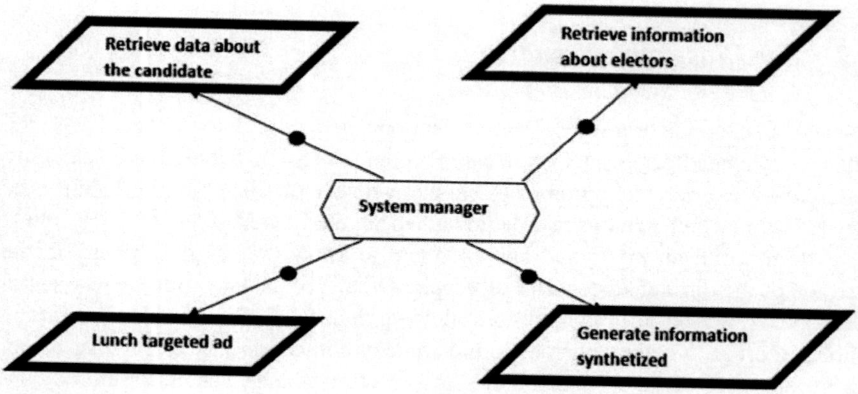

Fig. 2 The application of KAOS's responsibility model on the example

4 BKAOS: An Extension of KAOS for Big Data Projects

In this session, we present BKAOS (big data KAOS) which consists of an extension of KAOS method for big data projects. Prior to performing the BKAOS, we first clarify the need for this extension of the KAOS to support elicitation of the requirements for big data projects and then explain the concepts to be added.

4.1 The Needs for an Extension of KAOS

The use of RE in big data applications is a new development. There is a need for a better understanding (Madhavji et al. 2015). Errors and deficiencies can have disastrous effects on the subsequent development steps and on the quality of the resulting software product. Therefore, it is essential that RE be done with great care and precision (Dardenne et al. 1993). The elicitation is the most crucial step in RE (Horkoff et al. 2016), if it is not well done can lead to projects that do not respond well to the needs of the stakeholders. In the context of big data projects, the situation becomes even more complex. In order to meet a deadline, a big data project must process a large volume of data of a specific type in a short amount of time (structured, semi-structured, unstructured) (Madden n.d.).

Authors (Anderson 2015; Arruda 2018; Arruda and Madhavji 2018; Eridaputra et al. 2014; Madhavji et al. 2015; Noorwali et al. 2016; Otero and Peter 2015; Sangeeta and Kapil 2016) consider the big data characteristics as a challenge in the RE; also, the studies (Arruda and Madhavji 2018; Otero and Peter 2015; Sangeeta and Kapil 2016) confirmed that the big data software must include all three parameters (functional feature, time constraint, and verifiable during some period) to completely define the requirements specification for big data projects.

4.2 The Concepts Added to KAOS

The needs for extending RE method are different in every domain that explains why there is such amount of extension (Ali et al. 2014; Djeddi et al. 2018; Guzman et al. 2016; Lockerbie et al. 2012; Mazón et al. 2007; Morandini et al. 2017), because we always try to get the model closer to the reality by adding new concepts in purpose of improving the accuracy of the big data project that poses its specific challenges.

According to the literature's requirements for big data (Arruda 2018; Arruda and Madhavji 2018; Chen et al. 2014; Katal et al. 2013; Madden n.d.; Madhavji et al. 2015; Sangeeta and Kapil 2016), we've decided to include the ideas of execution time, data volume, data variety, and goal durability. We see clearly that to support big data projects by the KAOS method, we must ensure that the goals are attached

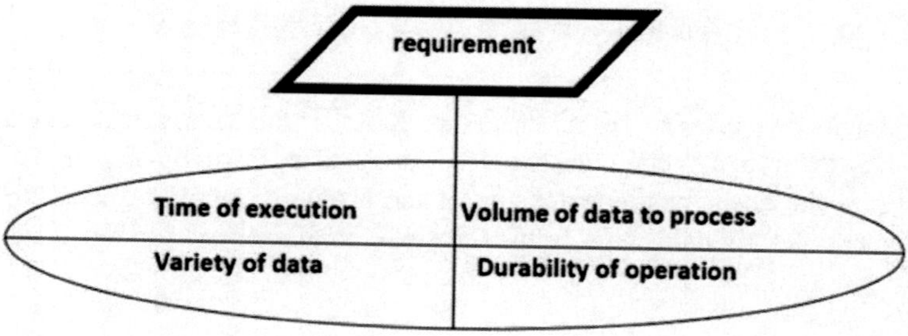

Fig. 3 The added concepts to the goal model and the responsibility model

to theirs (time, volume of data, variety of data, and goal's longevity are all factors to consider.).

Figure 3 shows the concepts added to the goal model and the responsibility model.

In the rest of this subsection, each concept is explained and clarified why adding these concepts is necessary.

4.2.1 The Volume of Data to Process

In big data projects, the volume is often large; we process data of hundreds of PB and technologies like Hadoop and NoSQL systems raised, but the volume remains a crucial point in big data projects. The current RE methods do not support this issue as an independent concept, but for eliciting the requirements for big data projects, specifying the volume of data to process seems evident.

4.2.2 The Execution Time

The execution time is important, a late result is considered as a wrong one, and the execution time must be exact. For big data it becomes more crucial due to the importance of the results. So, the RE methods must take into consideration the execution time.

4.2.3 The Variety of Data

The variety of big data is one of the most important properties of big data projects. Data have different nature structured (traditional databases), semi-structured (JSON, XML, etc.), and unstructured (images, emails, etc.), and its manipulation is quite different. The RE methods must undertake the nature of data to be process in the elicitation step.

4.2.4 The Durability of a Goal

In general, engineers develop a system to be functional during a specific time, and after that, it may become no more useful; it is the case in the context of big data projects. So, the RE method for big data must specify the durability of a goal from the beginning.

KAOS does not support the characteristics presented above, which do not allow a complete and refined elicitation of the requirements for big data. BKAOS came to overcome these issues and allow a better elicitation.

4.3 The Application of BKAOS on the Illustrative Scenario

However, we find in BKAOS that new concepts are linked to the goal "launch targeted advertising" which means that this goal must be done within 2 days, by analyzing 20 petabytes of unstructured and semi-structured data nature, and it must be in operation during the elections. The goal "Retrieve data about the candidate" must be done within 1 day, by analyzing 30 GB of structured data, and it must be functional during the elections. The goal "Generate information synthesized on the profiles of electors" must be done within 15 days, by analyzing 100 zettabytes of unstructured and semi-structured nature, and it must be functional during the elections. The goal "Retrieve information about elector" must be done within 1 day, by analyzing 30 structured data nature, and it must be functional during the elections.

BKAOS gives more completeness and refinement to the requirements, and we guaranty that there are no messing requirements.

Figure 4 shows the application of BKAOS responsibility model on the example of presidential elections.

Figure 5 shows the application of BKAOS goal model on the example of presidential elections.

The use of BKAOS allows to catch all the right specific requirements for big data projects.

5 Conclusion

The existing RE method doesn't support big data projects elicitation, and the specific characteristics of big data such as (volume, variety, etc.) are not undertaken by the existing RE methods; new methods must be developed to take into account big data projects. In this study, we propose BKAOS (big data KAOS) which is an extension of the KAOS method to properly elicit all the right requirements for big data projects. BKAOS takes into account the characteristics of big data projects in order to insure a proper elicitation of the requirements. We performed KAOS and

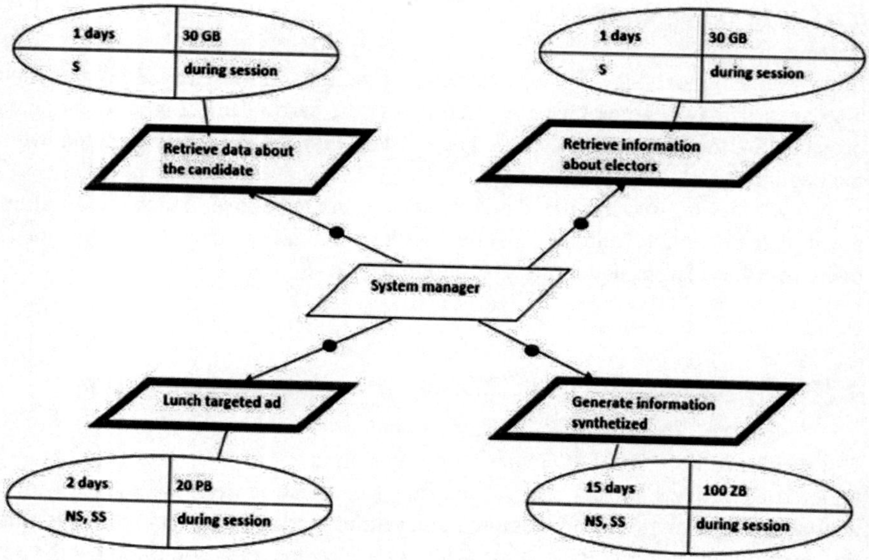

Fig. 4 The application of BKAOS responsibility model

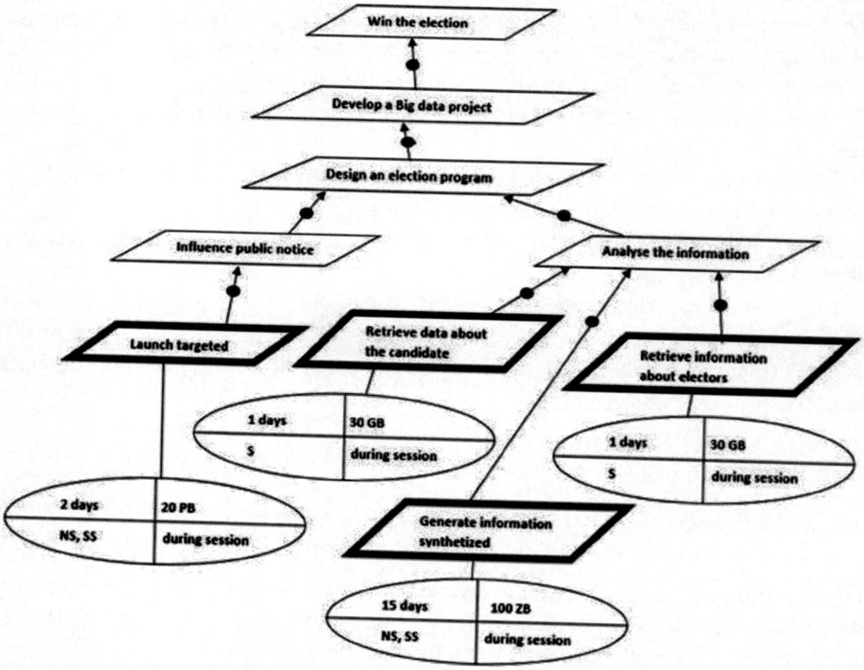

Fig. 5 The application of BKAOS goal model on the example of presidential elections

BKAOS on the same case study of the elections to illustrate the importance and the utility of BKAOS; with BKAOS we insure to catch all the right requirements for big data projects.

References

Ali, R., Dalpiaz, F., Giorgini, P.: Requirements-driven deployment: customizing the requirements model for the host environment. Softw. Syst. Model. **13**(1), 433–456 (2014). https://doi.org/10.1007/s10270-012-0255-y

Anderson, K.M.: Embrace the challenges: software engineering in a big data world. In: 2015 IEEE/ACM 1st International Workshop on Big Data Software Engineering, pp. 19–25 (2015). https://doi.org/10.1109/BIGDSE.2015.12

Arruda, D.: Requirements engineering in the context of big data applications. ACM SIGSOFT Softw. Eng. Notes. **43**(1), 1–6 (2018). https://doi.org/10.1145/3178315.3178323

Arruda, D., Madhavji, N.H.: State of requirements engineering research in the context of big data applications. In: Kamsties, E., Horkoff, J., Dalpiaz, F. (eds.) Requirements Engineering: Foundation for Software Quality, vol. 10753, pp. 307–323. Springer International Publishing (2018). https://doi.org/10.1007/978-3-319-77243-1_20

Attarha, M., Modiri, N.: Focusing on the Importance and the Role of Requirement Engineering 4. (n.d.).

Baig, M.I., Shuib, L., Yadegaridehkordi, E.: Big data adoption: state of the art and research challenges. Inf. Process. Manag. **56**(6), 102095 (2019). https://doi.org/10.1016/j.ipm.2019.102095

Brambilla, M., Cabot, J., Wimmer, M.: Model-driven software engineering in practice: second edition. Synth. Lectures Softw. Eng. **3**(1), 1–207 (2017). https://doi.org/10.2200/S00751ED2V01Y201701SWE004

Chen, M., Mao, S., Liu, Y.: Big Data: A Survey. Mob. Netw. Appl. **19**(2), 171–209 (2014). https://doi.org/10.1007/s11036-013-0489-0

Chung, L., do Prado Leite, J.C.S.: On non-functional requirements in software engineering. In: Borgida, A.T., Chaudhri, V.K., Giorgini, P., Yu, E.S. (eds.) Conceptual Modeling: Foundations and Applications, vol. 5600, pp. 363–379. Springer, Berlin Heidelberg (2009). https://doi.org/10.1007/978-3-642-02463-4_19

Dardenne, A., van Lamsweerde, A., Fickas, S.: Goal-directed requirements acquisition. Sci. Comput. Program. **20**(1–2), 3–50 (1993). https://doi.org/10.1016/0167-6423(93)90021-G

De Mauro, A., Greco, M., Grimaldi, M.: Understanding big data through a systematic literature review: the ITMI model. Int. J. Inf. Technol. Decis. Mak. **18**(04), 1433–1461 (2019). https://doi.org/10.1142/S0219622019300040

Djeddi, C., Charrel, P.-J., Laboratory, I., Jaurès, J.: Extension of iStar for Big Data Projects, p. 8 (2018)

Eridaputra, H., Hendradjaya, B., Danar Sunindyo, W.: Modeling the requirements for big data application using goal oriented approach. In: 2014 International Conference on Data and Software Engineering (ICODSE), pp. 1–6 (2014). https://doi.org/10.1109/ICODSE.2014.7062702

Guzman, A., Martinez, A., Agudelo, F.V., Estrada, H., Perez, J., Ortiz, J.: A methodology for modeling ambient intelligence applications using i* framework. CEUR Workshop Proc. **1674**, 6 (2016)

Horkoff, J., Aydemir, F.B., Cardoso, E., Li, T., Mate, A., Paja, E., Salnitri, M., Mylopoulos, J., Giorgini, P.: Goal-oriented requirements engineering: a systematic literature map. In: 2016 IEEE 24th International Requirements Engineering Conference (RE), pp. 106–115 (2016). https://doi.org/10.1109/RE.2016.41

IEEE Computer Society, ACM Sigsoft, IFIP Working Group 2.9: Classification of Research Efforts in Requirements Engineering. IEEE Computer Society Press (1997)

Katal, A., Wazid, M., Goudar, R.H.: Big data: issues, challenges, tools and good practices. In: 2013 Sixth International Conference on Contemporary Computing (IC3), pp. 404–409 (2013). https://doi.org/10.1109/IC3.2013.6612229
Lin, Y., Wang, H., Li, J., Gao, H.: Data source selection for information integration in big data era. Inf. Sci. **479**, 197–213 (2019). https://doi.org/10.1016/j.ins.2018.11.029
Lockerbie, J., Maiden, N.A.M., Engmann, J., Randall, D., Jones, S., Bush, D.: Exploring the impact of software requirements on system-wide goals: a method using satisfaction arguments and i* goal modelling. Requir. Eng. **17**(3), 227–254 (2012). https://doi.org/10.1007/s00766-011-0138-8
Madden, S.: From databases to big data. IEEE Internet Comput. **16**(3), 4–6 (n.d.)
Madhavji, N.H., Miranskyy, A., Kontogiannis, K.: Big picture of big data software engineering: with example research challenges. In: 2015 IEEE/ACM 1st International Workshop on Big Data Software Engineering, pp. 11–14 (2015). https://doi.org/10.1109/BIGDSE.2015.10
Mazón, J.-N., Pardillo, J., Trujillo, J.: A model-driven goal-oriented requirement engineering approach for data warehouses. In: Hainaut, J.-L., Rundensteiner, E.A., Kirchberg, M., Bertolotto, M., Brochhausen, M., Chen, Y.-P.P., Cherfi, S.S.-S., Doerr, M., Han, H., Hartmann, S., Parsons, J., Poels, G., Rolland, C., Trujillo, J., Yu, E., Zimányie, E. (eds.) Advances in Conceptual Modeling – Foundations and Applications, vol. 4802, pp. 255–264. Springer, Berlin/Heidelberg (2007). https://doi.org/10.1007/978-3-540-76292-8_31
Morandini, M., Penserini, L., Perini, A., Marchetto, A.: Engineering requirements for adaptive systems. Requir. Eng. **22**(1), 77–103 (2017). https://doi.org/10.1007/s00766-015-0236-0
Noorwali, I., Arruda, D., Madhavji, N.H.: Understanding quality requirements in the context of big data systems. In: Proceedings of the 2nd International Workshop on BIG Data Software Engineering – BIGDSE '16, pp. 76–79 (2016). https://doi.org/10.1145/2896825.2896838
Otero, C.E., Peter, A.: Research directions for engineering big data analytics software. IEEE Intell. Syst. **30**(1), 13–19 (2015). https://doi.org/10.1109/MIS.2014.76
Ramingwong, L.: A review of requirements engineering processes, problems and models. Int. J. Eng. Sci. Technol. **4**, 6 (2012)
Sangeeta, Kapil, S.: Quality issues with big data analytics. In: 3rd International Conference on Computing for Sustainable Global Development (INDIACom). IEEE (2016)
van Lamsweerde, A.: Goal-oriented requirements engineering: a guided tour. In: Proceedings Fifth IEEE International Symposium on Requirements Engineering, pp. 249–262 (2000a). https://doi.org/10.1109/ISRE.2001.948567
van Lamsweerde, A.: Requirements engineering in the year 00: a research perspective. In: Proceedings of the 22nd International Conference on Software Engineering – ICSE '00, pp. 5–19 (2000b). https://doi.org/10.1145/337180.337184
Werneck, V.M.B.: Comparing GORE Frameworks: I-star and KAOS. 12. (n.d.)
Yu, E. S.-K.: Modelling Strategic Relationships for Process Reengineering. 131. (n.d.)
Zowghi, D., Coulin, C.: Requirements elicitation: a survey of techniques, approaches, and tools. In: Aurum, A., Wohlin, C. (eds.) Engineering and Managing Software Requirements, pp. 19–46. Springer-Verlag (2005). https://doi.org/10.1007/3-540-28244-0_2

The Importance of the Internet of Things and Its Applications in the Field of Transport: Reference to Intelligent Transport Models in Some Countries

Nadia Soudani and Djamila Sadek

Abstract The study aims to investigate the extent of the importance of the Internet of Things and identify the most important fundamentals on which it is based, in addition to addressing the Internet, the means of transport, and the main pillars on which it is based. Some models in different countries will also be highlighted.

The study concluded with several results, the most important of which is the great development of the Internet of means of transport. Indeed, some countries and companies have become pioneers in this field. Countries also differ in terms of their dependence on intelligent transport, some have Internet of transport and traffic, and others are satisfied with a specific type of transport according to their needs and needs their capabilities.

Keywords The Internet of Things · Internet of transport (intelligent transport) · Intelligent transport systems · Sensors · Transport management

1 Introduction

Transportation is considered to be the economic and daily lifeline in all countries and for all business and tourism activities, etc., and it has experienced a great development similar to other economic fields, especially after the adoption of modern technology in these fields, so that the term Internet of Things emerged from the means of communication, the means of buying and selling, and the means of payment, until it became global companies specializing in this area and seeking to monopolize it and work to achieve leadership.

N. Soudani (✉)
Faculty of Economics, Business and Management Sciences, University Center of Tissemsilt, Tissemsilt, Algeria

D. Sadek
Faculty of Economics, Business and Management Sciences, University of Oran, Sénia, Algeria

S. Sedkaoui et al. (eds.), *International Conference on Managing Business Through Web Analytics*, https://doi.org/10.1007/978-3-031-06971-0_18

The transport sector has also seen great developments, so that the so-called intelligent transport (Internet of transport) has appeared, and the issue of the Internet of transport or intelligent transport requires the availability of platforms, sensors, and advanced communication technologies; and the Internet of transport brings great benefits to individuals and to the state economy as well as to the economy. This is what prompted countries to seek out its Internet and to work to continue to generalize the process by all means and to develop new ones.

There are countries that have become leaders in smart transportation, such as Dubai, France, and Canada, and this is what we will be discussing in this research paper.

Through this forward, we can pose the following question:

To what extent can the Internet of transportation be regarded as an inevitable inevitability?

We can also ask the following sub-questions:

What are the most important components of the Internet of transportation?
To what extent can the Internet of transportation be used?
What are the most important achievements in the field of intelligent transport?

The objective of the study: The objective of the study is to demonstrate the importance of intelligent transport and work on the Internet of transport means and to emulate the leading countries in this field.

The importance of the study: The importance of the study lies in showing the importance of the Internet of Things and the Internet of transportation, in addition to addressing the most important bases that must be available to be able to have the "Internet of transport"; then we will abandon the study on intelligent transport models in some countries.

The method adopted: The inductive approach is adopted with two analysis tools in data analysis.

Study divisions: The study was divided into the following axes:

Axis I: Introduction to the Internet of Things.
Axis II: Internet of means of transport (smart transport).
Axis III: Internet samples of means of transport.

2 Introduction to the Internet of Things

The Internet of Things started as a project for the US Department of Defense's Advanced Research Projects Agency in 1969 and was called in its day (ARPAnet), and it connected a group of sites that initially numbered four, and it was planned to link more than 50 billion devices by 2020 (Latif 2017). The first appearance of this term was in the beginnings of the twenty-first century precisely in 1999 at the hands of the British scientist Kevin Ashton, whose idea was to connect the digital devices around us, such as home appliances, in a way that allows us to know their specific cases and information without needing to be near them (Al-Marhaby and Al-Baz n.d.).

2.1 Definition

We can define the Internet of Things as:

A network comprising end networks or uniquely identifiable objects that communicate with each other via IP without human intervention. Basically, the Internet of Things takes into account sensors, smartphones, tablets, etc. (Kumar 2015).

Network: a term that has appeared recently and refers to the new generation of the Internet (the network) which enables understanding between devices interconnected to each other (via Internet Protocol). These devices include tools, sensors, and artificial intelligence tools. These devices allow the person to move around without needing their presence thanks to the control of these tools (Hassan 2019).

The Internet of Things: This is a vast network of connected devices connected to the Internet, including smartphones and tablets, and almost anything that contains a sensor, such as cars and machines in production plants, jet engines and oil well-drilling machines, and wearable devices is like wristwatches because objects collect and exchange data (Al-Marhaby and Al-Baz n.d.).

The term Internet of Things refers to the connection of devices and things to a private network or the Internet. These elements include sensors and actuators that help operate autonomous machines and intelligent systems. Internet-connected sensors allow objects to collect and exchange data in public spaces, workplaces and homes and to monitor the environment. Report their condition, receive instructions, and take appropriate action; in addition to connecting things, Internet technology enables a digital interconnection between components of the physical world, such as people, animals, air, and water (Economic and Social Commission for Western Asia, United Nations 2019).

Through the previous definitions, we can say that the Internet of Things connects things to programs and technological means with the aim of controlling them in a way that suits the party that uses them.

2.2 The Components of the Internet of Things and Its Basic Components

2.2.1 The Components of the Internet of Things

The Internet of Things is made up of four main components:

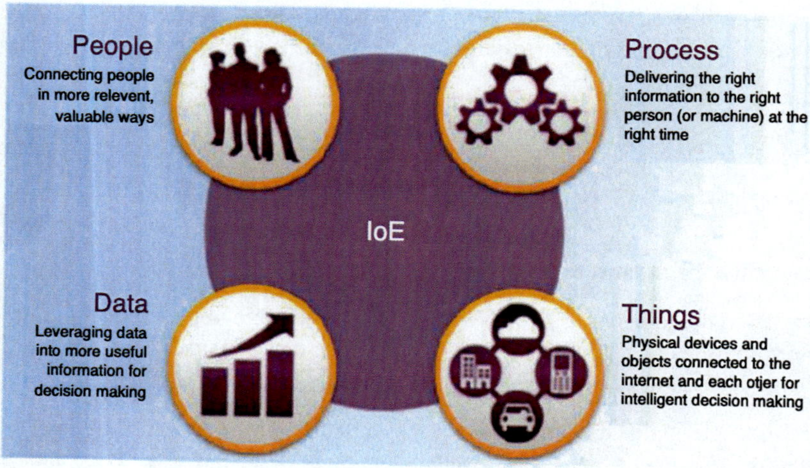

Fig. 1 The four elements of the Internet of Things. (Source: Mustafa Sadiq Latif, p. 50)

Things: What we hear here is everything around us, such as appliances, manufacturing, transportation, energy, nutrition, clothing, and even the human body and animals. In short, anything can be connected to the Internet through a small data chip that collects information without human intervention, and in the future there will be nothing that cannot be connected to the Internet.

- Communication networks: which connect them (Al-Marhaby and Al-Baz n.d.).
- Data: which is sent by things and people and received by the computer systems that process them.
- People (individuals) (Fig. 1).

2.2.2 The Basic Components of the Internet of Things

The components of the Internet of Things are:

- Mobility: its goal is to provide the ability to access the Internet from any device, anywhere and anytime.
- Cloud computing: this means providing IT services and resources distributed across the network so that they can be accessed from anywhere and anytime.
- Big Data: it is the ability to process and analyze huge data that steadily increases with the increasing number of devices and objects connected to the network.
- The new generation of Internet addresses: which will increase the number of addresses available to connect devices to the Internet from less than four billion addresses now to 10 × 4 ^ 38 addresses, which means connecting 50 billion Internet devices by 2020 will be easy and possible (Al-Marhaby and Al-Baz n.d.).

2.3 The Importance of the Internet of Things and the Areas of Use

2.3.1 The Importance of the Internet of Things (Atallah 2020)

The importance of the Internet of Things is:

- Help decision-makers make the best decision among the set of decisions available to them.
- Raise the level of performance and productivity strategy in factories.
- Provide better customer service.
- Increase productivity.
- Save effort, time, and money.
- Prevention and response to problems.

2.3.2 The Areas of Use the Internet of Things

The Internet of Things is used in the following areas:

- Smart homes.
- Smart cities.
- Health care.
- Automotive/transport.
- The field of agriculture.
- Retail trade.
- Smart clothing or electronic clothing (wearable).
- Energy management.
- Telecommunications.
- Factories and companies.

2.4 The World's Largest Pioneer in Internet of Things and Internet of Things Statistics

2.4.1 The World's Largest Pioneer in Internet of Things

Business Insider expects investments in Internet of Things technology to generate huge revenues totaling $ 13 trillion by 2025. Several global companies are competing for the Internet of Things, which has become an icon of business development in technology companies seeking to maximize their revenue, and these companies are:

- *Amazon*: It entered the Internet of Things by launching the Amazon electronic services platform, which controls nearly half of the world's cloud and is a huge central data market that allows businesses to access a vast base data center that enables economies of scale, eliminating the need for operational data centers. Amazon also dominates the smart home market, with the help of the voice assistant "Alexa" and "Echo Device" (Al-Nasser 2015).

- *Apple*: The American company is very present in the smart home sector thanks to its so-called "home" application, which integrates camera systems with door locks, as well as the management of lighting systems as well as kitchen appliances such as coffee machines.
- *Cisco*: When we talk about networking, Cisco has a say in this area. Cisco is one of the first companies to invest in providing solutions to the enterprise sector in the Internet of Things field, and it has very significant studies in this area (Al-Dubii 2018). Known for the expression "connect everything to the Internet," he has already embarked on the integration of the Internet of Things in infrastructure, television, and transport services (Al-Nasser 2015).
- *Google*: With its rich experience in the consumer sector, Big Data, and BI business intelligence, in addition to its hegemony in the mobile era in which we are currently living, Google has the tools needed to create smart solutions that benefit consumers and businesses alike (Al-Dubii 2018). Its approach focuses on self-driving car projects and unmanned aerial vehicle development, as well as the design of the Google cloud platform (Al-Nasser 2015).
- *Microsoft*: With an operating system (Windows) running on 1.5 billion devices and a cloud computing platform that is the best and strongest for the business sector, on top of that, Microsoft has launched a special version of Windows geared toward the Internet of Things, so Microsoft has what qualifies it to lead the software and cloud solutions industry in the Internet of Things market (Al-Dubii 2018). It also manages the "Azure IoT Suite" program, which allows users to remotely monitor devices, analyze their data, and predict maintenance needs to prevent outages.
- *Fitbit*: Is the leader in wearable device development, especially in fitness, and recently released Inspire and Versa devices that track individual athletic activities.
- *IBM*: Cognitive technology "Watson" is the main project of the American company in the field of the Internet of things, because this technology simulates the human method to answer questions and to acquire capacities of reflection and learning.
- *AT&T*: It connected 1.3 million cars to the Internet 3 years ago, drivers being exempt from paying the cost of creating an account on the network.
- *T-Mobile*: The company has set up an integrated hub to serve IoT customers by developing a platform that helps designers of IoT products benefit from modern technologies.
- *Comcast*: The media company acquired "I Control Networks" in 2017, which develops technology and platforms for connecting home security devices, as well as the development of the "Converge" software platform, which enables communicating with safety sensors and provides support for automation devices such as cameras and temperature measurements (Al-Nasser 2015).

- *Intel*: The number of devices that will shape the Internet of Things scene in 5 years will be twice as many as the mobile devices, computers, and tablets that work together today. Who will manufacture processors for this large number of devices? Simply, Intel will be the biggest participant in the hardware field, especially with its huge research and projects in this area after losing a significant share of the mobile device market (Al-Dubii 2018).

2.4.2 Internet of Things Statistics

According to some calculations, the number has reached between 15 and 25 billion connected devices, and the number is expected to increase from 50 to 212 billion by 2020, and some analyzes even predict that there will be one trillion connected devices by 2025 (United Nations, Government Summit 2015).

- *Personal clothing*, which is expected to reach 162.9 million units by the end of 2020.
- *The number of candidate smart home* devices grew from 83 million in 2015 to 193 million in 2020, including washing machines, refrigerators, dryers, security systems, and energy equipment such as lighting.
- *Small businesses* are among the areas that benefit the most from the Internet of Things, as Business Insider expects the volume of investments in devices and solutions for this technology to increase from $ 215 billion in 2015 to 832 billion dollars by 2020.
- *The transport sector*, in light of the expected increase in the number of passenger cars connected to it from 36 million cars in 2015 to 381 million cars in 2020, in addition to the trend of public transport from airplanes, trains and buses to use the Internet of Things to improve the travel experience, in parallel with the benefit of smart cities that will provide For intelligent car traffic and parking data (Al-Nasser 2015).

3 Internet of Means of Transport (Smart Transport)

3.1 The Definition of Internet of Means of Transport (Smart Transport)

It is the use of information and communication technologies to support and integrate transport and logistics systems, so-called intelligent movement (SMART MOBILITY), and it is possible to coordinate and "integrate" all levels of transportation into a unified virtual platform, including cars, trains, planes, and even bicycles and pedestrians (Sadiq and Sfour 2013).

The term intelligent transport is used to express the integrated applications of sensors, computers, communications, and electronics technologies and management strategies to provide people with the necessary information, increase the efficiency of transport systems, and strengthen road safety. Intelligent transport systems integrate information and communication technologies into existing transport management systems, with the aim of improving the services provided (Shakeri and Tal 2014a).

He is interested in the use of computer, electronics, communication, control, and development technologies to address the many challenges facing the public transport and traffic sector, through the use of modern technologies and their relation to public transport, and intelligent applications are characterized by their combination of the enormous capacity of information and control techniques for better transport management (Tunisian Ministry of Transport 2020).

Through the previous definitions, we can say that the Internet of transport or intelligent transport is based on computer technologies and through which all traffic movements, parking lots, traffic lights, etc. are controlled.

3.2 *The Characteristics of Internet of Means of Transport and Their Purpose*

3.2.1 The Characteristics of Internet of Means of Transport

The Internet of transport is characterized by the characteristics shown in the following figure (Fig. 2):

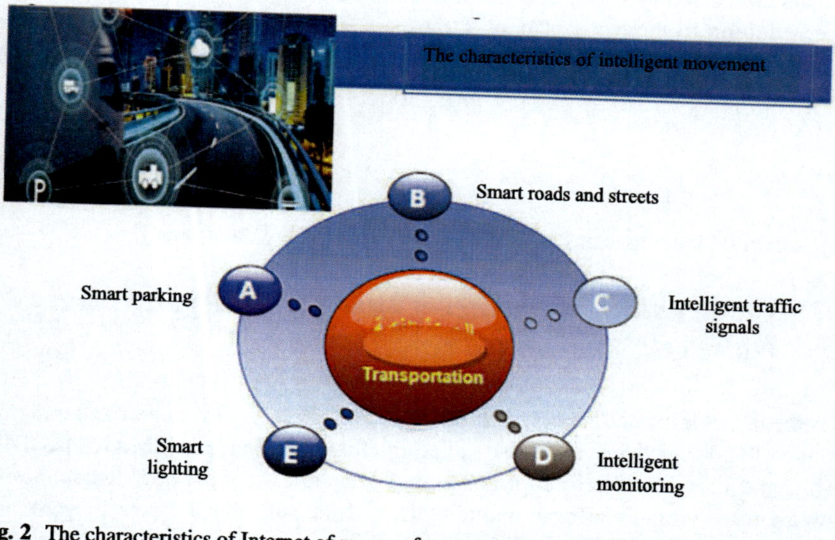

Fig. 2 The characteristics of Internet of means of transport. (Source: Djamel Subhi Hajeer)

3.2.2 The Objective of the Internet of Transport

The main objectives of intelligent transport systems, in particular to improve the current and future economic productivity of individuals, institutions, and the economy in general, are:

- Energy saving and environmental protection.
- Improve Golan standards and the well-being of travelers.
- Increase the operational efficiency and capacity of the transportation system (Al-Ghazi 2010).
- Ensure safety on the road network, reduce the frequency of road accidents, deaths and resulting injuries, and reduce their severity.
- Provide first aid services when such accidents occur, thanks to the rapid response and the raising of the level of rescue in emergency situations.
- Improve ease of movement and provide comfort and safety on the road network.
- Increase the efficiency of roads and increase the productivity of individuals, institutions, and the economic sector.
- Raise the level of road network management by adopting the efficiency of road network capacity.
- Shorten journey times and reduce delays.
- Save investments in the establishment and extension of road networks (Al Youssef and Hussein n.d.).

3.3 Types of Intelligent Transport Services and Their Components

3.3.1 Types of Intelligent Transport Services

Intelligent transport systems are generally classified into services whose hierarchy varies from country to country. For example, the National Engineering for Intelligent Transportation Systems in the USA includes 31 services grouped into 7 types of services, represented in the following table (Table 1):

3.3.2 Intelligent Transport Components

Intelligent transport systems consist of the following main components (Al Youssef and Hussein n.d.):

- Transport infrastructure such as road networks and public transport systems.
- Traffic control centers.
- Sensors and monitoring.
- Spatial identification and display devices in vehicles.
- Integrated communication systems.

Table 1 Types of intelligent transport services

Names of departments	Types of services
Information on transfers before departure Inform the driver on the road Traffic management Joint conduct and transgressions Information on services for travelers Modification of traffic Elimination of accidents Processing of mobility requests Sending and dilution of the experiment Railway junctions	1. Elimination of traffic and travelers
Carry out public transport Road notification Personalized public transport Safety in public transport	2. Provision of public transport
Electronic payment services	3. Online payment
Electronic verification of commercial vehicles Automated road safety checks Safety monitoring on board vehicles Administrative procedures related to commercial vehicles Intervention in the event of accidents related to hazardous materials Sale of the utility vehicle fleet	4. Trade in utility vehicles
Notification of urgent cases and personal safety Elimination of emergency vehicles	5. Act in an emergency.
Prevention of longitudinal collisions Prevention of side collisions Prevention of collisions at intersections Extreme vision to avoid collisions Security control Development of the tension device before impact Automatic driving of the vehicle	6. Security systems in vehicles
Processing of stored information	7. Information management

Source: Tunisian Ministry of Transport

- Basic information such as digital maps, safety information, and traffic.
- Intelligent transport systems represent the natural development of a country's transport infrastructure (Fig. 3).

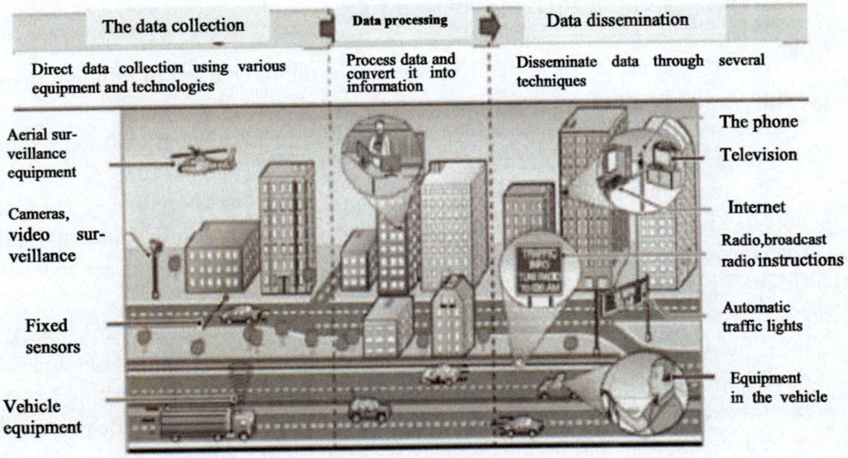

Fig. 3 The components of an intelligent transport system. (Source: Kholoud Sadiq)

3.4 Advantages of Intelligent Transport and Its Fields of Application

3.4.1 Advantages of Intelligent Transport

The Internet of transport leads to:

- Transportation systems are more efficient and intelligent.
- Take advantage of the efficient movement of vehicles, people, and goods, to reduce the load (United Nations, Governmental Summit 2015).
- Mobility application through a special application specializing in route planning and integration of means of transport (collective means, cars, bicycles, and parking lots).
- Provide clean, fast, and safe transportation.
- Avoid traffic problems and give better chances of finding parking or the like.
- Save time, improve mobility efficiency, reduce costs, and reduce carbon dioxide (CO_2) emissions.
- Provide traffic flow management, speed control, freight congestion, information systems, vehicle tracking, onboard security, and vehicle parking management (Shakeri and Tal 2014b).
- It is possible to get information on the performance of means of transport such as roads, streets, public transport and trains, as well as on transport demand (Balwan 2014).

3.4.2 Areas of Application of Intelligent Transport

- Transport demand management: This includes the movement of public transport vehicles and parking control.
- Management of emergency vehicles: Determining the location of the accident and prioritizing signals.
- Traffic management: Tracks the movement of cars, and immediately sends this data to a control center, allowing it to be managed by finding fast and unblocked roads in an emergency. as well as the ability to control traffic lights (Shaheen and Odeh 2016) (Fig. 4).
- *Parking lot management*, which is systems that have become commonplace in most public parking lots, especially shopping complexes (Princess Noura Bint Abdul Rahman University 2014), to know the number of vacant parking lots and reach the nearest one, as newer systems use ultrasonic sensors to detect vacant, occupied, and reserved parking spaces, as well as car parks for families. Special needs (Shakeri and Tal 2014a).
- *Reduce the level of pollution*: by installing sensors that monitor the general air quality.
- *Automated security check*: such as checking the validity of driving licenses, measuring vehicle weight.
- *Electronic tax collection services*: using electronic cards.
- *Underground platform*: the underground platform is represented by underground spaces, which have the capacity to alleviate urban problems and are exploited, as well as providing spaces and spaces for the growth and development of cities.
- *Smart traffic by imposing*: fluid insurance on the circulation of private vehicles and public transport in the city and by easing traffic jams and helping to reduce pollution rates.
- *Parking meters*: one of the most important tools for organizing parking in large cities.

Fig. 4 Priority systems. (Source: Mohsen Ali Balwan)

- *Car control systems*: as it became more sophisticated, it has become linked to communication networks and the Internet, where users can communicate remotely, sell their cars, and control them (Al-Saleh 2014).
- *Road lane control systems*: which are systems that make it possible to determine which lanes are sometimes closed or whose directions change during the day, as well as those which determine the lanes of trucks and modify the way they pass by their passage traffic (Al-Mashhadani 2019).
- *Notification systems*: that automatically notifies powerful people after an accident (Shaheen and Odeh 2016).

4 Internet Samples and Means of Transport

4.1 Driverless Transport in Dubai

In 2016, Dubai adopted a driverless transport technology strategy that will transform the transport system into a 25% intelligent system by 2030, and the Roads and Transport Authority and the Dubai Future Foundation are supporting this project, which focuses on people, technology, policies, legislation, and infrastructure (Economic and Social Commission for Western Asia, United Nations 2019). Dubai's smart transport strategy aims to improve several mobility indicators by 2030, including reducing carbon emissions by 12% and reducing parking demand by 50%, resulting in Dubai's interest for the construction of smart infrastructures which play a central role in the competitiveness of future cities. This strategy will increase road safety by 2030 by 12% and reduce transport costs by 44%.

4.2 The Driverless Metro at Princess Nora University in Saudi Arabia

An automated transport system is available at the university, which is a metro, which is the largest automatic university train in the world, as it does not require a driver and provides 24-h transport service for academics, administrators, and students. A station separated by a distance of about 400 m; as for the load capacity of each vehicle, it reaches 134 passengers, including 24 seated and 119 standing (Al-Saleh 2014).

4.3 The Mobile Phone Is Becoming the "Ticket" for Public Transport in France

Since 2008, three mobile telephone operators, in cooperation with public transport companies in France, have promoted offers that make it possible to transform the mobile telephone into a transport deposit. The traveler only has to pass his mobile phone a few centimeters from a special device (terminal) to confirm the reservation. Electronic tickets can be purchased on a mobile phone through a dedicated website, with the customer's credit card number inserted.

The aforementioned partners have agreed to set common technical specifications and to consider numerous economic examples centered on one basic principle: the customer pays the price of the ticket to the public transport company, which in turn pays a percentage to the price mobile operator (Al-Ghazi 2010).

4.4 Safe Scooters in Rwanda

The platform was launched in Rwanda in 2015 and aims to reduce road accidents and deaths and adopt the Uber model for motorcycles in two Africans. This platform collects driver data and analyzes it using the Global Positioning System and includes it in the classifications that customers assign to drivers to reward safe and responsible driving. Drivers with at least 3 years of experience can apply to join the platform, and this platform allows customers to pay the freight cost using an electronic wallet shipped by cellphone or credit card, and this platform uses the services of the Kigali incubator to interact with project owners, programmers, and developers. In Rwanda, the continuous improvement of the ICT infrastructure has been a major factor in the success of the platform (Economic and Social Commission for Western Asia, United Nations 2019).

4.5 CAMPAS Traffic Management System in Canada

The Canadian experience, through the CAMPAS traffic management system, in "Toronto," shows that the control of traffic accidents and overcrowding at certain points on the highway, number 401, has made it possible to reduce traffic. The duration of the intervention is between 86 and 30 min. It also reduced the response delay rate for each accident by 537 vehicle hours, in addition to forecasting around 200 accidents per year (by displaying information about the accident at the time of its occurrence), which has saved $ ten million.

4.6 GPRS Shelter System in Munich

Since April 2006, the city of Munich has implemented a system produced by Siemens and Citrix to guide drivers when looking for free space in stacked shelters. This system works by sending all shelter information to a central distributor via the Internet (GPRS). This distributor broadcasts this information to the public after having analyzed it on digital screens in the street and on the radio link.

This system allows the city of Munich to earn 500 thousand euros by considering only public works and is characterized by a low operating cost due to its use of the Internet as a means of sending information (Al-Ghazi 2010).

5 Conclusion

The means of transport are the main artery of daily life and foreign trade, as they are exported and imported, and they differ according to the needs, and the need for their Internet has become an unavoidable necessity, as many countries have benefited from their services which were previously deprived of it.

Results
Through the study, the following results were achieved:

- *The Internet of Things appeared with the emergence of the Internet and the development of communication technologies. It depends on specialized techniques and programs and has become a source of profit for many companies working in the field of technologies and services.
- *Smart transport is considered an important and necessary means in daily life and international trade, due to its characteristics that distinguish it from other traditional means of transport.
- *Internet and means of transport have developed considerably, and some countries and companies have become pioneers in this field.
- *Countries differ in terms of their dependence on intelligent transport; some of them have internet transport means, traffic, etc.; and there are those who are satisfied with one type of specific transport, according to their needs and capacities.

Recommendations
We can make the following recommendations:

- Countries should emulate pioneering and successful experiences in intelligent transport.
- Countries must work on the Internet and the means of transport, take advantage of the advantages it offers, and overcome the difficulties they may encounter to achieve this.

- Sign agreements with leading countries in the field of intelligent transport, and benefit from a transfer of technology within the framework of exchanges of experiences or investments in this field.

References

Al Youssef, I.J., Hussein, M.M.: Smart and sustainable cities: perspectives and aspirations in the footsteps of 21st century cities

Al-Dubii, Y.: How does the Internet of Things Work?, Posted on 08-09-2018 (2018)

Al-Ghazi, S.: The Importance of ITS Intelligent Transport Systems in Transport Problem Solving and Crisis Control, Date Posted: 09/18/2010 (2010)

Al-Marhaby, K.A., Al-Baz A.M.: Internet of Things and Cities

Al-Mashhadani, B.A.H.: The Role of Sustainable Transportation and Intelligent Transportation in Allevising Transportation Problems in the Emirate of Dubai, paper presented at the tenth international conference on geophysical, social, human and natural challenges in a changing environment, July 25–26, 2019, Istanbul (2019)

Al-Nasser, N.: What Do You Know About the Internet of Things ?, Date Posted: March 4, 2015 (2015)

Al-Saleh, M.: Information Security in Smart Cities. J. Sci. Technol., Issue 111, May 2014 (2014)

Atallah, S.: What is the Internet of Things, Posted: May 22, 2020 (2020)

Balwan, M.A.: Smart Traffic Systems in Future Cities, J. Sci. Technol., number 111, May 2014 (2014)

Economic and Social Commission for Western Asia, United Nations.: Innovation and Technology for Sustainable Development Promising Prospects in the Arab Region for the Year 2030 (2019)

Hassan, H.H.: Smart Cities and their Role in Solving Community Services Problems in Cities (Bagdad City) for example, Reflecting the Flow of Literature, Special Issue for Conferences 2018–2019 (2019)

Kumar M.: Building Smart Cities from Smart Data, October 2015 (2015)

Latif, M.S.: Introduction to the Internet of Things - Part 1, February 7 (2017)

Princess Noura Bint Abdul Rahman University.: Smart Cities, J. Sci. Technol., Number 111, May 2014 (2014)

Sadiq, K., Sfour, M.H.: Smart Cities and their Role in Finding Solutions to Urban Problems (Case Study: Transportation Problems in the City of Damascus), Damascus University Journal of Engineering Sciences, Volume twenty-ninth, number two (2013)

Shaheen, B.R., Odeh, M.J.: The Role of the Information Environment and Building a Smart City. J. Eng., issue July 22, 2016 (2016)

Shakeri, A., Tal, M.M.: Public transport is the artery of the city. J. Sci. Technol., Issue 111, Rajab 1435 AH, May 2014 (2014a)

Shakeri, A., Tal, M.M.: Public transport, artery of medina. J. Sci. Technol., Issue 111, May 2014 (2014b)

Tunisian Ministry of Transport.: Tunisia Intelligent Transport Systems (2020)

United Nations, Government Summit.: Government Summit Research Series, Smart Cities: A Regional Perspective, February 2015 (2015)

United Nations, Governmental Summit.: Government Summit Research Series, Smart Cities: A Regional Perspective, February 2015 (2015)

An Adaptive Medical Advisor to Improve Diabetes Quality of Life

Abdelouahab Belazoui, Abdelmoutia Telli, and Chafik Arar

Abstract This paper aims to improve the diabetes quality of life (DQOL) by making it easier to control the disease anytime and anywhere. There are so many factors affecting diabetes that self-management has become a complex task. Moreover, the therapeutic interventions differ according to the patient's profile, making this process similar to solving a multi-objective optimization problem. In this context, we propose a new solution to the problem of diabetes control shaped an adaptive tool that assumes the medical adviser's role for those suffering from diabetes. The proposed tool exploits mobile technology to help individuals manage their diabetes independently, anywhere, anytime. It allows patients to avoid the complications of the disease and encourages them to become actively involved in their own care processes.

Keywords Diabetes quality of life · Adaptation · Medical advisor · Mobility

1 Introduction

The medication discovery is an innate behavior practiced by all living things in a subliminal way that is mainly aimed at preserving their lives. Several treatment techniques and strategies have been gradually developed by humans since the beginning of history where they previously depended on the use of crops provided by nature and human experiments in the curing process.

A. Belazoui (✉) · C. Arar
LAMIE Laboratory, University of Batna 2, Fesdis, Batna, Algeria
e-mail: a.belazoui@univ-batna2.dz; chafik.arar@univ-batna2.dz

A. Telli
LAMIE Laboratory, University of Batna 2, Fesdis, Batna, Algeria

Computer Science Department, University of Biskra, Biskra, Algeria
e-mail: a.telli@univ-biskra.dz

S. Sedkaoui et al. (eds.), *International Conference on Managing Business Through Web Analytics*, https://doi.org/10.1007/978-3-031-06971-0_19

At present, the facilities that technological advancements have introduced into human life have hampered his physical activity, which has led to the emergence of diseases related to obesity and diabetes. The latter is a syndrome characterized by a metabolic disorder and abnormally high blood sugar caused by insulin deficiency, decreased sensitivity of tissues to insulin, or both. Diabetes leads to serious complications and even premature death. However, there are precautions people with diabetes can take to control the disease and reduce the risk of complications. These precautions are tightly linked with the patient's blood sugar level. Practically, it must satisfy the influencing factors in the glucose equilibrium process to have controlled diabetes which is considered a goal for every patient who suffers from this disease.

This paper aims to provide an informatics solution that makes life with diabetes easier under the best circumstances. The proposed approach enables self-management of diabetes using an interactive mobile application as a medical advisor which provides adaptive counsels to the patient according to his introduced parameters.

After giving an overview of related work on recently invested technologies in the field of patient's self-care and prove its importance, Sect. 3 presents the influencing factors in controlled diabetes to know the patient's needs that we help in the modeling phase of the proposed medical advisor. Section 4 shows a general architecture of the proposed system by expressing the functionalities of each module and their communication with each other. Section 5 exhibits the scenarios of the use of the proposed system and captures the dataflow under its components. Section 6 provides the software configuration necessary to develop the proposed system. Finally, the last section shows the results of the research described in this paper.

2 Literature Review

Improving diabetes healthcare has recently attracted the interest of researchers that has led to the emergence of several approaches and informatics tools to make life easier for individuals with this chronic disease. Indeed, most of the research undertaken in this area has mainly aimed at supporting disease self-management and improving quality of life which considered the main goal of its early diagnosis and treatment (Trikkalinou et al. 2017).

Artificial intelligence (AI) techniques have revolutionized diabetes care, and they have been very helpful in making medical decisions. For more information, it's recommended to review the work done in (Contreras and Vehi 2018); this investigation has contributed to know recent works aimed at using AI techniques to support the diabetes management process. Also in relation to that, the work presented in (Dovc and Battelino 2020) has tried to catch the evolutionary line of diabetes technology. This work targets type 1 diabetic patient who relies on periodic insulin injections. That's why the research focused on the development of injection technology.

The first studies that raised the issue of computer intervention in diabetes management focused on helping healthcare professionals choose the best monitoring procedure by predicting blood glucose profiles resulting from alternative control policies. Deutsch and his team in (Deutsch et al. 1990) developed a complex computer system architecture to support the management of diabetes, which depends primarily on patient blood glucose monitoring. Researchers have gone one step further by developing an automated insulin dosing adviser (Lehmann and Deutsch 1993; Lehmann et al. 1994); they have focused on insulin adjustment for the insulin-dependent diabetic patient. Their proposed solution differs from traditional purely algorithmic and model-based methodologies generally applied for treatment planning by combining rule-based reasoning and model-based reasoning to select appropriate control actions. In addition, the adaptation was introduced into diabetes management by (Juhasz et al. 1994) where it was applied in its control techniques. Multi-agent systems (MAS) have brought about a paradigm shift in medical decision support. In this context, the authors in (Greenwood et al. 2003) attempt to solve the problem of self-adaptation of the agent-based user interface to enable real-time decision support.

Fuzzy logic techniques were inspired in the form of a glucose-regulating counseling algorithm for a type 1 diabetic patient under insulin intensive treatment based on a multiple daily dosing regimen (Femat et al. 2006). Moreover, the reasoning techniques and mechanisms have also been exploited in the same context; the authors in (Ahmed et al. 2015) have shown that the case-based reasoning paradigm is the best methodology of reasoning technique and provides an optimal solution for diabetic expert systems. The adaptation based on reinforcement learning was also proposed by (Sun et al. 2019) where they have encouraging results for diabetic patient under multiple daily injections treatment; this approach increases the time spent in the target range while simultaneously reducing risk factors.

After studying the most interesting work done in the computerization of diabetes management, it appeared that the majority of studies were focused especially on the pharmaceutical side through taking of insulin doses. They helped to facilitate the application of the instructions of the attending physician outside of medical consultation times. In contrast, the solutions proposed did not take account of all the factors affecting the control of diabetes and spared patients from serious complications of the disease.

To address this problem, the authors identify in the following section the main factors affecting diabetes control that will be the main pillars of the solution proposed in this paper.

3 Factors Influencing Controlled Diabetes

Diabetes is a chronic disease that cannot be cured. Its treatment aims to improve the patient's quality of life and prevent complications. Therefore, its treatment differs

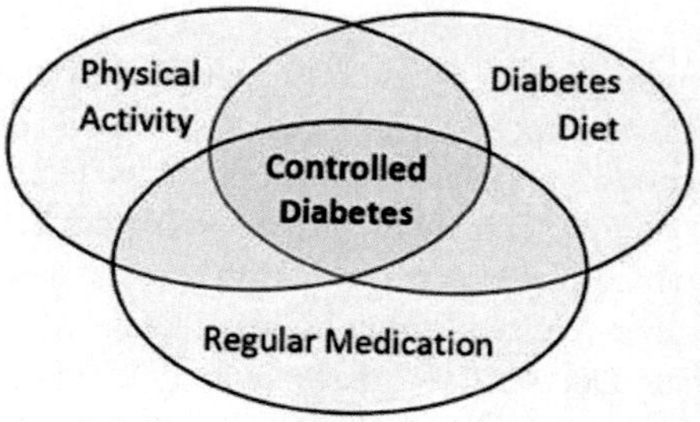

Fig. 1 Major factors in controlling diabetes

from other diseases and depends on the satisfaction of a few factors. Figure 1 illustrates these factors involved in controlling diabetes.

There may be other factors that affect controlled diabetes, but the authors save three main factors, namely, physical activity, diabetes diet, and regular medication. These factors form a framework for the proposed tool in the next section.

Firstly, the physical activity enables its practitioners to have good fitness and avoid many diseases. Physicians strongly recommend involving patients in physical activities that help burn fat to keep blood sugar levels moderate. The World Health Organization (WHO) experts recommend brisk walking for 20 min a day as a minimum physical activity for individuals with diabetes.

Secondly, the diabetic diet is used by diabetes mellitus or high blood sugar to reduce the symptoms and dangerous consequences of the disease. There are diet educators tasked with indoctrinating healthy eating habits for elderly and ill-read patients in particular. For knowing the recent nutritional recommendations for diabetics, it is recommended to review the work presented in (Gray and Threlkeld 2019).

Finally, the regular consumption of medications prescribed by the attending doctor contributes to improving the quality of life of the patient; this matter therefore should not be tolerated because it can lead to catastrophic consequences for the health of the patient. It is appropriate to provide a mechanism that helps patients remember when to take their medications in specific doses.

4 Proposed System Architecture

In this paper, the authors have attempted to help individuals with diabetes with a clever tool to improve their quality of life and prevent them from problems related to poor disease management. However, its intelligent aspect lies in the fact that the patient's profile guides the advice given. As well as their mobile aspect, patients can enjoy the benefits of self-management of the disease anytime and anywhere thanks to intelligent devices that have become widely available recently.

This section presents a detailed architecture of the proposed approach based on four main components, namely, the therapeutic model, the monitoring agent, the

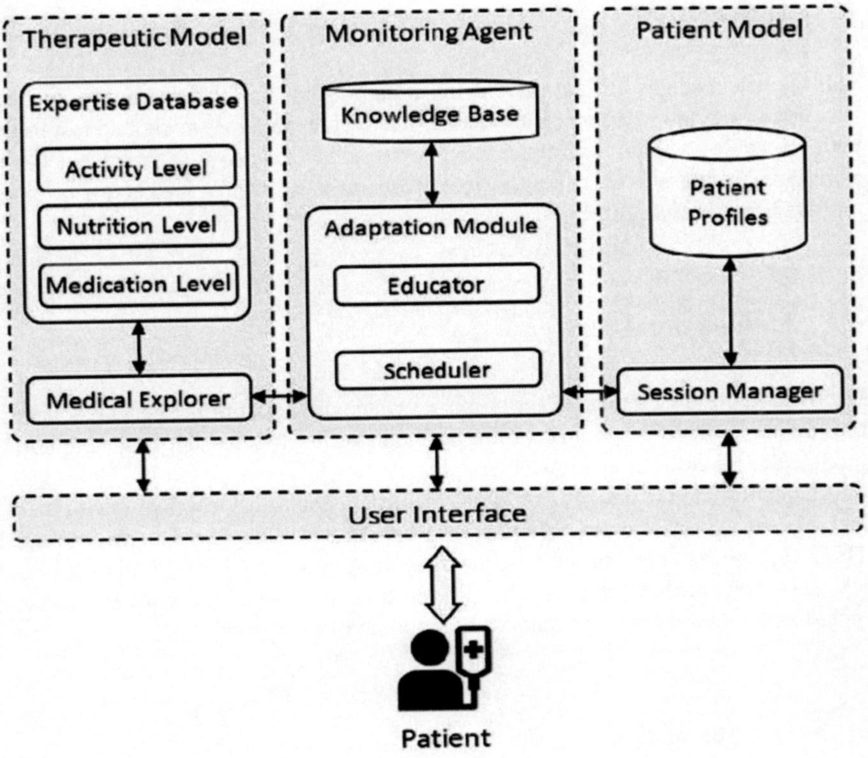

Fig. 2 Detailed architecture of the proposed system

patient model, and the user interface. A detailed description of this architecture is illustrated in Fig. 2.

4.1 Therapeutic Model

It coordinates with the monitoring agent to ensure that medical advice is provided to patients according to their profile. Conceptually, this component is indeed composed of a layered database of diabetes knowledge and a medical explorer.

The layered database performs logistical support to the exploration module by supplying relevant medical knowledge. It was designed by extracting and organizing the knowledge of a human expert in a layered database at three levels, namely, physical activity, nutrition, and medication. Moreover, the exploration module bears the task of the disease area expert. It acts on the medical knowledge extracted from the database and provides advice for the best management of diabetes.

4.2 Patient Model

This model manages critical patient information that directly affects the overall behavior of the proposed medical advisor. It is responsible for saving and restoring patient profiles at each consultation of this medical advisor. To carry out such a mission, it consists of two parts that work in coordination, namely, a session manager and patient profiles.

4.3 Monitoring Agent

The most important element of the proposed system is that it offers an escort adapted to individuals with diabetes according to their profile. It consists of an adaptation module and an associated knowledge base.

Adaptability to the patient's profile is the intelligent aspect of the proposed medical advisor, while its main objective is to benefit from adaptive self-care service. There are two features provided by the adaptation module which are diabetes diet education and medication reminder. In addition, the associated knowledge base includes data that enables the smooth functioning of the system.

5 Scenarios of Use

To facilitate the implementation of the proposed tool in this paper, it is appropriate to illustrate the data flow across their components. Indeed, the proposed tool offers two essential functionalities from which patients can benefit after their identity verification or registration: the deposit of their values concerning the factors affecting their diabetes or the receipt of adapted advice on their real state of health through their mobile device. The authors describe in the following how their medical advisor functions using sequence diagrams. They have adopted UML (Uniformed Markup Language) to describe the information exchanges within this proposed tool.

5.1 Setting Mode

The proposed tool either requires users to preregister or must register and open a new account. In all cases, the setting mode allows the creation and enrichment of the patient's profile which will affect the provided advice nature. There are two ways the user can use the proposed application in setting mode, namely, creating a new account or entering the new parameters in a previously existing one.

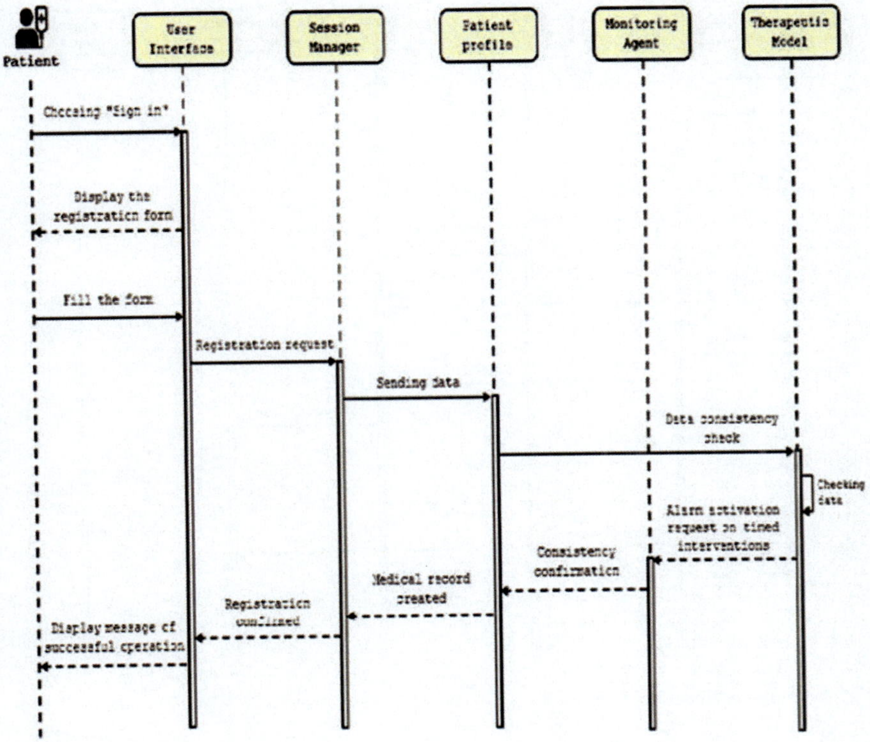

Fig. 3 Registration process for a new patient

When creating a new account, the user must choose "sign in" after running the application on his mobile device. A registration form must then be filled out. Critical information regarding healthcare metrics must be checked before being saved as a new medical record. On the other hand, saving parameter values in an existing account does not require data checking. Figure 3 shows a sequence diagram describing how a new patient is registered under the proposed application.

5.2 *Advising Mode*

With the help of the proposed application, the user can use it in an advisory mode such as an alarm clock to consume the medication prescribed by the attending physician or a diet educator adapting the diet according to the patient's profile. Below in Fig. 4, a sequence diagram described an example of the provision of advice by the proposed application.

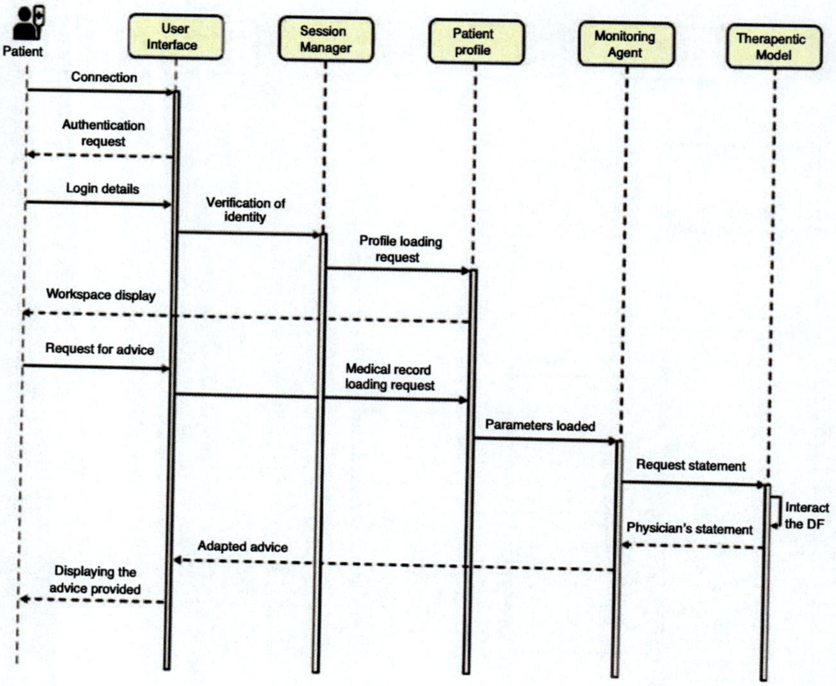

Fig. 4 Adaptive advice process

6 Implementation

The implementation phase in software engineering encompasses all the processes involved in the proper functioning of an element in its environment. To achieve this objective, it must identify in this phase the various programs and appropriate tools that correspond to the specificities of the desired software to be developed, followed by translating one's own model using the chosen set of tools as a usable product.

The authors have chosen to use the Java and XML languages under Android Studio as IDE (integrated development environment) based on a software package (libraries, tools) called SDK Android and ADT (Android developer tools). They also took advantage of the AVD (Android Virtual Device) emulator to run and test the application to be developed on Windows. Moreover, SQLite was adopted for database creation and manipulation.

A prototype has been developed using the software range chosen for the implementation phase. The proposed tool is a mobile application called AMAD (Adaptive Medical Advisor for Diabetics). It is currently under the software validation phase of its development cycle and will soon be placed on the Google Play platform for free use by people suffering from diabetes.

7 Conclusion

As the factors involved in the stabilization of diabetes have multiplied, disease surveillance has become a tiring process. It requires a good appreciation of therapeutic intervention strategies which vary according to the patient's profile. In this context, the purpose of this paper is to improve the quality of life of people with diabetes by enabling them to live better with their disease. To reach this aim, the authors introduce a medical advisor for individuals with diabetes by integrating of adaptive and mobility techniques.

In light of the key factors influencing the diabetes stability identified at the beginning of this paper, the authors presented the architecture of the proposed medical advisor by specifying these different components. Thereafter, they have clarified the working mechanisms of the proposed tool through sequence diagrams which capture the data flow among their different components. Moreover, they have specified the appropriate software configuration for such programming kind. A prototype has been developed under the name AMAD which will be made available to patients after having completed the software validation phase. Arguably, the tool proposed in this paper will be a good partner for diabetics through its adaptive consultations according to the disease situation in every place and time.

In future work, the authors will place their medical advisor for free use across the Google Play platform. After a trial period, they will conduct a field survey with a sample of diabetic patients to see how the proposed tool will affect their quality of life.

References

Ahmed, I.M., Alfonse, M., Aref, M., et al.: Reasoning techniques for diabetics expert systems. Procedia Computer Science. **65**, 813–820 (2015). https://doi.org/10.1016/j.procs.2015.09.030

Campos-Delgado, D.U., Hernández-Ordoñez, M., Femat, R., et al.: Fuzzy advisor algorithm for glucose regulation in type 1 diabetic patients on a multi doses regime. IFAC Proceedings. **39**(18), 309–314 (2006). https://doi.org/10.3182/20060920-3-FR-2912.00057

Contreras, I., Vehi, J.: Artificial intelligence for diabetes management and decision support: literature review. J. Med. Internet Res. **20**(5) (2018). https://doi.org/10.2196/10775

Deutsch, T., Carson, E.R., Harvey, F.E., et al.: Computer-assisted diabetic management: a complex approach. Comput. Methods Prog. Biomed. **32**(3–4), 195–214 (1990). https://doi.org/10.1016/0169-2607(90)90102-F

Dovc, K., Battelino, T.: Evolution of diabetes technology. Endocrinol. Metab. Clin. N. Am. **49**(1), 1–18 (2020). https://doi.org/10.1016/j.ecl.2019.10.009

Gray, A., Threlkeld, R.J.: Nutritional recommendations for individuals with diabetes. In: Feingold, K.R., et al. (eds.) Endotext. MDText.com, Inc., South Dartmouth, MA (2019) Available from: https://www.ncbi.nlm.nih.gov/books/NBK279012/

Greenwood, S., Nealon, J., Marshall, P.: Agent-based user Interface Adaptivity in a medical decision support system. In: Moreno, A., Nealon, J.L. (eds.) Applications of Software Agent Technology in the Health Care Domain Whitestein Series in Software Agent Technologies and Autonomic Computing. Birkhäuser, Basel (2003). https://doi.org/10.1007/978-3-0348-7976-7_4

Juhasz, C., Asztalos, B., Lerner, B., et al.: AdASDiM: application of adaptive control technique to diabetic management. In: Proceedings of 16th Annual International Conference of the IEEE Engineering in Medicine and Biology Society, Baltimore, USA, pp. 928–929 (1994). https://doi.org/10.1109/IEMBS.1994.415217

Lehmann, E.D., Deutsch, T.: AIDA^2: a Mk. II automated insulin dosage advisor. J. Biomed. Eng. **15**(3), 201–211 (1993). https://doi.org/10.1016/0141-5425(93)90116-G

Lehmann, E.D., Deutsch, T., Carson, E.R., et al.: AIDA: an interactive diabetes advisor. Comput. Methods Prog. Biomed. **41**(3–4), 183–203 (1994). https://doi.org/10.1016/0169-2607(94)90054-X

Sun, Q., Jankovic, M.V., Mougiakakou, S.G.: Reinforcement learning-based adaptive insulin advisor for individuals with type 1 diabetes patients under multiple daily injections therapy. In: 41st Annual International Conference of the IEEE Engineering in Medicine and Biology Society, Germany, pp. 3609–3612 (2019). https://doi.org/10.1109/EMBC.2019.8857178

Trikkalinou, A., Papazafiropoulou, A.K., Melidonis, A.: Type 2 diabetes and quality of life. World J. Diabetes. **8**(4), 120–129 (2017). https://doi.org/10.4239/wjd.v8.i4.120

The Role of Data Bank Algeria as a Big Data Service Provider in Evaluating the Lending Policy of Public Banks Using the Capital Asset Pricing Model for the Period (2010–2016)

Ilifi Mohamed, Belghalem Hamza, and Serir Abdelkade

Abstract This study aimed to evaluate the lending policies of public banks in the Algerian banking system for the period 2010–2016, and to achieve this, the capital asset pricing model was built based on the big data collected on the Bank of Algeria website as a source of open data, through which we came to consider that the lending policy of all public banks. It is defensive in line with the stagflation state that the Algerian economy is going through during the studied period, but it remains in line with the capital asset pricing model.

Keywords Lending policy · Capital asset pricing model · Open government data · Big data

1 Introduction

Big data has become an important variable in the digital environment, due to its great role in building plans and anticipating the future. Therefore, in light of the spread of this data, especially through social media sites, governments focus on collecting them within their official sites in the form of open government data to draw and build macroeconomic policies. These data are stocks of information that are processed in modern ways to store, process, and distribute them that help in the process of making various decisions.

Within this context, the Bank of Algeria collects this data to assist banks in building and drawing their lending policies through the data provided by many preventive centers specialized in this field, the most important of which is the centralization of risks, to achieve the safety and stability of the banking system as a whole and avoid systemic risk.

I. Mohamed (✉) · B. Hamza · S. Abdelkade
University Khemis, Miliana, Algeria
e-mail: m.ilifi@univ-dbkm.dz

S. Sedkaoui et al. (eds.), *International Conference on Managing Business Through Web Analytics*, https://doi.org/10.1007/978-3-031-06971-0_20

With the help of the data available on the Bank of Algeria website on the Internet and using the capital asset pricing model, it is possible to evaluate the effectiveness of public banks in using the available resources within their various lending policies and their ability to finance the economy in a banking environment that knows rapid transformations to extract strengths to enhance them and highlight shortcomings and weaknesses to find appropriate solutions to overcome them, as this model works on a brief and comprehensive evaluation by calculating the required rate of return on investment in the bank's assets and comparing it to the actual rate and thus judging the efficiency of the lending decision included in the lending policy of public banks. Accordingly, from the above, we can ask the following main question: *Can the data collected in the Bank of Algeria report From extrapolation of the effectiveness of the lending policies of Algerian public banks?*

Sub-questions Asking the main question contributes to asking the following set of sub-questions:

- What is the role of open government data in light of big data?
- What is the theoretical basis of the capital asset pricing model?
- What is the advantage of the lending policy of Algerian public banks?

Hypotheses This research is based on the main hypothesis: that the lending policies of public banks summarized in the data of the Bank of Algeria are acceptable within the capital asset pricing model.

The Importance of the Research The research is of great importance because the effectiveness of the lending policy of commercial banks has a direct impact on the optimal allocation of financial resources to serve economic development projects, and since bank financing is dominant in financing the Algerian economy, the open government data has become the Bank of Algeria provides important data capable of giving the ability to read. These policies are analyzed and evaluated in a brief and useful manner using some of the entrances in this, which helps in modifying them in a way that serves to achieve economic development in Algeria.

Research Objective Through this research, we aim to identify the content of big data and its role in evaluating the lending policies of Algerian public banks according to the capital asset pricing model.

Study Boundaries The time frame for this study is limited to the period from 2010 to 2016, while the spatial framework concerns the six Algerian public banks.

The Approach Used Given the nature of the research topic and the attempt to answer the main question and test the validity of the hypotheses, we will rely on the deductive approach in order to describe and analyze the various dimensions of the study.

2 A Conceptual Framework for Big Data and Open Government Data

The importance of data has increased in light of the development in modern technology, and it has become so great that it has necessitated the adoption of methods to collect and analyze it to facilitate the process of its use, especially in decision-making, and given this, governments seek to collect them in the form of open government data:

2.1 *The Concept of Big Data*

In fact, it is not possible to give any specific definition of big data, as it is a complex multiform term; its definition differs between the societies that care about it as a user or service distributor, etc.; and accordingly some concepts can be given related to the term big data.

Big data is defined as a stock of information that is characterized by great size, speed, and diversity that requires innovative and effective forms of processing it that differ from ordinary data processing so that it enables its users to improve visibility, decision-making, and the automation process (Suleiman Rashwan 2018, page 27).

It is also known as a huge amount of complex data that achieves high levels of distribution and huge quantitative data sources; its speed and diversity are great; its size exceeds the ability of traditional software and computers to store, process, and distribute it; and it is often available in its time and takes various forms if understood in depth. It was used better in the decision-making process (Ghobeiry and Hassan Hassan 2019, page 35), as defined by the International Organization for Standardization as a group or groups of data that have characteristics such as size, speed, diversity, variance, data validity, etc., and cannot be efficiently processed using current and traditional technology to take advantage of them (Suleiman Rashwan 2018, page 27).

According to the previous definitions, big data can be defined as a group of large and complex data that is difficult to process using traditional techniques and tools, enabling its users to improve visibility and make appropriate decisions.

2.2 *Characteristics of Big Data*

Big data has the following characteristics (Suleiman Rashwan 2018, page 28) (Sakri et al. 2019, page 45) (Shukla et al. 2015, p. 6):

- Volume: refers to the amount of data generated, which may reach a large number of data and determine the size of its value.

- Speed: it means the speed of production and extraction of data to cover the demand for it, as speed is a crucial element in making a decision based on this data, and it is the time that we spend from the moment this data arrives at the moment the decision is made based on it.
- Diversity: it is the extracted data that helps users, whether they are researchers or analysts, to choose the appropriate data for their field of research and includes structured and unstructured.
- Data such as images, clips, photo, video recordings, SMS, call records, and map data and requires time and effort to prepare them in a suitable form for processing and analysis.
- Value: After discussing all the previous characteristics, there is one characteristic that must be taken into account when looking at big data, and it is valued; everything is good in accessing data, but unless we can convert it into a value, it is useless.
- Honesty: It is related to the quality of the data that are obtained, and this requires a careful analysis of it in terms of its usefulness with an investigation of its source and validity.
- Of great value: To benefit from big data, we need specialists with sufficient experience and skills to deal with this data and analyze it with appropriate analysis, in which case the information is considered valuable.
- Variable value: meaning that the same information or the same data can mean several things, and based on the context in which they are mentioned, their true value can be determined and analyzed appropriately.
- Multiple appearances: When using big data, it must be analyzed and presented in different forms commensurate with the nature of its use, and it takes multiple forms such as statistics, numbers, geometric shapes, etc.

2.3 Types of Big Data

Big data is divided into structured (or structured) data, but it represents a small part less than 10% and irregular (or unstructured) data and represents the largest part of the data, and therefore it is represented as follows (Latabi, 2019, p. 59):

- Structured data: It is the big data stored in database fields, so it can be searched easily.
- Unstructured data: it is everything that cannot be categorized easily, such as images and graphs, audio and video clips, website clicks, PDF files, emails, and social media posts. Although these types of files have a special internal structure, they are considered (non-organization) because its data is not coordinated into uniform columns suitable for a database.
- The data is semi-structured: it is a mixture between the two, but it lacks a regular structure, such as word-processing programs.

2.4 Definition of Open Government Data

The term open government data consists of two components: the first is government data, which is any data produced or collected by government agencies (Al-Saadani 2015, p. 49), and the second is open data that can be freely used, reused, and redistributed by anyone (Vasarhelyi and Alzamil 2019, p. 178) and when bringing the two elements together, can be interpreted as government-produced data that can be freely accessed, used, or modified, and thus, this is support for increased transparency and public participation (Rahmat et al. 2019, p. 136).

In other words, it is any content that government agencies publish on informational networks, and the data may be a simple text or a statistical file, an image, or an audio file (Talib, May 2020, p. 4).

For government data to be described as open, several conditions must be met, and these conditions are a set of principles that were formulated in 2007 (Rodrigo et al. 2018, p. 7) (Al-Saadani 2015, pp. 49–50 -51) Completeness: all working data is made available to the public so that the available data sets are as complete as possible, reflecting a complete picture of what is recorded on a particular topic, and public data are those data that are not subject to privacy, security, confidentiality, or discriminatory restrictions.

- Nondiscriminatory: must be made available to any person without precedence to register or obtain a specific permit.
- Nonproprietary: must be available in a manner that does not make any party have the right to control or dispose of it absolutely.
- The license is open: the data is not subject to copyrights to patents or trademarks, except that some restrictions may be imposed for privacy and information security purposes.
- Supporting scientific research: obtaining open government data enables the researcher and research institutions to conduct accurate research by making use of available data.
- Encouraging innovation: the flow of data and information provides citizens with wide opportunities for innovation and creativity by developing complex applications to integrate and synthesize information from different sources, which may provide some useful tools for conducting market analysis, forecasting trends, etc.
- Helping new projects: the benefits of open government data are not limited to already existing projects but also potential projects and business sectors that plan to establish their activities. Accordingly, the corporation can decide to select the appropriate location for its commercial activity.
- Promote citizen participation: this data creates a basis for citizen participation in governance and makes them more aware and effective.
- Promoting good governance and reducing corruption: the essence of open government data initiatives is data on government expenditures and performance, and thus publishing such data may allow civil society to uncover government misconduct and help it curb corruption.

– Improving the effectiveness of public service delivery: displaying data on service providers correctly may help citizens make informed decisions about the best services to be provided, and it also helps service providers improve their performance. For example, the United States of America provides applications developed based on the open government data initiative. It provides data on automotive products and security, airline performance, and mortgage plans, helping consumers make the right choice and helping service providers improve their performance.

2.5 *The Role of the Bank of Algeria as a Provider and Provider of the Big Data Service*

Big data is of great importance to many parties, especially government agencies and macroeconomic policymakers, since it has become available information that is currently a valuable and economic wealth that contributes to preparing plans and policies that work to achieve economic and social development for any country; hence it can be said big data is a developed industry like other industries related to information, and it contributes to developing the effectiveness of many industries and fields, perhaps the most prominent of which is the banking system (the banking sector), through the benefits of analyzing big data at the level of the banking system as a whole or the level of the banking unit alone (Suleiman Rashwan 2018, page 29) (Prasad and Balachandran 2017, p. 1118).

– Identify the areas of shortcomings and weaknesses, and improve operations in all financial and management units.
– Make better decisions based on the information generated by analyzing big data for all financial and administrative units.
– Discover untapped opportunities and potential weaknesses in all businesses.
– Enabling stakeholders to find solutions to potential problems that are revealed when analyzing big data in some operations.
– Increased chance of making clear and correct decisions.

Since commercial banks play the role of financial intermediation despite the technological and information development that the banking industry knows, they are forced to collect data from the outcome of daily interactions with digital products or services, including cell phones, credit cards, and social media platforms to attract more deposits and grant loans, The latter requires a lot of information to grant loans to categories of acceptable creditworthiness to avoid the occurrence of the problem of bad debts, that is, to help them build an acceptable lending policy. Within this direction, the Bank of Algeria, as a provider of big data service and in its capacity as a provider of government data, provides information and data on credit activity in Algeria to commercial banks, as the Bank of Algeria works within the framework of a group of preventive centers to collect the necessary information that aims to help the banking system in reducing the credit risk. These preventive centers are represented in the centralization of risks, centralization of the contingent of payment, centralization of budgets, and the financial stability committee.

2.5.1 Centralization of Risks

This department was established according to Article 106 of Law No. 10-90 on cash and loan, which was called the risk center, and currently it is regulated and operated by the Regulation No. 12-01 of February 20, 2012, which includes the regulation of centralization of risks for institutions and families and their work, which is called centralization of risks. According to this last system, the centralization of risk is divided into two (02), namely (Bank of Algeria 2012a, page 45):

- **Corporate risk centrality:** in which the data related to loans granted to legal and natural persons are recorded. Those who engage in unpaid professional activity.
- **Household risk centrality:** it collects data related to loans granted to individuals.

The centralization of risk is a risk centralization interest that is assigned to each bank and financial institution in particular (called authorized institutions) to collect the identity of the beneficiaries of the loans, the nature and ceiling of the loans granted, the number of uses, the number of unpaid loans, as well as the guarantees taken about each class of loans. The institutions declared in this regard must declare the following to the centrality of risks according to the nature of the data in their section devoted to institutions and in their section devoted to families (Bank of Algeria 2012b, page 45).

- The data related to the definition of the beneficiaries of the loans and the ceiling of the loans granted to the customers, regardless of the amount, in the title of the operations carried out at the level of their windows, as well as the guarantees are taken, whether in kind or person, about each class of loans, and this information is called positive data.
- Unpaid amounts from loan lists and this information are called negative data.

The Central Department of Risk in Algeria is one of the models that fall within the centers of return and managed by central banks (government agencies), and these systems are characterized by the mandatory provision of banking institutions with data and credit information and then to ensure greater responsiveness by these institutions, in addition to what the supervisory authority enjoys. There are credibility and transparency in dealing with this data and information, which leads to a greater guarantee of the safety and proper use of this data, but on the other hand, its main defects are the information gathered in itself, as it is mostly negative information about default, bankruptcy, and liquidation cases, and does not include the information positivity that has an impact on building creditworthiness is something that is corrected by System No. 01–12 (Muhammad 2012, pages 17–18).

2.5.2 Centralization of Payment Incidentals

The Bank of Algeria, for the precaution and protection of more than the risks associated with banking operations, established the centralization of the incidentals of payment in accordance with Regulation No. 92-02 of March 22, and this centralization, for each payment method and/or loan, undertakes the following (Muhammad 2012, pages 17–18).

- Organize a central index of payment barriers and the follow-ups that may arise from them, and then manage and organize this index.
- Periodically informing the financial intermediaries and every other designated authority of the list of payment impediments and the consequent follow-ups.

This centralization was supported by the issuance of Regulation No. 08-01 of January 20, 2008, relating to arrangements for preventing and combating the issuance of checks without balance. This procedure was based on a system of centralizing information related to the symptoms of payment of checks due to the absence or shortage of the balance and publishing it at the level of banks, the public treasury, and the financial interests of Post Algeria, in order to be informed and exploited, especially upon the delivery of the first checkbook to its customer, and as soon as there is a payment counterpart due to a lack of a shortage in the balance, the drawee must, in accordance with the provisions of the commercial law, authorize the centralization of the payment incident within 4 days for the work following the date of presenting the check, and he must prepare and deliver or assign the delivery of a nonpayment certificate to the beneficiary, and the drawee is also required to send to the check issuer within the stipulated time an order indicating that the offeror of payment has been authorized to centralize the incident of payment. Failure to settle within 10 days will prevent the issuance of checks during a period of 5 years at all authorized establishments, starting from the date of ordering the order and criminal prosecutions in the absence of a settlement (Bank of Algeria 2008, pages 21–22).

2.5.3 Centralization of Budgets

This centralization was created at the Bank of Algeria according to Regulation No. 96-07 dated July 03, 1996, which includes organizing the centralization of budgets and their flow. In this regard, banks, financial institutions, and rental credit companies are required to provide the centralization of budgets with accounting and financial information related to the last 3 years to their customers of the borrowing economic institutions according to a unified model developed by the Bank of Algeria, and the financial and accounting information according to the concept of this system includes the budget, the results accounts table, and the attached data (Bank of Algeria 1996, October 27, page 23).

2.5.4 Financial Stability Committee

A committee affiliated to the Bank of Algeria appeared in 2009, which allows early detection of weaknesses through continuous monitoring of the performance of Algerian banks and financial institutions, by reviewing a set of financial solidity indicators estimated at 11 indicators called minimum indicators with targeting other indicators called the proposed indicators. In 2010, great importance was given to the periodic evaluation of the stability of the banking and financial system by means of rigidity tests introduced since 2007, in addition to its more interest in the structural liquidity surplus that characterizes the Algerian banking system, and the committee found that the risk of nonpayment remains the banking risk. The main one is at the level of the Algerian banking system, from various sources in the different preventive centers (by creating data and metadata that describe the borrowing institutions) and presenting it through analyzing the big data. And to provide the necessary infrastructure by searching data sources and collecting data by direct request, storing and integrating them, so that Algerian banks can benefit from them in building a good or acceptable lending policy.

3 Evaluating the Lending Policy of Algerian Banks

Algerian banks represent the cornerstone of the financial system as the dominant source of financing the national economy, and therefore the performance of their lending policy is directly reflected in economic development, and, accordingly, with the help of the data of the Bank of Algeria as a big data service provider and the capital asset pricing model as a performance evaluation model, we will try to find this out as shown as follows:

3.1 Introducing the Capital Asset Pricing Model (CAPM)

The capital asset pricing model is considered one of the most important models for evaluating and calculating the required rate of return on investment, and since the investment decision emanating from investment policies depends on two important elements, namely, the return and risk, which the capital asset pricing model paid attention to, as it provided a measure of the systematic risks surrounding the investment assets as provided, the investor has the minimum return that the investment asset should achieve in order to compensate him for the risks that cannot be avoided by diversifying the investment (Abd Rabbo 2019, page 310).

3.1.1 The Genesis of the Capital Asset Pricing Model (CAPM)

This model represents the essential extension of the portfolio theory founded by Harry Markowitzin 1952, the model provides the practical framework for balancing the expected return with the associated risk, and it is also called the Beta model, The one factor model for its dependence on the rate of return of the market portfolio as the only factor that affects the required rate of return for an investmentasset (Al-Amiri 2012, pp. 69–70). Harry Markowitz – as mentioned above – was the first to introduce the idea of using the single factor model to solve the investment problem of choosing the investments in which they would like to invest, stating that they want to choose the assets with the lowest risk. Known expected return or greater expected return to a known risk (these two conditions are known as the principle of dominance or control) as they seek to avoid risk. This idea was developed by William Sharpe in 1964 and Lintner in 1965, followed by Maussin in 1966 for a model combining risk and return required in At the same time, which assumes that the expected return on any investment asset is linked to a positive linear relationship with systemic (undistributed) risks for this asset, that is, assets with higher returns bear higher degrees of risk (Hatab and Jameed 2008, p. 33), so that the evaluation of financial decisions becomes more objective, and this model was used in many areas such as differential decisions between financial structures, estimating the cost of financing, the cost of private capital., the weighted average of the cost of capital and in investment spending decisions the capital budget (Al-Amiri M., 2012, p. 70).

3.1.2 The Basic Assumptions of the Capital Asset Pricing Model (CAPM)

Preparing a mathematical model to evaluate the price of an investment asset is difficult because of the different elements affecting the price, and among these factors are the amount of commission, short selling, speculation, competition, and the difference in the concept of return, so the asset pricing model capitalism is based on multiple assumptions, combining these factors with the aim of calculating the required rate of return for investment. They can be summarized as follows (Kamel 2010, page 14):

- Returns follow a normal distribution, or utility functions are quadratic (Badroni and Stripe 2017, p. 55).
- All investors are competent, and the choice depends on the personality of each investor and the risks he is willing to accept in light of the equity curve.
- The cost of executing deals is zero (excluding commission and fees on deals).
- There are no taxes on income resulting from investment, regardless of the return achieved or its source.
- The investor can enter the market in any amount, whatever the size of the invested capital.

Providing an element of perfect competition in the financial market and having no influence on prices for any investor;

The investor can borrow and lend on the basis of an interest rate equal to the risk-free rate of return.

Mathematical formula of the model: the mathematical formula of the capital asset pricing model is given as follows:

$$\overline{R_i} = R_f + \left(R_m - R_f\right)\beta_i \quad (1)$$

whereas

$\overline{R_i}$: required rate of return on the investment asset i
R_f: risk-free rate of return
R_m: market rate of return
β_i : beta factor of the investment asset i

It is noted from Formula No. 01 above that the model depends on the systematic risk measured by a factor β_i that can only be borne in return for a return that compensates the investor for bearing this risk, while the erratic risks associated with the investment can be controlled through effective diversification (Al-Momani 2014, p. 194), and it turns out that the expected return of an investment asset depends on three elements (Ghoneim 2005, p 488):

- The time value for money $R_f$which expresses the consideration given to the investor as a result of delaying spending without bearing any risk.
- The counterpart to bearing a moderate degree of systemic risk measured by the market risk premium ($R_m - R_f$).
- The systematic risk value of the investment asset (β_i) as a percentage of the average market risk.

3.1.3 Capital Asset Pricing Model (CAPM) Rules

This model is based on a set of rules that we refer to in the following (Al-Ali 2019, page 252):

The capital asset pricing model is market-based to reflect self-contained risks and as such is a useful way to think about risks for assets.

When the capital asset pricing model is applied in practice, it provides accurate answers to important questions about risks and required rates of return.

As the capital asset pricing model is logical in the sense that it represents the way in which investors behave avoiding risk, and therefore this model is a useful tool for a large company of investors.

It is appropriate to think about many financial problems within the framework of the capital asset pricing model, but it is important to be aware of the limitations imposed on this model when using it in practice.

3.1.4 The Ability of the Capital Asset Pricing Model to Assess Lending Policy

This model can be used in evaluating investment policies on the asset side of the Islamic bank's budget, as the model is based on the assumption that the required rate of return on any investment asset equals the rate of return on vacant investments of the risks plus the risk premium, that is, the amount of risk in any investment should be reflected in the return, so if the risk increases, that means the return will rise, and vice versa. To clarify this hypothesis, we know that the investor maintains investment assets as part of a diversified investment portfolio (such as the Islamic bank that maintains a group of diverse assets within its budget) in order to obtain a total return (total) on all the components of the portfolio in light of the overall risk associated with it, not the return and individual risk for each investment asset, and this assumption is not limited to investment portfolios only but also includes the asset's aspect of Islamic banks, because the latter is trying to obtain a return commensurate with the degree of risk within the total assets of the investment policy (Gharaibeh 1997, p.67).

3.2 *Methodology for Assessing Lending Policy*

The method of evaluating the lending policy of public banks in the Algerian banking system is summarized in knowing the contribution of their financing to the total financing directed to the national economy and then knowing the direction of the macroeconomic activity in order to evaluate the lending policy according to the systemic risk perspective to know the type of policy and judge it.

3.2.1 Method of Data Enumeration and Sample Selection

We used the sampling method by selecting a sample from a statistical population, represented in the non-probability control sample by taking data from specific units capable of providing the required information (Taherdoost 2016, p. 23). The statistical community in our research of 20 banks operating in the Algerian banking market during the period 2010–2016. For the purpose of studying the assumptions given and answering the main question of the research, we used the controlling sample represented in all six public banks, given that they represent the largest market share in financing the national economy, estimated at 86.96% during the studied period, as the sample banks dominate financing the national economy according to the maturity period, whether short, medium- and long-term loans, as well as seniority in practicing banking activity in the Algerian banking market that extends for more than 54 years, enabling us to generalize the evaluation of its lending policy and to circulate the results of the research to the statistical community

Table 1 Study sample

Bank	Date of establishment	Seniority
Foreign Bank of Algeria (BEA)	October, 1967	53 years
National Bank of Algeria (BNA)	June 13, 1966	54 years
Popular loan Algeria (CPA)	May 14, 1967	53 years
Local development Bank (BDL)	April 30, 1985	35 years
Bank of Agriculture and Rural Development (BADR)	March 13, 1982	38 years
National Savings and reserve fund (CNEP)	August 10, 1964	56 years

Source: Shaker Al-Qazwini, Lectures on Banking Economics, University Press Office

as a whole. Most of the private banks represent branches of foreign banks whose lending policy is focused on financing with mostly short-term loans during the studied period.

3.2.2 Presentation of Algerian Public Banks

The study sample consists of the following banks (Table 1).

3.2.3 Method of Data Collection

Special data were collected to assess the lending policies of public banks from the reports of the Bank of Algeria on the economic and monetary development of Algeria from 2010 to 2016 published on the website www.bank-of-algeria.dz.

3.2.4 The Statistical Tools Used

The statistical and standard tools used are summarized as follows:

- **The arithmetic mean** is used to determine the level of data during the search period and is measured by the following formula:

$$\overline{R} = \frac{1}{n}\sum\nolimits_{i=1}^{n} R_i$$

- Jarque-Bera test: to check that the residuals of the estimation follow the normal distribution:
- Coefficient of covariance $\mathrm{COV}(\mathrm{R_i}, \mathrm{R_j})$: it measures the sensitivity of the change in returns of an asset i as a result of changing returns of an asset j, given by the following formula:

$$\mathrm{COV}\left(\mathrm{R_i}, \mathrm{R_j}\right) = \sum\nolimits_{t=1}^{n} \left(\mathrm{R_{(i,t)}} - \overline{\mathrm{R_i}}\right)\left(\mathrm{R_{(j,t)}} - \overline{\mathrm{R_J}}\right)$$

- Variance: it measures the overall risk score, calculated with the following relationship:

$$\sigma_i^2 = \frac{1}{n}\sum_{I=1}^{n}(R_i - \overline{R})^2$$

- Beta: it measures the systematic risk of the asset, according to the following:

$$\beta_i = \frac{COV\,(R_i, R_M)}{\sigma_M{}^2}$$

whereas $\sigma_M{}^2$ It represents the variance of market returns (R_M).
- General trend vehicle detection test: the least squares method is used in the estimation.
- Risk-free rate of return R_f: it represents the rate of return on Algerian public treasury bonds with maturities of only 1 year.

3.2.5 Evaluating the Contribution of Public Bank Financing to the Overall Financing of the Algerian Economy

Table 2 includes the volume of financing for public banks and the total loans directed to the national economy.

It is evident from Table 2 and the figure on the next page that the Algerian public banks dominate financing the national economy at an average rate of 86.96% during the period 2010–2016. The period, bearing in mind that medium and long loans constitute an important percentage in the total loans granted by it to the national economy, therefore, is evident that in the absence of a developed financial market, the public banking sector remains the main source of financing economic development in Algeria, while private banks remain overly restrictive of financing the national economy, whose financing is concentrated in some profitable activities only with the dominance of short-term financing. Competition between it and the public banking sector is considered the most important item for reforming the Algerian banking system within the Monetary and Loan Law.

Table 2 The volume of financing in the banking sector in Algeria for the period 2010–2017

Bank	Year						
	2010	2011	2012	2013	2014	2015	2016
Public banks	2.84	3.19	3.72	4.46	5.71	6.37	6.93
Private banks	4.30	5.31	5.7	6.98	7.91	9.1	9.83
Share of public banks (%)	86	85	86	86	87	88	88
Share of private banks (%)	13	14	13	14	12	13	12

Source: Prepared by researchers, based on the reports of the Bank of Algeria
Unit: 10^{12}DZD

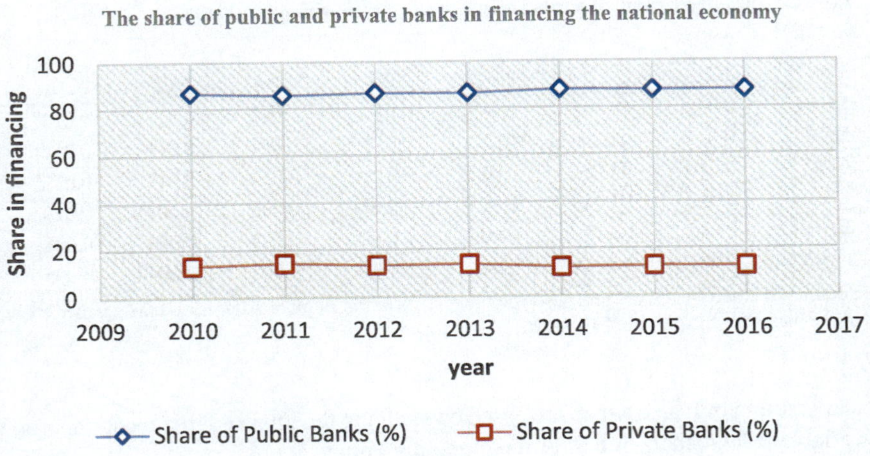

Source: Prepared by researchers according to Table 2

3.2.6 Defining the Capital Asset Pricing Model for the Lending Policy of Algerian Public Banks

In this regard, we rely on the capital asset pricing model to calculate the total required rate of return for the lending policy of each public bank and then compare it with the average actual returns of the bank's lending policy during the studied period, where if it is the average, the actual returns are greater than the total required rate of return, so the lending policy is good, but in the case of the opposite, it is unacceptable, and if they are equal, it is acceptable, and to achieve this we will be exposed to the following.

3.2.6.1 Calculating the Effective Rate of Return on Total Assets (ROA)

This ratio is included in the profitability indicators, and it expresses the efficiency of the bank's management in generating profits from its assets, i.e., the share of each unit of assets from the net profit (Karoumi 2016;p: 138), measured by dividing the net annual result on the total assets of the bank, and Table 3 shows this rate for the public banks during the studied period.

3.2.6.2 Calculate the Beta Coefficient (β_i)

The returns of the lending policy of banks are affected by the systemic or regular risk measured by the beta coefficient, which means that the sensitivity of the returns of the lending policy to the change in the systemic risk, and given that the credit analysis models in the lending policy focus on economic conditions in the analysis

Table 3 The effective rate of return on total assets for the period 2010–2016

Year	Bank 2010	2011	2012	2013	2014	2015	2016
BEA	0.008	0.0115	0.015	0.01	0.012	0.013	0.013
BNA	0.023	0.022	0.013	0.014	0.011	0.011	0.011
CPA	0.016	0.013	0.013	0.012	0.013	0.017	0.015
BDL	0.002	0.004	0.005	0.004	0.003	0.007	0.019
BADR	0.013	0.011	0.006	0.005	0.004	0.004	0.008
CNEP	0.001	0.001	0.001	0.0002	0.002	0.004	0.005

Source: Prepared by researchers based on the reports of the Bank of Algeria and the balance sheets of the banks for the period 2010–2016

and granting of loans, which are based on studying the impact of the most important variables Macroeconomic Return on Lending Policy.

Perhaps among the most prominent is the real economic growth, which represents an important factor in describing systemic risk. Macroeconomic instability (instability of real economic growth) resulting from imbalances resulting from successive changes in the structure of the national economy, such as fluctuations in terms of trade exchange and fluctuations in the rate of inflation, is considered an element. Decisive in the ability of the banking system to play the role of mediation, especially granting credit and providing liquidity and other fluctuations at the macroeconomic level (marginalization of Elevy's doctoral thesis), accordingly we will use the real economic growth rate as a variable to calculate the beta coefficient of the returns of the lending policy of Algerian public banks. We will calculate both the variance of the real economic growth rate ($\sigma_M{}^2$). And coefficient of covariance ($COV(R_i, R_M)$) shows the actual returns on total assets of each public bank and the real economic growth rate (R_{TCR}). It represents market returns (R_M) as follows:

- Calculating the variance of the real economic growth rate ($\sigma_M{}^2$): we calculate the average real economic growth rate in Algeria($\overline{R_{\text{TCR}}}$):

$$\overline{R}_{\text{TCR}} = ((0.036 + 0.028 + 0.033 + 0.028 + 0.038 + 0.037 + 0.033))/7$$
$$= 0.03329$$

Based on that, we calculate ($\sigma_M{}^2$) as follows:

$$\sigma_M{}^2 = \Big((0.036 - 0.03329)^2 + (0.028 - 0.03329)^2 + (0.033 - 0.03329)^2$$
$$+ (0.028 - 0.03329)^2 + (0.038 - 0.03329)^2 + (0.037 - 0.03329)^2$$
$$+ (0.033 - 0.03329)^2\Big)/7$$
$$= 1,42 \times 10^{-5}$$

- **Coefficient of covariance ($COV(R_i, R_M)$) among the actual returns on the total assets of the public banks:** we denote the actual returns on the total assets of the public bank by (R_i), while we denote the average actual returns on the total assets of the public bank b–$\overline{R}_i$. And it computes $\overline{R}_i$ for each public bank the following:

$$\overline{R}_{BEA} = (0.0081 + 0.01148 + 0.01541 + 0.00991 + 0.011556 + 0.01284 + 0.01343)\ /7 = 0.01182$$

$$\overline{R}_{BNA} = (0.02293 + 0.02148 + 0.01319 + 0.01384 + 0.01137 + 0.01086 + 0.01105)\ /7 = 0.01496$$

$$\overline{R}_{CPA} = (0.01582 + 0.01339 + 0.0134 + 0.01226 + 0.0129 + 0.01718 + 0.01541)\ /7 = 0.01434$$

$$\overline{R}_{BDL} = (0.00192 + 0.00437 + 0.00458 + 0.00387 + 0.00252 + 0.00865 + 0.01978)\ /7 = 0.00653$$

$$\overline{R}_{BADR} = (0.01338 + 0.01099 + 0.00573 + 0.0046 + 0.00445 + 0.00435 + 0.00767)\ /7 = 0.00731$$

$$\overline{R}_{CNEP} = 0.00116 + 0.00131 + 0.00126 + 0.00022 + 0.00193 + 0.00422 + 0.00508)/7 = 0.00217$$

Hence, the covariance factor of the actual returns on the total assets of the public banks is calculated according to the following:

$$COV\,(R_{BEA}, R_{TCR}) = ((0.0081 - 0.01182)\,(0.036 - 0.03329) + (0.01148 - 0.01182)\,(0.028 - 0.03329) + (0.01541 - 0.01182)\,(0.033 - 0.03329) + (0.00991 - 0.01182)\,(0.028 - 0.03329)) + (0.01155 - 0.01182)\,(0.038 - 0.03329) + (0.01284 - 0.01182)\,(0.037 - 0.03329) + (0.01343 - 0.01182)\,(0.033 - 0.03329) = 4 \times 10^{-7}$$

$$COV\,(R_{BNA}, R_{TCR}) = (0.02293 - 0, 01496)\,(0.036 - 0.03329)$$

$$+ (0.02148 - 0,01496)\,(0.028 - 0.03329) + (0.01319 - 0,01496)$$
$$(0.033 - 0.03329) + (0.01384 - 0,01496)\,(0.028 - 0.03329)$$
$$+ (0.01137 - 0,01496)\,(0.038 - 0.03329) + (0.01086 - 0,01496)$$
$$(0.037 - 0.03329) + (0.01105 - 0,01496)\,(0.033 - 0.03329)$$
$$= -5,35 \times 10^{-6}$$

$$COV\,(R_{CPA}, R_{TCR}) = (0.01582 - 0.01434)\,(0.036 - 0.03329)$$
$$+ (0.01339 - 0.01434)\,(0.028 - 0.03329) + (0.01340 - 0.01434)$$
$$(0.033 - 0.03329) + (0.01226 - 0.01434)\,(0.028 - 0.03329)$$
$$+ (0.01226 - 0.01434)\,(0.038 - 0.03329) + (0.01718 - 0.01434)$$
$$(0.037 - 0.03329) + (0.01541 - 0.01434)\,(0.033 - 0.03329) = 3,39 \times 10^{-6}$$

$$COV\,(R_{BDL}, R_{TCR}) = (0.00192 - 0.00653)\,(0.036 - 0.03329)$$
$$+ (0.00437 - 0.00653)\,(0.028 - 0.03329) + (0.00458 - 0.00653)$$
$$(0.0333 - 0.03329) + (0.00387 - 0.00653)\,(0.028 - 0.03329)$$
$$+ (0.00252 - 0.00653)\,(0.038 - 0.03329) + (0.00865 - 0.00653)$$
$$(0.037 - 0.03329) + (0.01978 - 0.00653)\,(0.033 - 0.03329)$$
$$= -1,86 \times 10^{-7}$$

$$COV\,(R_{BADR}, R_{TCR}) = (0.01338 - 0.00731)\,(0.0167 - 0.03329)$$
$$+ (0.01099 - 0.00731)\,(0.0193 - 0.03329) + (0.00573 - 0.00731)$$
$$(0.021 - 0.03329) + (0.0046 - 0.00731)\,(0.0167 - 0.03329)$$
$$+ (0.00445 - 0.00731)\,(0.0198 - 0.03329) + (0.00435 - 0.00731)$$
$$(0.0183 - 0.03329) + (0.00767 - 0.00731)\,(0.0186 - 0.03329)$$
$$= -1,82 \times 10^{-6}$$

$$COV\,(R_{CNEP}, R_{TCR}) = ((0.00116 - 0.00217)\,(0.036 - 0.03329)$$
$$+ (0.00131 - 0.00217)\,(0.028 - 0.03329) + (0.00126 - 0.00217)$$
$$(0.033 - 0.03329) + (0.00022 - 0.00217)\,(0.028 - 0.03329)$$

Table 4 The value of the beta parameter (β_i)

Bank	Beta $\beta_i = \frac{COV(R_i, R_{TCR})}{\sigma_M^2}$
BEA	$\beta_{BEA} = {4 \times 10^{-7}} / {1,42 \times 10^{-5}} = 0.02816$
BNA	$\beta_{BNA} = {-5,35 \times 10^{-6}} / {1,42 \times 10^{-5}} = -0.37714$
CPA	$\beta_{CPA} = {3,39 \times 10^{-6}} / {1,42 \times 10^{-5}} = 0.23902$
BDL	$\beta_{BDL} = {-1,86 \times 10^{-7}} / {1,42 \times 10^{-5}} = -0.01312$
BADR	$\beta_{BADR} = {-1,82 \times 10^{-6}} / {1,42 \times 10^{-5}} = -0.12821$
CNEP	$\beta_{CNEP} = {2,57 \times 10^{-6}} / {1,42 \times 10^{-5}} = 0.18095$

Source: Prepared by researchers

$$+ (0.00193 - 0.00217)(0.038 - 0.03329) + (0.00422 - 0.00217)$$
$$(0.037 - 0.03329) + (0.00508 - 0.00217)(0.033 - 0.03329)$$
$$= 2,57 \times 10^{-6}$$

Based on the results obtained above, the value of the beta parameter (β_i). For each public bank, it is given as in Table 4.

- **Interpretation of the beta coefficient of (β_i) public banks:** It is noted that the beta coefficient of all public banks is less than the correct one, so it can be said that Algerian public banks follow a defensive (investment) lending policy, which means that banks during the studied period give in this policy – also known as a conservative policy – to the security element priority over the return component because it is the portfolios of loans formed according to this policy which are called defensive or cautious portfolios that seek to reduce losses associated with invested capital, as they are mostly composed of loans whose returns are insensitive to market changes (beta coefficient Small(β_i)) (Kapoor 2014, p. 1368), that is, the lending policy consists mostly of high collateral loans (as the Algerian state guarantees through the public treasury, the loan guarantee fund for small and medium enterprises and the investment loan guarantee fund) and resorted to in times characterized by the presence of indicators that clearly reflect the cases of economic downturn and also notes that the beta coefficient (β_i). The Algerian National Bank, the Local Development Bank, and the Agriculture and Rural Development Bank are negative, which is attributed to government interference in directing loans to the sectors financed by the aforementioned banks.

Table 5 The general trend detection test for the series of real GDP growth rate

Dependent variable: TCVA				
Method: Least squares				
Date: 01/07/20 time: 16:58				
Sample: 2000Q1 2018Q4				
Included observations: 76				
Variable	Coefficient	Std. error	t-statistic	Prob.
C	0.047843	0.003848	12.43270	0.0000
@trend	−0.000360	8.86E-05	−4.064666	0.0001

Source: Prepared by researchers based on Eviews 09 program output

3.2.6.3 Building a Capital Asset Pricing Model

After determining the value of the beta parameter (β_i), for the lending policy of public banks, and calculating the average real economic growth rate $\left(\overline{R_{TCR}}\right)$ which represents average market returns R_M, the risk-free rate of return has to be determined R_f which expresses the average yield of 1-year public treasury bonds of 0.0284 in 2016, and therefore the capital assets pricing model for public banks according to Relationship No. (01) is:

$$RRR_i = 0.0284 + (0,0192 - 0.0284)\,\beta_i = 0,0284 - 0,0092\beta_i \tag{2}$$

Note that the relationship between systemic risk measured by (β_i) in the Algerian banking market and the rate of return on demand on the lending policy of public banks in Algeria is a reverse policy, given, as we mentioned earlier, the large government interference in directing these banks' loans in Algeria.

To apply the model within Eq. (2), it is necessary to verify the actual returns on total assets during the studied period following the normal distribution, and by using the Jarque-Bera test within the Eviews 9 program on the data of Table 3, it is evident from Appendix Fig. 1. Each of the public banks follows the normal distribution, because the probability value of the test is greater than 05%, so we accept the null hypothesis, i.e., the hypothesis of the normal distribution of the remainder of the estimate. On the other hand, we should know the direction of economic activity in Algeria for the period studied, by studying the trend of both real GDP growth and the rate of inflation, and by using the detection test of the year trend using the least squares method in Eviews 9 for the series of real GDP growth rate, we find Table 5.

It is evident from Table 4) that the value of the general trend component is statistically significant at the level of significance 5%, meaning that there is an effect of time on the studied series, and therefore we say that a series contains the general direction component, and the general direction complex sign is negative, and this is what it indicates that the series GDP is decreasing over time, as confirmed by Appendix Fig. 2. By using the same test on the series of inflation rate in Algeria, we get Table 6.

Table 6 The general trend detection vehicle test for the series of real GDP growth rate

Dependent variable: TINF				
Sample: 2007 M01 2018 M12				
Included observations: 144				
Variable	Coefficient	Std. Error	t-Statistic	Prob.
C	4.299933	0.251648	17.08712	0.0000
@trend	0.007569	0.003043	2.487567	0.0140

Source: Prepared by researchers based on Eviews 09 program output

Table 7 Features approximately 2010–2016

Bank	$\overline{R}_i$	*RRR*	β_i	Evaluation
BEA	0.01182	0.0284 + 0.02816(0.03329 – 0.0284) = **0.02853**	**0.02816**	Not good
BNA	0.01496	0.0284 – 0.37714(0.03329 – 0.0284) = **0.02656**	**–0.37714**	Not good
CPA	0.01434	0.0284 + 0.23902(0.03329 – 0.0284) = **0.02957**	**0.23902**	Not good
BDL	0.00653	0.0284 – 0.01312(0.03329 – 0.0284) = **0.02834**	**–0.01312**	Not good
BADR	0.00731	0.0284 – 0.12821(0.03329 – 0.0284) = **0.02777**	**–0.12821**	Not good
CNEP	0.00217	0.0284 + 0.18095(0.03329 – 0.0284) = **0.02928**	**0.18095**	Not good

Source: Prepared by researchers

It is evident that the value of the general trend component is statistically significant at a level of 5% significance, meaning that there is an effect of time on the series of inflation rate in the Algerian economy, and therefore we say that a series contains the general trend component, and a complex sign of the general direction is a wave, and this is what it indicates that the inflationary chain is increasing over time (see Appendix Fig. 3). Therefore, it can be concluded that during the studied period the Algerian economy suffers from the phenomenon of stagflation.

3.2.6.4 Comparison of Actual and Requested Returns

As mentioned above, to evaluate the lending policy of public banks, the actual return should be compared to the required return calculated on the basis of the capital asset pricing model included in Eq. (3) and the values of the coefficient (β_i) as shown in Table 7.

We note by the output of the Table 7 that the defensive lending policy of public banks is not good because the required rate of return is greater than the actual or achieved the rate of return, but it is consistent with the case of stagflation in Algeria during the studied period of the External Bank of Algeria (BEA) and the Algerian Popular Loan (CPA), the National Savings and Reserve Fund (CNEP), and the National Fund for Savings and Reserve (CNEP), but it is more precautionary than necessary because the banks greatly overcome the element of safety over the element of profitability; in other words, their lending policy focuses on financing in projects guaranteed, especially by the guarantee funds approved in this regard by the state, and the lending policy of each of the National Bank of Algeria

(BNA), the Local Development Bank (BDL), and the Bank for Agriculture and Rural Development (BADR) is in contrast to the stagflation that accompanied the Algerian economy for the period studied, and this is due, as we mentioned above, to government intervention in directing loans toward sectors that the government attaches importance to in raising from the level of real economic growth. Therefore, despite the decline in real economic growth, the required rate of return increases for the lending policy of these banks as a result of the state's guarantee of these loans. The reasons for the ineffectiveness of the lending policy of public banks according to the capital assets pricing model can be referred to as follows:

- A high share of the average volume of its activities with a low return.
- The increase in the volume of nonperforming loans, especially the loans granted under the (ANGEM, CNAC, and ANSEJ) program, whose maturities are too long.
- -Weak management in the banking field, as we find that the Bank of Algeria, in its annual report for the year 2010, stated that the reason for the weak return on the assets of public banks is due to the decrease in the margin of mediation and the margin outside the mediation.
- Significant government interference in the work of public banks, as they use the banks they own to finance projects of dubious return and profitability, and view them as the main financier of the public treasury by lending the public sector at large rates from its own and non-self-resources, which leads to the emergence of Major difficulties and problems resulting from the poor performance of the financing companies, which will affect the future of these banks.
- The weakness of the performance of the Algerian economy under the weight of what is known as the resource curse, which in its narrowest form lies the inverse proportion between the increase in dependence on natural resources on the one hand and the rates of economic growth on the other hand, that is, it shows the decrease between the increase in nonrenewable natural resources that leads to diminishing growth, the economic development, and the emergence of negative results for economic development (Salman 2015, page 04). It is evident to us in the case of Algeria that the dependence of its overall economy on oil revenues has a negative impact on the rate of economic growth and the heavy dependence on oil collection and its revenues contributed to the spread of corruption in public banks, especially by granting loans and concentrating them in limited and unproductive investment categories.

4 Conclusion

Through the study we conducted on public banks in Algeria, which focused on evaluating their lending policy, we can come up with the following results:

- Algerian public banks dominate the volume of loans distributed to the national economy, as they guarantee full financing for the public sector while largely financing the private sector, and medium- and long-term loans constitute a significant share, and therefore we find that these banks remain the only way to finance the national economy in light of the reluctance of private banks stock market stalemate.

- The Algerian economy is undergoing during the studied period the phenomenon of stagflation, which was reflected in the performance of the lending policies of the public banks in Algeria.
- The lending policy of public banks is characterized as a defensive policy aimed primarily at achieving a basic goal, which is banking security, despite being consistent with the situation of the Algerian economy during the period studied, but this policy during the mentioned period was not acceptable from the perspective of the capital asset pricing model, given that the rate and the actual yield were less than the required rate of return.
- The performance of the lending policy of public banks within the capital asset pricing model summarizes the lack of independence of these banks in banking management and their inefficiency, especially in terms of granting loans that are known as government interference in granting and directing loans, which contributed to the high volume of corruption in these banks and thus the high volume of bad debts and so on, accompanied by a decrease in its returns.

Given the results presented above, we can present recommendations that will help us increase the effectiveness of the lending policy of public banks, which remain the dominant source of financing the national economy, as follows:

- Increasing the volume of competition in the Algerian banking market through measures that stimulate private banks to increase their contribution to financing the national economy, in order to distribute risks to all units of the Algerian banking system.
- Activating the Algerian stock market to relieve pressure on public banks in financing the national economy.
- Work to diversify the national economy outside the hydrocarbon sector.
- Reducing government intervention in directing and granting loans to expand the degree of independence of public banks in this field.
- Activating the practice of banking operations related to participatory banking in all Algerian public banks, with the possibility of issuing banking laws that facilitate the transformation of some traditional public commercial banks into Islamic banks.
- The Bank of Algeria should address the phenomenon of inflation by following the deflationary monetary policy, as the main reason for this phenomenon is due to the expansion or excess of the money supply.

Appendices

Appendix 1

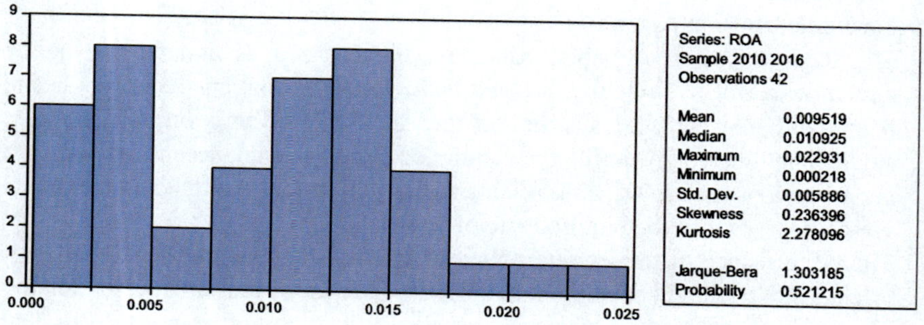

Fig. A.1 Results of the Jarque-Bera test on the residual estimation of the series of actual returns on the total assets of the Algerian banking system. (Source: Prepared by researchers based on Eviews 9)

Appendix 2

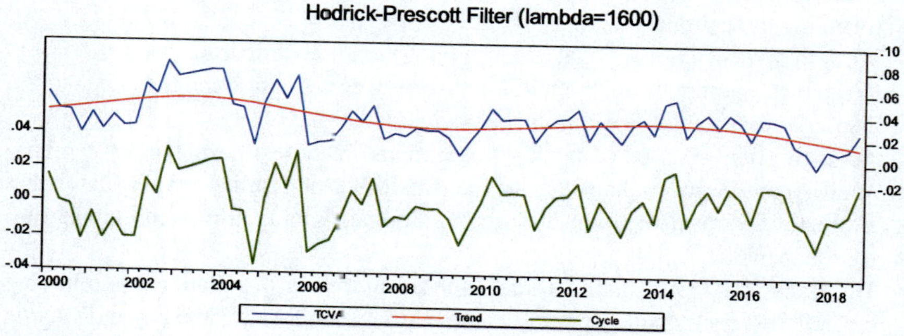

Fig. A.2 Evolution of the real GDP growth rate, the general trend, and the cyclical growth rate. (Source: Prepared by researchers based on Eviews 9)

Appendix 3

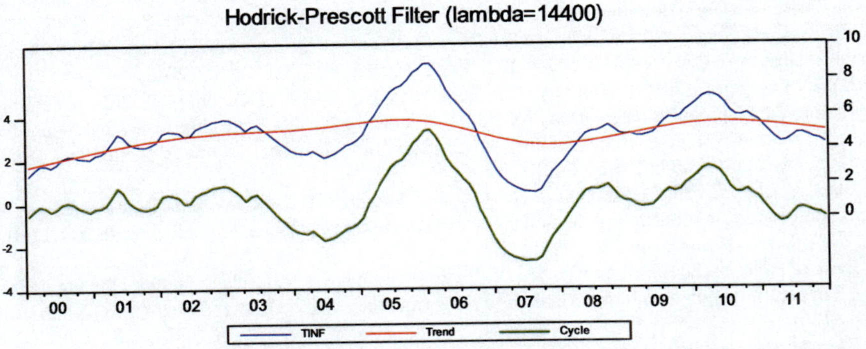

Fig. A.3 Evolution of the inflation rate, the general trend and the cyclical growth rate. (Source: Prepared by researchers based on Eviews 9)

References

Books

Kamel, A.S.D.: Portfolio Management, 1st edn. Circuit of the March, Amman (2010)

Al-Amiri, M.: Managing Investment Portfolios, 1st edn. Ithraa House, Amman (2012)

Al-Momani, G.: Modern Investment Portfolio Management, 1st edn. House of Curricula, Amman (2014)

Ghoneim, H.: Studies in Finance. The Academic Library, Cairo (2005)

Prasad, S., Balachandran, B.: Challenges and Benefits of Deploying Big Data Analytics in the Cloud for Business Intelligence. In: International Conference on Knowledge Based and Intelligent Information and Engineering. France (2017)

Hattab, S., Jumaid, S.: Examination of the Capital Assets Pricing Model on the Amman Stock Exchange (An Applied Study on the Amman Stock Exchange). Amman Stock Exchange, Amman (2008)

Articles

Abd Rabbo, M.: Measuring the efficiency of a modified model for pricing capital assets in the Egyptian Stock Exchange during the period 2005–2010. Sci. J. Bus. Environ. Stud. **10**(N 1, Part 1), 321–322 (2019)

Abdul Hamman Mohammed Suleiman Rashwan. (2018). The role of big data analysis in rationalizing financial and administrative decision-making in Palestinian universities – a field study. J. Econ. Finan., vol 11, N 1, p 27.

Al-Ali, H.H.: The use of the international capital asset pricing model to estimate the financial returns required for the international investment portfolio: a case study in Zain International. Univ. Sharjah J. Human. **16**(1), 252 (2019)

Badroni, A., Stripe, H.: Realistic study of the capital assets pricing model on the Algiers stock exchange "NCA Rouiba as a model". J. Econ. Manag. Bus. Sci. **18**, 55 (2017)

Bank of Algeria. (2012a, February 15). Regulation No. 11-07 of October 19, 2011 amending and supplementing Regulation No. 08-01 of January 20, 2008 relating to arrangements for preventing and controlling issuance of checks without balance. Official Gazette, The Forty-Ninth Year (Issue 08), pp. 21–22.

Bank of Algeria. (2008). Regulation No. 08-01 of January 20, 2008, relating to arrangements for preventing and controlling the issuance of checks without balance. Official Gazette, The Forty-Fifth Year (Issue 8), pp. 21–22.

Bank of Algeria. (2012b, June 13). Article 01, Regulation No. 12-01, which includes the regulation of centralization of enterprise and family risks and their work. Official Gazette, forty-ninth year (36), p : 45.

Bank of Algeria. (1996, October 27). For more details, see: Regulation No. 96-07 of July 03, 1996, which includes organizing the centralization and functioning of budgets, the official newspaper of the Algerian Republic. Official Gazette, year thirty-three, 23.

Gharaibeh, H. (1997). Capital assets pricing model an applied study on the Amman international market. Yarmouk Res. J. Human. Soc. Sci. Ser., vol 13, N 3, 67.

Ghobeiry, M., Hassan Hassan, A.: Big data and its impact on realizing the Kingdom of Saudi Arabia vision 2030 an empirical study. J. Strat. Dev. **3**(Part 1), 35 (2019)

Kapoor, N. (2014). Financial portfolio management: overview and decision making in investment process. Int. J. Res. (IJR), Vol 1, N 10, 1368.

Karoumi, A.: Evaluating the performance of commercial banks by means of financial ratios, an applied study during the period (2005–2014). Al-Bashaer Econ. J. **2**(5), 138 (2016)

Latabi, M.: Big data and the information industry. Al-Hikmah J. Media Commun. Stud. **6**(4), 59 (2019)

Al-Saadani, M.A.R.: Open government data in the world is a survey study with a systematic vision proposal. Inform. Peer Rev. Sci. J. (15), 49–51 (2015)

Rahmat, A., Nurmandi, A., Kusuma Dew, D.: Does the government effectively in optimizing open data? analysis. J. Govern. **4**(2), 136 (2019)

Rodrigo, H., Deisy, C., & Barbiero. (2018). Open government data: concepts approaches and dimensions over time. E&G Economia E Gestão, vol 18, N 49, 7.

Sakri, A., Nashad, A., Brahimi, A.: Exploiting big data for the purposes of sustainable development in Arab countries – opportunities and challenges. Arsad J. Econ. Admin. Stud. **2**(2), 45 (2019)

Salman, H.A.: The role of Dutch disease and the resource curse in the spread of corruption in Iraq. Gulf Econ. J. **31**(25), 4 (2015)

Shukla, S., Kukade, V., & Mujawar, S. (2015). Big data: concept, handling and challenges: an overview. Int. J. Comput. Appl., vol 114,N 11: 6.

Taherdoost, H.: Sampling methods in research methodology; How to choose a sampling technique for research. Int. J. Acad. Res. Manag. **5**(2), 23 (2016)

Vasarhelyi, M., Alzamil, Z.: A new model for effective and efficient open government data. Int. J. Disclos. Govern. **16**(4), 174–187 (2019). https://doi.org/10.1057/s41310-019-00066-w

Other Reports

Muhammad, Y. (2012). Development of credit information systems and risk concentrations in Arab countries, the Arab Monetary Fund. Récupéré sur www.amf.org.ae/sites/default/files/econ/amdb/Committee%20%20Publications/en/2.pdf.

A Privacy Guard Mechanism for Cloud-Based Home Assistants

Khaoula Mahdjar, Radja Boukharrou, and Ahmed-Chawki Chaouche

Abstract Using smart personal assistants (SPA) for smart homes provides valuable service and an easy control of connected devices, despite allowing the emergence of privacy threats and disclosing sensitive data about users. In this paper, the proposed idea is to associate with any SPA a privacy guard gateway system (PGG for short) in order to mitigate privacy issues related to cloud-based devices, like profiling and linkage. In an original way, the PGG system is based on a noise addition mechanism that interrogates smart speakers within some dummy requests. The PGG system is adapted to any SPA and has the advantage of strengthening privacy protection without the user's activities being stored during the operation.

1 Introduction

The emergence of the Internet of Things (IoT) provides an opportunity to connect isolated devices in order to assist the user in a better way. IoT facilitates the automation of many tasks and contributes to the emergence of new innovations and services with new features. With the growth of intelligent and connected devices like smartphones and smart wearables, the interaction of humans with computers has changed due to the intelligence embedded in these devices that act proactively using the knowledge gathered about the user (Alshohoumi et al. 2019). Moreover, other devices have emerged, offering voice interaction with the user by means of natural languages, such as Smart Home Personal Assistants (SPA) (Abdi et al. 2019).

Nowadays, security and privacy are considered the topmost challenges of IoT (Alshohoumi et al. 2019; Ziegeldorf et al. 2014), especially in home, because user activities are captured all the time, often without asking the permission of the user. Thereby, the data collected on a user can harm his privacy, especially sensitive data.

K. Mahdjar · R. Boukharrou · A.-C. Chaouche (✉)
MISC Laboratory, University Abdelhamid Mehri – Constantine 2, Constantine, Algeria
e-mail: khaoula.mahdjar@univ-constantine2.dz; radja.boukharrou@univ-constantine2.dz; ahmed.chaouche@univ-constantine2.dz

S. Sedkaoui et al. (eds.), *International Conference on Managing Business Through Web Analytics*, https://doi.org/10.1007/978-3-031-06971-0_21

Privacy of users may be exposed as a result of serious breaches of users' sensitive information, which may occur in several levels: devices, storage, during communication, and at processing too. According to Abdi et al. (2019), most of the smart device owners are oblivious to their collected data being stored in device manufacturers' clouds. With the larger adoption of the cloud-based home assistants, several threats are accrued; mobile applications and clouds collect both wanted and unwanted data about the users including their locations, names, addresses, purchases, vocation, etc. These hubs can collect user's habits that can be used and learned by some data analytics engines that analyze user routines by means of machine learning techniques (Ahmed et al. 2020; Edu et al. 2019). They can go much further and analyze the recorded voices by smart speakers to get profiles about the user concluding much data about his health, whether he is suffering from chronic disease, and so on. In addition, their data analysis is able to understand the context of interactions between users (Edu et al. 2019).

In any home assistant, data analytics is an essential concept, allowing the system to conclude fine-grained decisions and to offer appropriate services, although it compromises users' privacy. The huge amount of data gathered by the home assistant can be stored locally or remotely in cloud servers. Untrusted cloud providers can perform privacy attacks against their clients, by exploiting the collected data for profiling or by selling them to third parties to gain profit. Therefore, it is hard to construct software solutions for such devices that look like a black box and use encrypted data (Gao et al. 2018). Indeed, common security threats and privacy concerns need to be studied and addressed in depth.

For this end, different solutions have been proposed. A countermeasure is proposed in Acar et al. (2018) based on generating spoofed network traffic to hide real activities of devices. The authors show how machine learning can be exploited by passive attackers, emphasizing that if we use encryption techniques for transmissions, much information can be inferred without using advanced techniques. Works as (Chandrasekaran et al. 2018; Gao et al. 2018) propose obfuscation techniques that prevent the voice assistant from listening and recording private conversations, by using ultrasound signals. These signals jam the assistant's microphone with inaudible obfuscation sound sent from the obfuscator, when listening to the user's hot-word lift the jamming. However, analyzing the user utterances for inferring the hot-word increases the time of responding to the user request.

A solution based on remote-controlled plugs or built-in mute buttons has been proposed (Abdi et al. 2019) allowing to safely control the functioning of voice assistants. Nevertheless, the participants in the survey proposed in this work find the latter solutions inconvenient because they degrade the usability of SPAs. The authors of (Chandrasekaran et al. 2018) suggest hardware solutions for application layers of IoT architectures using signals in an environment containing a SPA. In fact, our approach is gone in that way since we propose noise based techniques allowing the obfuscation by performing dummy requests and speeches that introduce uncertainty about the true data.

In this paper, we aim at strengthening the privacy of SPA within IoT devices use cases. In fact, we propose a novel system based on a user-centered approach, called Privacy Guard Gateway (PGG). This system is physically associated with any SPA, in order to mitigate privacy concerns which can be violated by it. Our proposed solution consists to collect the user speech the same as the assistants do and to perform thereafter a noise addition based mechanism. The main role of this mechanism is to fix the profiling concerns and thus ensure data scrambling, as said "too much information is equal to no information."

The present paper is organized as follows: Sect. 2, enumerate the privacy issues that SPAs can cause. In Sect. 3, we present the architecture of our privacy guard system, which strengthens the privacy of IoT devices and SPA users. In Sect. 4, we experiment the functioning of our system. The proposed scrambling mechanism is well-developed. The last section concludes and brings out our next perspectives.

2 Smart Personal Assistants and Privacy Issues

The smart assistants are increasingly used in homes making them smart, often based on cloud platforms, such as Google Home and Amazon Echo. They interact with the user through mobile apps, or hardware Smart Personal Assistants (SPA), knowing that SPAs are considered more simple and user-friendly (Abdi et al. 2019). A SPA generally uses voice recognition skills to perform user's commands (Edu et al. 2019). Through connected devices, SPAs provide easy and remote control of the smart home. In fact, they can support different manufacturers IoT devices and connect to their cloud services to trigger pertinent services. Standard protocols are defined to ensure compatibility of SPAs with different IoT devices. Actually, there are two types of SPAs:

SPA without interfaces: allows the user to perform commands by using voice only, usually involving smart speakers, such as Google Home and Amazon Echo, and cloud-based voice personal assistants such as Amazon Alexa and Google Assistant. Thus, speech recognition and analysis are based on cloud services. By taking Google Home as an example, this one can respond to all the commands of home users that start their command with the expression "Ok, Google." Google Home allows us to manage emails, ask questions, and control different IoT devices of the home.

SPA with interfaces: allows users to perform commands from touchable interfaces, like Samsung SmartThing and Hubitat Home Hub. Completely managed using screens, they connect different devices in a smart home for easy automation. Thanks to additional dongles, Hubitat can find and control all the Z-wave and Zigbee devices in the smart home. Some SPAs without interfaces can operate offline, thus not compromising user privacy.

2.1 Privacy Concerns in SPA

In Alshohoumi et al. (2019), different smart-environment architectures proposed in the literature are discussed in terms of security and privacy, and as it is expected privacy was the last interest.

Whereas security is concerned about the way that the data appears and transmitted, privacy concerns about the data itself, by answering to many questions: What kind of data is collected? How is it managed? Who is allowed to store thereby use it? And what is the purpose it is used for? In the context of smart environments, it is also important to know the used IoT devices and their impact and scope.

In this paper, we focus on Cloud-based SPA which raises many privacy issues and threats, as speech is a rich source of sensitive acoustic and textual information about the users and their environment. According to Atlam and Wills (2020); Edu et al. (2019); Ziegeldorf et al. (2014), the main privacy threats to which SPA suffer are:

- **Weak authentication.** The authentication is done just with wake-up words, so the SPA accepts any command preceding the wake-up keyword from anyone;
- **Weak authorization.** The absence of a pertinent method in SPA for permissions and access control which defines to the users their role in smart home, allows to any user to control any device, and/or modify the SPA set-up;
- **Identification.** IoT devices in home can be related with private data about the owners such as (name, addresses, location, ...);
- **Profiling.** Being attentive to the wake-up keyword, SPA is always on and always listening. This would allow us to process data about users' activities and actions during a long period of time, to classify users according to some features. Usually, the results are sent to cloud and exploited as advertisements;
- **Inventory and linkage.** Data can be illegitimately gathered about the existence and characteristics of IoT devices in specific places, opening the opportunity to other types of attacks such as profiling and tracking. In addition, separate data sources can be linked and combined to conclude new facts that are not perceived;

To preserve user privacy, any device should be designed to interrogate users about any sensitive data dissemination and even for software updates and also allow him to reset and review the permissions and data policies. However, viewed as a black box for final users, cloud-based platforms do not ensure the user's privacy because they always depend on the manufacturer cloud which can gain access to the user's data without his knowledge. Indeed, users never have the ability to filter their voice commands, even if the manufacturers promise that they clear all the recorded voices.

Actually, the data collection process from IoT devices is more passive, pervasive, and less intrusive (Abdi et al. 2019). To better protect the SPA users, collected data should be untraceable and unlinkable, making it difficult to identify hiding information about the relationship between any device (AL-mawee et al. 2012). Moreover, the anonymity of users is very important, by hiding information about the user who performed a given action and by using pseudonyms instead of using real identifiers.

The more important threat faced by the SPA user is profiling. The collected user data includes his personal information, his location, and his habits that can be learned by some data analytics engines that analyze user routines. The analytics engines are generally deployed in the cloud. Thereby, the cloud services already support a number of speech processing functions like speaker identification, vocal recognition and text analysis such as topic modeling, document categorization, sentiment and relations analysis and identity detection that can extract sensitive information. Applying these functions can significantly undermine the user's privacy. Most of the time, data is exploited as advertisements, but this might be a great problem for users with sensitive positions (like governments, commercial parties, etc.), or could be used for account settlement between enemies. Moreover, another problem that disturbs smart home owners is related to home control, which could be used by thieves.

3 Privacy Guard Architecture

In this paper, we propose a privacy guard system, dedicated to SPA operating in smart environments. It is physically associated with any SPA, in order to mitigate privacy concerns caused by the use of IoT devices. The privacy guard system collects the user speech the same as the SPA does and thereafter performs a noise addition based mechanism. The main role of this mechanism is to fix the profiling concerns and thus ensure data scrambling.

Figure 1 depicts the main physical entities composing the proposed architecture and the interactions between them.

SPA and cloud-based platform. The majority of SPAs use voice recognition skills to perform user requests. Viewed as a black box for the final users, SPAs are cloud-based assistants always depending on the manufacturer cloud which can gain access to the user's data without his knowledge.

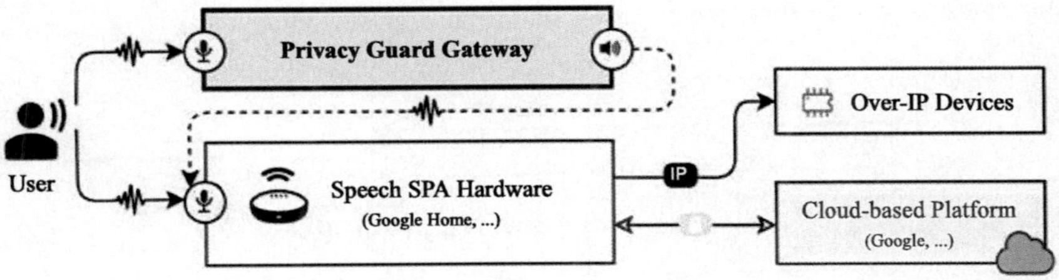

Fig. 1 Privacy guard architecture

Usually, SPAs are based on over-IP protocols to communicate with IoT devices, therefore, they do not allow to control non-IP devices, requiring an interoperability gateway. Moreover, communication between the SPA and cloud platform is over-IP encrypted, i.e., no request can be interpreted or analyzed. Although this ensures the confidentiality of the communication, Thereby it does not allow any intermediary privacy guard system to filter the interaction between the SPA and the cloud platform.

Privacy Guard Gateway (PGG). In order to provide a privacy guard, a physical gateway called PGG is placed beside the SPA and embeds a microphone and a speaker, as for SPA. Its role is to collect the same speech forwarded to the SPA in order to apply a scrambling mechanism after processing the user requests. First, through the microphone, PGG retrieves the speech and proceeds to the extraction of the request features. Then, an analysis of the request is performed allowing the scrambling mechanism which consists of producing some dummy requests by using the speaker.

4 Privacy Guard Process

As stated by Fig. 2, the PGG system operates in three processing phases:

Phase 1: Features Extraction

Firstly, PGG captures the speech of the user request and translates it into text by using any Speech to Text technique, like in SpeechRecognition 3.8.1 (2020). Then, it proceeds to the extraction of the so-called *request features*. These features are keywords and synonyms generated from the user request that are sorted according to several types of context features, like devices or actions.

Definition 1 (Request Feature) Let $\mathcal{N}$ be the set of all possible feature names, and $\mathcal{T}$ be the set of possible types of request features, such as: Device, Action, Space, and Time. A request feature is a pair $\langle name, type \rangle$, where $name \in \mathcal{N}$ is the name of the feature and $type \in \mathcal{T}$ is the corresponding type. The mapping

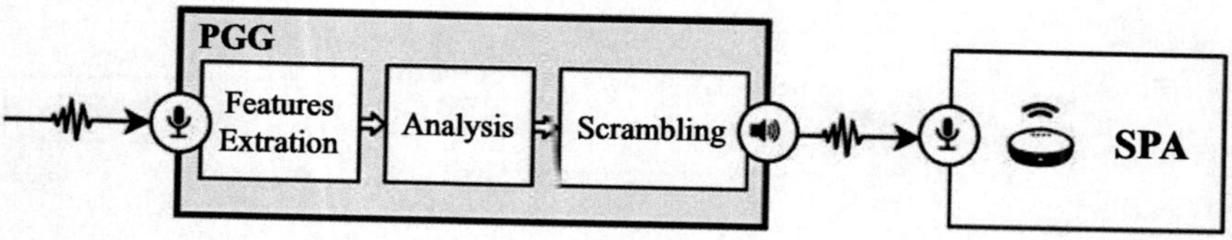

Fig. 2 Process of the privacy guard system

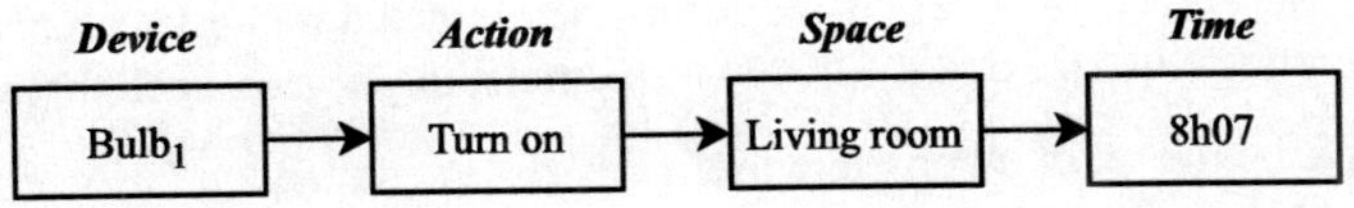

Fig. 3 An example of user request

$\lambda : \mathcal{T} \rightarrow 2^{\mathcal{N}}$ yields the set of feature names corresponding to a given type, whereas $\lambda^{-1} : \mathcal{N} \rightarrow \mathcal{T}$ gives the type of a feature name.

In Fig. 3, the request features are represented by rectangles, and the arrows connecting them constitute the request patterns. For example, the request pattern "$Bulb_1 \rightarrow Turn\ on \rightarrow Living\ room \rightarrow 8h07$" means that this is a request to turn on the primary bulb in the living room at 8h07.

Let $\mathcal{I}$ be the set of possible interpreted sentences of the user, and $\mathcal{F}$ be the set of all possible request features. The function $extract\ :\ \mathcal{I} \rightarrow 2^{\mathcal{F}}$ allows us to extract a set of request features from an interpreted sentence given by the speech to text technique. In this phase, the words of a sentence that cannot correspond to any type are ignored and do not be considered in the next phases. In addition, some features can be implicitly extracted without appearing in the sentence of the request, such as the temporal context of it (i.e., Time features). For instance, $extract("Turn\ on\ the\ living\ room\ bulb") = \{\langle Turn\ on, Action\rangle, \langle Living\ room, Space\rangle, \langle Bulb, Device\rangle\}$. The time feature is deduced from the current time at which the request is launched.

Phase 2: Feature Analysis

In this phase, the features previously extracted are classified and matched to some predefined *request pattern*. The request patterns represent the possible and expected requests that the user can launch.

Definition 2 (Request Pattern) A request pattern P ($P \in 2^{\mathcal{F}}$) is a set of request features, such that $\forall f_1, f_2 \in P$, $f_1 = \langle name_1, type_1\rangle$ and $f_2 = \langle name_2, type_2\rangle$, then $name_1 \neq name_2 \wedge type_1 \neq type_2$.

All possible requests constitute the so-called *Request Feature Tree*, where nodes correspond to request features and paths to request behaviors. The matching of extracted features with some request pattern yields a *user request*. It is based on a matching threshold, allowing to determine if the extracted features are eligible or are to be discarded.

Definition 3 (User Request) Let $D \in [0, 1]$ be the matching threshold used to determine the eligibility of the extracted features. A user request is a triplet

$\langle EF, P, d \rangle$, where $EF \in 2^{\mathcal{F}}$ is the set of the extracted request features, P is the matched request pattern and $d \in [0, 1]$ is the matching degree, such that $d \geq D$.

Let $\mathcal{UR}$ be the set of all possible user requests. We define the function $analyze : 2^{\mathcal{F}} \times [0, 1] \rightarrow \mathcal{UR}$, to yield a user request from a set of extracted features and some matching threshold.

Phase 3: Scrambling Mechanism

Now that the user request has been analyzed and approved, PGG can apply the scrambling mechanism. Based on a speaker and on some Text to speech technique, the proposed scrambling mechanism consists of forwarding dummy requests in order to frustrate the profiling and linkage problems of SPAs. This technique enables the interference of the information that the cloud can extract from the orders forwarded to the over-IP devices, adopting an engineering saying "too much information is equal to no information."

Thereby, two types of request features are considered, *real* and *dummy features*. A dummy feature can be introduced at any type by implementing it in PGG and this one can be configured with SPA to respond to it. For instance, dummy features can be dummy devices, actions, spaces, or times. These dummy features are added when configuring real features. In Fig. 5, the real features are represented by solid line rectangles, whereas dummy ones are depicted by dotted line rectangles.

Definition 4 (Scrambling Request) A scrambling request R ($R \in 2^{\mathcal{F}}$) is a request whose features can be real or dummy. Let $\mathcal{SR}$ be the set of all possible scrambling requests. We define a mapping $scramble^{-1} : \mathcal{SR} \rightarrow \mathcal{UR}$, which yields the user request from which a given scrambling request originated.

As the scrambling mechanism is dedicated to IoT device controls, if the device feature of a scrambling request is real, then all other features must be real. Indeed, it is not possible to perform a dummy action in a real device, because this one provides only predefined (real) actions.

As shown in Fig. 4, two kinds of scrambling requests are possible: (a) A scrambling request based on a real device, with its own actions and deployed in a well-defined space. Therefore, the scrambling can only be done on a dummy time; (b) A scrambling request based on a dummy device offers more possibilities for scrambling, because other features (actions, spaces, and times) can also be dummy.

Let $\mathcal{S}$ be the set of scrambling strategies. The function $scramble : \mathcal{UR} \times \mathcal{S} \rightarrow 2^{\mathcal{SR}}$ is used to give a set of scrambling requests from the pattern of a given user request based on a scrambling strategy. A simple strategy would be to generate $n \in \mathbb{N}^*$ scrambling requests for each user request. The choice of dummy and real (device) features in scrambling requests would be based on some probabilistic law.

The Algorithm 1 synthetically highlights the PGG process for reinforcing the privacy guard of SPAs. PGG starts by extracting features (EF) from the interpreted

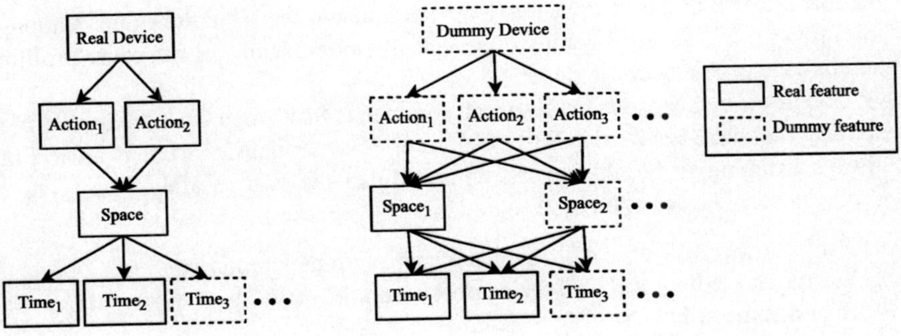

(a) (b)

Fig. 4 Scrambling request structures **(a)** Scrambling requests based on a real device **(b)** Scrambling requests based on a dummy device

Algorithm 1 PGG process

```
Require:
D := 0.5 ∈ [0, 1]  /* Matching threshold */
S ∈ S  /* Scrambling strategy */
while speech_in := listen() do
    I_in := speechToText(speech_in)
    EF := extract(I_in)
    ur := analyze(EF, D)
    if ur ≠ Null then
        SR := scramble(ur, S)
        for sr ∈ SR do
            I_out := sentence(sr)
            speech_out := textToSpeech(I_out)
            emit(speech_out)
        end for
    end if
end while
```

sentences (I_{in}) of the user request. Then, the features are analyzed to determine if the user request (ur) is eligible or not. In the third phase, PGG produces a set of scrambling requests (SR), where for each of them a sentence is made (I_{out}) and emitted using the PGG speaker.

An Illustrative Scenario

Now, we experiment our approach within the most widely used SPA on the market, Google Home.

In this paper, we take the most widely used SPA on the market, Google Home as a case study for our proposed approach. Indeed, Google Home is a cloud-based platform, which is able to collect IoT device data and send it to the cloud. Although this allows to enhance the quality of service of Google Home, it can cause profiling and linkage problems for the user.

Google Home responds to all speech commands forwarded from the users which start by the wake-up word "Ok, Google" (or "Hey Google"). So, it is always on, always listening to be able to react instantly to the user. Mainly, five types of commands can be accepted by Google Home:

- **Useful information:** like "What's the latest with the coronavirus?"
- **Phone and calls:** like "Call the nearest hospital."
- **Broadcasting:** like "Broadcast, Dinner is ready!"
- **Time, timers, and alarms:** like "Set alarm for 7:30 AM."
- **Smart home control:** like "Turn off the living room primary bulb."

Here, we focus on smart home control commands, which allow the user to control several types of IoT devices, like turn on/off lights or open doors. As mentioned in this section, the extracted features in smart home control requests concern mainly the requested device and the performed action on it.

We experiment our approach by developing a hardware project embedding the proposed PGG system. The hardware used is: Google Home Mini, Raspberry Pi 3 Model B, Logitech USB Mic, Rokit Boost Orbit Speaker, and Horsky WiFi Outlet. In this project, we opted for Raspberry Pi 3 Model B (RPi3) as a nano-computer, due to its interesting hardware features, such as the performance of CPU, RAM, and peripherals.

For efficiency reasons of the PGG system, the RPi3 with its microphone and speaker should be beside Google Home with a distance between 10 cm and 50 cm. Moreover, the closer PGG is to the Google Home, the greater the noise nuisance caused by the volume of the speaker.

In our approach, some scrambling requests include dummy device (and action) features, therefore, it is necessary to virtually implement them, like on real device firmware. As for real devices, dummy devices must be accessible from Google Home, and configured with it. Although each one of them can be implemented in a separate connected hardware, they all can be integrated in one same hardware. Moreover, to avoid using additional hardware, the RPi3 can embed all dummy devices, in addition to containing the PGG system. The communication between Google Home and dummy devices in RPi3 is done over-IP, like for real devices.

The PGG process is triggered when the user requests are listened to by the microphone of PGG. After the achievement of the three phases Extraction-Analysis-Scrambling, this one starts the emission of the generated scrambling requests through the speaker plugged into the RPi3.

In order to test the functioning of PGG, we launch a voice request consisting of turning on the primary bulb in the living room at 8h07. This request is simultaneously listened to by both Google Home and PGG and this later generates a set of scrambling requests according to some strategy (giving four scrambling requests in this example). Figure 5 depicts the features of both user request ur and generated scrambling requests sr_1, sr_2, sr_3, and sr_4. In this example, the *Device*

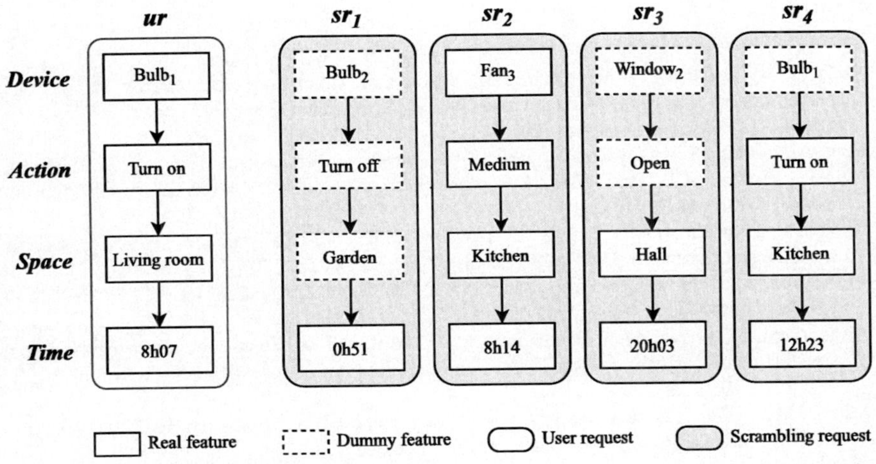

Fig. 5 Examples of user and scrambling requests

layer contains many IoT devices, such as $Bulb_1$, Fan_3, and $window_2$, whereas *Space* one contains room names of a given smart home.

5 Conclusion

Smart home Personal Assistants (SPAs) are interesting connected devices for home users, however, they suffer from profiling and linkage issues, mainly in the case of IoT device remote controlling. In this paper, we have proposed an efficient privacy guard system, called PGG, that is adapted to any SPA, as long as the PGG and SPA are next to each other. In order to guard the user activities, PGG is based on a noise addition based mechanism that injects dummy requests according to some strategy.

The main advantage of PGG is that it protects the user's privacy while being transparent. Indeed, no user request is rejected or lost by PGG, because all requests reach the SPA and the PGG in real time. Moreover, it does not store any activity of the user during its operation.

In order to enhance the proposed approach, different scrambling strategies can be explored. Furthermore, in addition to handling smart home control, the PGG system can be expanded to other types of SPA commands (like useful information and broadcasting commands).

References

Abdi, N., Ramokapane, K.M., Such, J.M.: More than smart speakers: security and privacy perceptions of smart home personal assistants. In: Symposium On Usable Privacy and Security. USENIX Association, Berkeley (2019)

Acar, A., Fereidooni, H., Abera, T., Sikder, A.K., Miettinen, M., Aksu, H., Conti, M., Sadeghi, A.R., Uluagac, A.S.: Peek-a-boo: I see your smart home activities, even encrypted! Preprint. arXiv:1808.02741 (2018)

Ahmed, S., Chowdhury, A.R., Fawaz, K., Ramanathan, P.: Preech: a system for privacy-preserving speech transcription. In: {USENIX} Security Symposium, pp. 2703–2720 (2020)

AL-mawee, W.: Privacy and Security Issues in IoT Healthcare Applications for the Disabled Users a Survey. In: Master's Theses, vol. 651, (2012)

Alshohoumi, F., Sarrab, M., AlHamadani, A., Al-Abri, D.: Systematic review of existing IoT architectures security and privacy issues and concerns. Int. J. Adv. Comput. Sci. Appl. **10**(7), 232–251 (2019)

Atlam, H.F., Wills, G.B.: IoT security, privacy, safety and ethics. In: Digital Twin Technologies and Smart Cities, pp. 123–149. Springer, Berlin (2020)

Chandrasekaran, V., Linden, T., Fawaz, K., Mutlu, B., Banerjee, S.: Blackout and obfuscator: an exploration of the design space for privacy-preserving interventions for voice assistants. Preprint. arXiv:1812.00263 (2018)

Edu, J.S., Such, J.M., Suarez-Tangil, G.: Smart home personal assistants: a security and privacy review. Preprint. arXiv:1903.05593 (2019)

Gao, C., Chandrasekaran, V., Fawaz, K., Banerjee, S.: Traversing the quagmire that is privacy in your smart home. In: Workshop on IoT Security and Privacy, pp. 22–28 (2018)

SpeechRecognition 3.8.1 (2020). Accessed August 25, 2020. https://pypi.org/project/SpeechRecognition/. https://ifttt.com/

Ziegeldorf, J.H., Morchon, O.G., Wehrle, K.: Privacy in the internet of things: threats and challenges. Secur. Commun. Netw. **7**(12), 2728–2742 (2014)

A Lightweight Phishing Detection System Based on Machine Learning and URL Features

Alaa Eddine Belfedhal and Mohammed Amine Belfedhal

Abstract Over the last few years, the number of phishing attacks has been increasing rapidly on the Web. Those attacks are used by cybercriminals to steal sensitive information, which threats users and enterprises' confidentiality and costs a lot of money. Therefore, detecting phishing webpages automatically in real time is becoming more crucial than ever. In this paper, we propose a machine learning (ML)-based lightweight system to detect phishing webpages from URL features. To choose the best ML models and features, we experiment with eight different models on two sets of features. We evaluate the performance of each combination of model/features by using a benchmark dataset of 73,575 phishing/legitimate web URLs. Experimental results show that the multilayer perceptron (MLP) model can achieve very high accuracy of 98.76% with a false negative rate of only 1.35%.

Keywords Web security · Phishing detection · Machine learning · URL features

1 Introduction

Nowadays, many useful electronic services are available on the web, ranging from e-banking, e-commerce, e-health, and e-learning to social media and email platforms. To access this type of services, the user must generally provide some authentication credentials like a username/password via a webpage form, so the system can verify his identity and give him the adequate access rights. Phishing is a social engineering technique that tries to trick the user by providing a fake webpage that looks similar to a legitimate one of a given service (a bank or a social media website for instance). The URL of the malicious webpage is necessarily different from the URL of the

A. E. Belfedhal (✉)
EEDIS Laboratory, Ecole Superieure en Informatique, Sidi Bel Abbes, Algeria
e-mail: a.belfedhal@esi-sba.dz

M. A. Belfedhal
EEDIS Laboratory, Djillali Liabes University of Sidi Bel Abbes, Sidi Bel Abbès, Algeria

S. Sedkaoui et al. (eds.), *International Conference on Managing Business Through Web Analytics*, https://doi.org/10.1007/978-3-031-06971-0_22

original webpage, but the attackers try to generate and register a URL that also looks similar to the legitimate one to fool the user. If the user is not prudent, he can reveal his confidential information by attempting to use the service. This malicious technique is used by cybercriminals and hackers to steal sensitive and private information of a victim web user. Examples of such information include user's credentials (login/password), credit card number, bank account, social security number, etc. (Feng et al. 2018). After stealing the information, the malicious website can even redirect the victim to the original legitimate website, so the user will not even be aware of the phishing. The URL of the fake site is usually sent to the victim using emails (SPAMs), popups, ads, social media, SMS, or blogs.

The problem with fishing is not only about confidentiality and privacy issues caused by information leakage but also about phishing costing companies and individuals a lot of money. According to (Retruster 2021), the total amount of money lost due to phishing can reach billions of dollars in 1 year.

Attackers are also known for seizing the opportunities presented by events and trends like New Year, discounts period, and even disasters and pandemics like the coronavirus, for the sake of cyberattacks. According to the US Department of Homeland Security (DHS (Department of Homeland Security) 2020), cybercriminals have taken advantage of the COVID-19 crisis to lunch increasingly more complex phishing and malicious attacks. Starting in March 2020, cybercriminals have launched various phishing and malware attacks – with COVID-19 as "phishing theme" – against remote workers, healthcare institutions, and those who have recently lost their jobs (APWG (Anti-Phishing Working Group) 2020). An example of such attacks is sending phishing emails of fake Zoom video-conferencing meeting notifications. The URLs in those emails redirect users to webpages designed by the phishers to steal Zoom accounts credentials. The attackers targeted Zoom because of the increasing number of users, including businesses (because of the pandemic businesses sent their staff to work from home).

As a result of what we mentioned above, creating robust and accurate real-time phishing detection systems is becoming more important than ever before. In this paper, we present a machine learning (ML)-based system for real-time phishing detection. The proposed system uses a multilayer perceptron (MLP) model and a set of features extracted from URL. We used two categories of features: the first category represents lexical features extracted from the URL string. The second one represents tokens frequency using the TF-IDF (term frequency-inverse document frequency) algorithm. Unlike many other works that use webpages content or third-party information as features, we used URL-based detection to speed up the detection time. We performed many experiments, using eight different ML algorithms on a large dataset of phishing and legitimate URLs from (Ebbu2017 n.d.). The experimental results showed that our system achieved very good results with an accuracy of 98.76%, a true positive rate of 98.64%, and a very small false negative rate of 1.35%.

The rest of this paper is organized as follows: in Sect. 2, we present a literature review of phishing detection techniques and systems. In Sect. 3, we detail our proposed system, and we show the principal components of our solution. Tests and results are presented in Sect. 4. The last section of this paper is a conclusion in which we present perspectives and future work.

2 Related Work

In this section, we present an overview of various phishing detection systems proposed in the literature. The presented systems can be categorized according to the type of used features or the algorithmic approach used for phishing detection.

2.1 Type of Features Used for Phishing Detection

According to the type of used features, phishing detection approaches can be mainly divided into three categories: content-based detection, third-party information-based detection, and URL-based detection.

2.1.1 Content-Based Detection

In this detection category, features are extracted from the code source and other downloadable resources of the webpage associated with a particular URL. Those features include webpage text, icon, images, hyperlinks, HTML code, CSS style sheet, JavaScript code, and so on (Mao et al. 2019). The major issue with content-based detection is its slowness. Before determining whether or not a webpage is phishing, the detection system must download the webpage code and files and then extract necessary features.

2.1.2 Third-Party Information-Based Detection

Some research works, which are based on third-party information, use search engines (like Google or Bing) to get ranking information. This information is useful to websites classification under the assumption that phishing websites have very low probability of being well ranked (Huh and Kim 2011). Other works use WHOIS services to get information on the domain, like the domain owner, the hosting server, the IP address, etc. Reference (Sahoo et al. 2019) showed that this domain information can be used to detect phishing websites. The problem with this kind of detection is that it is very consuming in terms of time and resources and depends on the availability of the third-party services.

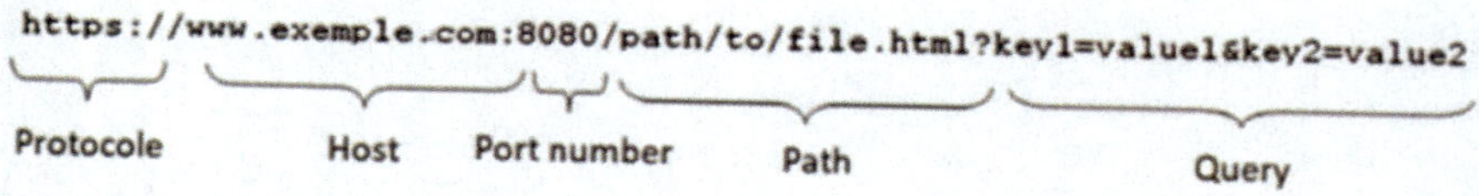

Fig. 1 Syntax of a URL

2.1.3 URL-Based Detection

URL-based detection systems use information extracted from the URL to classify webpages. A URL (Uniform Resource Locator) is a string of characters that represents a global address to locate resources (like a webpage) on the Internet.

A webpage URL is composed of up to five components, delimited by special characters (IBM Knowledge Center 2018) (an example is given in Fig. 1):

1. A *Protocol*. It identifies the application layer protocol used to access the webpage content.
2. A *Hostname*. Also called the domain name; the host name identifies the server that accommodates the webpage, for example, "www.website.com." A port number can be specified after the host name, for example: "www. website.com:80."
3. A *Path*. It identifies the specific file in the server that the browser wants to access, for example, "/folder/exemple.html."
4. A *Query script*. It provides a number of name and value pairs that are used by the server to respond to a certain query (e.g., to perform a search or process some data). The query string begin after the character "?," for example,? name1 = value1&name2 = value2.
5. A *Page fragment*

URL-based features can be directly extracted from the URL string (length of URL, the number of special characters, etc.) or by using a text vectorization technique like n-grams (Joshi et al. 2019) or word2vec (Yuan et al. 2018). The main advantage of URL-based detection is that it is lightweight (very fast in terms of both training and detection) in comparison to content or third-party-based detection. URL features are discussed further in Sect. 3.3.

2.2 *Algorithmic Approach Used for Phishing Detection*

For the three categories of features mentioned, detecting malicious webpages can be done either by lists, heuristics, or machine learning techniques. In the following paragraphs, we discuss these approaches and provide examples of state-of-the-art works using them.

2.2.1 List-Based Approaches

In this situation, a list is a database of URLs or other content-based features (like hyperlinks or images). List-based techniques mainly use one of two types of lists: lists that contain phishing features ("blacklists") or legitimate features ("white lists"). Blacklist-based detection is very efficient for detecting known phishing web sites, but it cannot detect previously unseen, new ones. Thus, the false negative rate can be very high. On the other hand, white list-based detection allows only known legitimate websites and blocks all others. In this case, all new websites are to be blocked even if they are legitimate. Therefore, the false positive rate can be very high when it comes to white list detection.

Lists can be created manually – by experts or based on user rating (crowd sourcing) – or automatically by other approaches like heuristics. Examples of approaches that use list-based detection include using database of (legitimate or phishing) URLs, images, DOMs, and hyperlinks (Rao and Ali 2015). Presented below are some list-based systems.

PhishNet is a blacklist-based system proposed by (Prakash et al. 2010). It uses two components to detect phishing websites. The first component is based on heuristics that generate combinations of known phishing URLs to detect new ones. The second is a matching method used to calculate URLs similarities against entries in the blacklist. Experimental results on PhishNet gave 3% of false positives and 5% of false negatives.

Y. Zhou et al. (2014) used snapshot images extracted from webpage to detect phishing. Their approach was based on "visual similarity" between a known legitimate website and a new suspected one. The experimental results showed that the proposed system achieved 90% true positive rate and 97% true negative rate. Jain and Gupta (2016) proposed a system using white list of legitimate hyperlinks (extracted from the webpage content) and a "domain IP address" matching model. When a browser tries to access a website which is not present in the white list, the system block this potentially phishing website. The experimental results showed that the proposed system achieved 86.02% of true positive rate and 1.48% of false negative rate.

Using DOM (Document Object Model) to detect phishing was proposed by many researchers (Rosiello et al. 2007; Cui et al. 2017), for instance (Cui et al. 2017), presented a system called DOMAntiPhish that uses layout similar information to distinguish between malicious and benign webpages. The similarity is calculated using DOM-Tree representation of the website. If the website is similar to a legitimate one – that uses the same credentials – but with different domain, it is considered as phishing website.

2.2.2 Heuristics-Based Approaches

Heuristics are approximate algorithms "that do not guarantee a correct solution, but typically yield a reasonable one" (Todd 2001). In the context of phishing detection, heuristics are generally rules based on the researcher's expertise of what constitutes a phishing webpage; and how it should look like in comparison with a legitimate one. Examples of heuristic features include (Zhang et al. 2007) age of domain, known images and logos, suspicious URL, number of dots in URL, etc.

For instance, CANTINA (Zhang et al. 2007) detects phishing pages, using "term frequency-inverse document frequency" (Wu et al. 2008) (TF-IDF) algorithm, and combines it with other heuristics to reduce false positives. It is a content-based system that uses text and hyperlinks from webpages to classify them as phishing or legitimate. CANTINA is capable of correctly labeling approximately 95% of phishing sites. Another example is "Lightweight Phish Detector" (LPD) proposed by (Varshney et al. 2016). LDP uses Google search engine to search for strings extracted from the webpage URL and title. If the domain name of the specified URL is present in the top-K search results, the webpage is labeled as legitimate. If there is no match, it is regarded otherwise.

The major problem with heuristics-based approaches is that a cybercriminal can circumvent the heuristic once he finds out the method used in the detection system (Rao and Ali 2015).

2.2.3 Machine Learning-Based Approaches

To overcome the weaknesses of lists and heuristics-based approaches, researchers have proposed the use of machine learning (ML) techniques. The advantage of ML methods is their ability to detect new (previously unseen) phishing webpages (Orunsolu et al. 2019). In the following sections, we present some of the ML-based phishing detection research works.

Authors of (Chiew et al. 2015) proposed a machine learning technique to search and extract the logo from all downloadable images of a webpage. The extracted logo is then used as a search image with "Google image search service" to find the real domain name associated with the logo (considered as the real identity). The founded domain name is subsequently compared with the domain name of the webpage to decide whether it is phishing or legitimate.

Opara et al. (2020) presented a system called HTMLPhish, which is based on HTML analysis and deep learning. The authors used word and character embeddings with Convolutional Neural Networks (CNNs) to extract the semantic relations in the text content of the webpage. They reported results of over 93% accuracy.

Sahingoza et al. (2019) experimented with seven different machine-learning algorithms and natural language processing-based features to create an anti-phishing system. To measure the performance of their system, they constructed a dataset of 73,575 URLs, which contained 36,400 legitimate URLs and 37,175 phishing URLs. The authors reported that random forest algorithm with only NLP-based features gave the best performance, at an accuracy rate of 97.98%.

Another interesting ML-based detection method is the one proposed by (Chatterjee and Namin 2019) which uses deep reinforcement learning for URLs classification. The "Q-learning" (Sutton and Barto 2018) algorithm was applied to Ebbu2017 (n.d.) dataset and gave a 90.10% of classification accuracy. Ghalati et al. (2020) used URL N-grams as features and experimented with three different ML models. The best result of 96.78% accuracy was obtained by the random forest classifier.

3 Proposed System

In this section, we present our phishing detection system. We first detail the system architecture, the used dataset, the features extraction, and the used models. Then, we show experimental some results and we discuss them.

3.1 System Architecture

Figure 2 shows the architecture of the proposed system. Our system takes as input a URL string and then extracts two types of features from it. The first type of features is lexical, extracted directly from the URL (like the size of the URL). The second type represents token features, extracted from the URL and a previously constructed token matrix obtained through the training dataset. After predicting, the used MLP (multilayer perceptron) model outputs one of two possible labels: either "phishing" or "legitimate."

Before choosing the MPL model, we trained eight different machine-learning algorithms, using the same dataset. The used models are logistic regression, decision tree, random forest, k-nearest neighbors (KNN), support vector machine (SVM), MLP, AdaBoost, and XGBoost. We experimented with two sets of features; the first set is composed of lexical features only; however, the second set is composed of lexical features associated with token frequency features. Experimental results (showed in Sect. IV) suggest that using the second set of features gives much better performances.

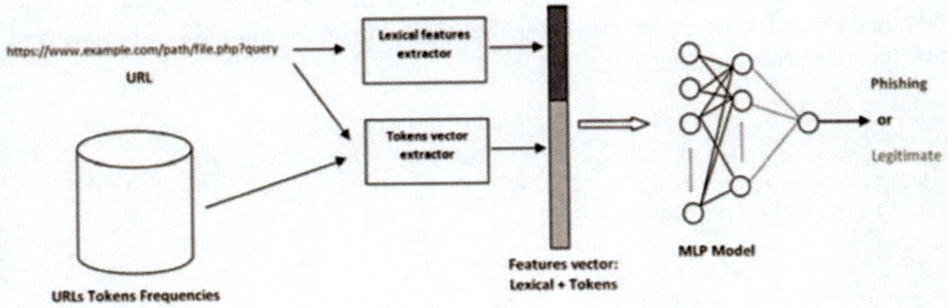

Fig. 2 Architecture of our system

3.2 Dataset

To train and evaluate the models and compare the obtained results with other literature works, we used the Ebbu2017 Phishing Dataset (Ebbu2017 n.d.). This dataset was generated and published by Sahingoza et al. (2019) in 2018. It is composed of 73,575 different URLs, among which 37,175 are phishing, and 36,400 are legitimate. To collect the dataset, Sahingoza et al., developed a script to query the Yandex Search API[1] to get the highest-ranking webpages, which would be labeled as "legitimate." They also used PhishTank[2] service to obtain a list of "phishing" URLs.

3.3 Features Extraction

In our work, we use URL features, extracted from URLs without downloading any further content or code, nor using a third-party service (like WHOIS service or search engine), which makes our system very fast. We extracted two types of features, namely, lexical features and token-based features:

3.3.1 Lexical Features

Lexical features are features directly extracted from the URL string. Examples of such features include URL length, parts of URL lengths, number of special characters in different parts of the URL, etc. We extract a total of 40 different lexical features based on our literature review (IBM Knowledge Center 2018; Yuan et al. 2018). In Table 1, 15 examples of the lexical features used in this work are presented with their min, max, and mean values in both legitimate "legit" and phishing "phish" parts of the used dataset.

Principal component analysis (PCA) was then used to eliminate highly correlated features and to extract the 15 most relevant ones. The features' values are then normalized, using the min-max scaling method.

To see whether lexical features are sufficient to distinguish between phishing and benign URLs, we experimented with different machine learning classifiers. The results are showed in Sect. 4.2.

[1] https://yandex.com/

[2] https://www.phishtank.com/

Table 1 15 Lexical features extracted from URLs

	Min phish	Min legit	Max phish	Max legit	Mean phish	Mean legit
URL length	45	53	88	88	84.07	83.82
Host part length	9	10	77	11	10.21	10.36
Path part length	0	0	71	71	64.76	65.37
Query part length	0	0	58	56	2.02	1.04
Parameters part length	0	0	41	37	0.007	0.001
Host to URL ratio	0.11	0.11	0.91	0.20	0.12	0.12
Path to URL ratio	0	0	0.80	0.80	0.76	0.77
Query to URL ration	0	0	0.66	0.64	0.023	0.01
Number of dots in host part	0	0	9	0	0.006	0
Number of digits in URL	0	0	34	27	6.68	6.89
Number of digits host part	0	0	10	1	1.01	1
Number of "-" in URL	0	0	25	7	0.40	0.74
Number of "?" in URL	0	0	15	1	0.05	0.02
Number of "*" in URL	0	0	7	0	0.0004	0
Number of "@"	0	0	2	1	0.002	0.0001

3.3.2 Token Frequency Features

In order to extract token frequency features, we used the TF-IDF (term frequency inverse document frequency) algorithm. TF-IDF has been originally used by research engines and information retrieval systems to index documents based on important keywords. The main idea behind this algorithm is to give a term a certain weight based on its frequency in a corpus of documents. The weight indicates the importance of a term in a corpus of documents and is calculated using the formula (1) below (Wu et al. 2008):

$$\text{tfidf}\,(t, d, D) = tf\,(t, d) \times idf\,(t, D)\,. \tag{1}$$

Where t is a term, d is a document and D is a corpus of documents. To calculate (1), we have to calculate *tf(t,d)* and *idf(t,D)*.

tf(t,d): It represents the term frequency; it's calculated as follows:

$$tf\ (t, d) = \log\left(1 + \text{freq}\ (t, d)\right). \tag{2}$$

where *freq(t,d)* is the frequency of the term t in the document *d*.

idf(t,D): inverse document frequency. It represents the importance of a term t in a corpus D:

$$idf\ (t, D) = \log\ \left(N/\left(\text{count}\ \left(d \in D : t \in d\right)\right)\right) \tag{3}$$

where *N* is the number of documents d in the corpus *D*.

Instead of using natural language "terms" and documents corpus, we used "tokens" and a dataset of phishing/benign URLs. In our case, a token is a substring of the URL string. Tokens are separated by special characters (like: "/," "." and "-"). The number of tokens depends on the size of the dataset and its diversity. In our case, the total number of tokens was 106,173. Unlike CANTINA (Wu et al. 2008), we applied TF-IDF method on the URL string (not the webpage content).

4 Tests and Results

This section describes the experimental tests that we performed in order to select the best model and set of features to be used in our final system. This section also presents the results of comparison between our system and three other research works that used the same dataset.

4.1 Test Environment and Metrics

Experiments were executed on a laptop with 2.6 GHz Core-i7 processor and 16 GB of DDR4 RAM with Windows 10 as operating system. To perform tests, we used python with Scikit-Learn (version 0.23.1) and Pandas (version 1.0.4).

To evaluate the eight used machine-learning algorithms on the two sets of features, we used the "confusion matrix (true positive "TP," true negative "TN," false positive "FP," and false negative "FN")" (Orunsolu et al. 2019) to calculate five traditional machine-learning metrics, namely, accuracy, precision, recall, F-measure, MCC (Matthews Correlation Coefficient), and FNR (Ghalati et al. 2020).

4.2 Results and Discussion

For the first series of tests, we used only the lexical features for training and prediction. We experimented with different hyper-parameters to tune the models. The best results from training the different ML algorithms are given in Table 2.

Table 2 Evaluation of machine learning algorithms on lexical features

Model	Accuracy (%)	Precision (%)	Recall (%)	F-measure (%)	MCC
Logistic regression	75.95	76.51	75.95	75.74	0.5231
Decision tree	82.07	82.12	82.07	82.04	0.6413
KNN (K = 3)	79.42	79.42	79.42	79.42	0.5881
Random forest	82.73	82.85	82.73	82.69	0.6551
SVM	70.68	76.55	70.68	68.63	0.4633
MLP	82.12	82.31	82.12	82.07	0.6436
AdaBoost	73.89	75.65	73.89	73.27	0.4924
XGBoost	**82.84**	**82.95**	**82.84**	**82.81**	**0.6573**

Table 3 Evaluation of machine learning algorithms on lexical features combined with token features

Model	Accuracy (%)	Precision (%)	Recall (%)	F-measure (%)	FNR(%)
Logistic regression	97.39	97.40	97.39	97.38	03.26
Decision tree	95.80	95.81	95.80	95.80	05.06
KNN (K = 3)	96.86	96.90	96.86	96.86	04.61
Random forest	96.79	96.83	96.79	96.79	04.58
SVM	71.09	74.14	71.16	69.45	25.84
MLP	**98.76**	**98.77**	**98.76**	**98.76**	**01.35**
AdaBoost	78.43	78.44	78.43	78.43	22.50
XGBoost	93.63	93.64	93.63	93.64	07.07

From Table 2, we can see that the XGBoost algorithm gives the best results. XGBoost is an ensemble machine-learning algorithm that combines the results of many simple classifiers. An accuracy rate of 82.84% is acceptable, but not sufficient for cyber security applications. As a conclusion of our experiment, we consider that lexical features "only" applied to the used dataset are not sufficient for good phishing detection system.

In the second series of tests, we combined the lexical features with token frequency features, and we retrain the eight algorithms using the new set of features. Results are showed in Table 3.

From Table 3, we can see that the MLP algorithm achieved a very high accuracy and F-measure on the new set of features. We can also see that the false negative rate is very low, which indicates that the MLP model can be used for real-life phishing detection systems. We notice also that all models gave better results when applied to the new set of features. This indicates that this set of features characterizes more effectively the difference between phishing and benign URLs.

After choosing the best set of features and the best classification model, we implemented our final system using MLP. We then compared our system with three other research works that use the same dataset. The results are showed below in Table 4. From the comparison table, we can see that our system showed better results compared with the three recent phishing detection works, namely, (Sahingoza et al. 2019; Chatterjee and Namin 2019; Ghalati et al. 2020).

Table 4 Comparison between our system and three other research works

Approaches	Accuracy (%)	Precision (%)	Recall (%)	F-measure (%)
Sahingoza et al. (2019)	97.98	97.00	**99.00**	98.00
Chatterjee and Namin (2019)	90.10	86.70	88.00	87.30
Ghalati et al. (2020)	96.78	90.10	95.8	94.45
Our system (MLP)	**98.76**	**98.77**	98.76	**98.76**

5 Conclusion

In this paper, we have presented a lightweight system for phishing detection based on MLP model applied on URL features. To choose the best set of features and the best training model, we experimented with eight different ML algorithms applied on two sets of features. Experimental results showed that our system achieved very high accuracy compared to other literature works. The false negative rate was also very low, which is suitable for real-life phishing detection systems. Another advantage of our system is its lightness, which gives it the possibility to detect phishing URLs in real time without downloading any webpage content, nor using any third-party service. As a future endeavor, we intend to experiment with deep learning techniques on a larger dataset and implement our system as a browser plug-in for fishing detection.

References

APWG (Anti-Phishing Working Group).: Phishing Activity Trends Report, 1st Quarter 2020. https://docs.apwg.org/reports/apwg_trends_report_q1_2020.pdf

Chatterjee, M., Namin, A.S.: Detecting phishing websites through deep reinforcement learning. In: IEEE 43rd Annual Computer Software and Applications Conference (COMPSAC) (2019). https://doi.org/10.1109/COMPSAC.2019.10211

Chiew, K.L., Chang, E.H., Sze, S.N., Tiong, W.K.: Utilisation of website logo for phishing detection. Comput. Secur. **54**, 16–26 (2015). https://doi.org/10.1016/j.cose.2015.07.006

Cui, Q., Jourdan, G.V., Bochmann, G.V., Couturier, R., Onut, I.V.: Track in phishing attacks over time. In: Proceedings of the 26th International Conference on World Wide Web (2017)

DHS (Department of Homeland Security).: Alert (AA20-099A) COVID-19 Exploited by Malicious Cyber Actors. (2020). https://us-cert.cisa.gov/ncas/alerts/aa20-099a. Accessed: 10/06/2021

Ebbu2017.: Phishing Dataset (n.d.). https://github.com/ebubekirbbr/pdd/tree/master/input

Feng, F., Zhou, Q., Shen, Z., Yang, X., Han, L., Wang, J.: The application of a novel neural network in the detection of phishing websites. J. Ambient. Intell. Humaniz. Comput. (2018). https://doi.org/10.1007/s12652-018-0786-3

Ghalati, N.F., Ghalaty, N.F., Barata, J.: Towards the detection of malicious URL and domain names using machine learning. In: Technological Innovation for Life Improvement – 11th IFIP DoCEIS 2020, Proceedings, pp. 109–117. Springer (2020) 9 p

Huh, J.H., Kim, H.: Phishing detection with popular search engines: simple and effective. In: International Symposium on Foundations and Practice of Security. FPS 2011: Foundations and Practice of Security, vol. 6888, pp. 194–207. LNCS (2011)

IBM Knowledge Center.: The components of a URL (2018). https://www.ibm.com/support/knowledgecenter/SSGMCP_4.2.0/com.ibm.cics.ts.internet.doc/topics/dfhtl_uricomp.html. Accessed: 04/06/2021

Jain, A.K., Gupta, B.: A novel approach to protect against phishing attacks at client side using auto-updated whitelist. EURASIP J. Inf. Secur., 1–11 (2016). https://doi.org/10.1186/s13635-016-0034

Joshi, A., Lloyd, L., Westin, P., Seethapathy, S.: Using lexical features for malicious URL detection-a. Machine Learning Approach. (2019). arxiv.org. https://arxiv.org/ftp/arxiv/papers/1910/1910.06277.pdf

Mao, J., Bian, J., Tian, W., Zhu, S., Wei, T., Li, A., Liang, Z.: Phishing page detection via learning classifiers from page layout feature. EURASIP J. Wirel. Commun. Netw. **43** (2019). https://doi.org/10.1186/s13638-019-1361-0

Opara, C., Wei, B., Chen, Y.: HTMLPhish: enabling accurate phishing web page detection by applying deep learning techniques on HTML analysis. arXiv, 1909.01135v3 [cs.CR] (2020)

Orunsolu, A.A., Sodiya, A.S., Akinwale, A.T.: A predictive model for phishing detection. J. King Saud Univ. Comput. Inf. Sci. (2019). https://doi.org/10.1016/j.jksuci.2019.12.005

Prakash, P., Kumar, M., Kompella, R.R., Gupta, M.: PhishNet: predictive blacklisting to detect phishing attacks. In: 2010 Proceedings IEEE INFOCOM (2010). https://doi.org/10.1109/INFCOM.2010.5462216

Rao, R.S., Ali, S.T.: PhishShield: a desktop application to detect phishing webpages through heuristic approach. In: Proceedings of the Eleventh International Multi-Conference on Information Processing 2015 (IMCIP-2015) (2015). https://doi.org/10.1016/j.procs.2015.06.017

Retruster.: Learn About Phishing Email Statistics For 2019. Online: https://retruster.com/blog/2019-phishing-and-email-fraud-statistics.html. Accessed: 02/06/2021

Rosiello, A.P.E., Kirda, E., Kruegel, C., Ferrandi, F.: A layout-similarity-based approach for detecting phishing pages. In: Proceedings of the Third International Conference on Security and Privacy in Communications Networks and the Workshops, SecureComm. IEEE (2007). https://doi.org/10.1109/SECCOM.2007.4550367

Sahingoza, O.K., Buberb, E., Demirb, O., Diric, B.: Machine learning based phishing detection from URLs. Expert Syst. Appl. **17**, 345–357 (2019). https://doi.org/10.1016/j.eswa.2018.09.029

Sahoo, D., Liu, C., Hoi, S.C.H.: Malicious URL detection using machine learning: a survey. arXiv:1701.07179v3. (2019) 37 pages

R.S. Sutton, and A.G. Barto. "Reinforcement learning: an introduction". MIT Press. 2018

Todd, P.M.: Heuristics for decision and choice. In: International Encyclopedia of the Social & Behavioral Sciences (2001). https://doi.org/10.1016/B0-08-043076-7/00629-X

Varshney, G., Misra, M., Atrey, P.K.: A phish detector using lightweight search features. Comput. Secur. **62**, 213–228 (2016). https://doi.org/10.1016/j.cose.2016.08.003

Wu, H.C., Luk, R.W.P., Wong, K.F., Kwok, K.L.: Interpreting TF-IDF term weights as making relevance decisions. ACM Trans. Inf. Syst. **26**(3), 1–37 (2008)

Yuan, H., Yang, Z., Chen, X., Li, Y., Liu, W.: URL2Vec: URL modeling with character Embeddings for fast and accurate phishing website detection. In: IEEE International Conference on Big Data and Cloud Computing (BdCloud) (2018). https://doi.org/10.1109/BDCloud.2018.00050

Zhang, Y., Hong, J.I., Cranor, L.F.: Cantina: a content-based approach to detecting phishing web sites. In: Proceedings of the 16th International Conference on World Wide Web, pp. 639–648. ACM, Banff (2007)

Zhou, Y., Zhang, Y., Xiao, J., Wang, Y., Lin, W.: Visual similarity based anti-phishing with the combination of local and global features. In: Proceedings of the 2014 IEEE 13th International Conference on Trust, Security and Privacy in Computing and Communications, TrustCom. IEEE (2014). https://doi.org/10.1109/TrustCom.2014.28

Advances in Search Engine Optimization Through Web Analytics Development: GuinRank's Web Analytics Case Study

Keltoum Bentameur and Isma Belmihoub

Abstract The amelioration of website's relationship with search engines is one of the most advanced Internet marketing strategies in today's businesses, through the improvement of the sites content. Search engine optimization (SEO) interests in how search engines work and computer-programmed algorithms, which determine the behavior of the search engine, and what people are searching. As soon as the site appears in higher ranking on the search results page, the increasing number of visitors coming to the site via the search engine may be converted into customers. For business web analytics, GuinRank is one of new and innovative websites designed to help marketers managing content. This paper aims to clarify how to work with sites analysis tools and to demonstrate their importance in improving the sites relationship with search engines. Therefore, by adopting the descriptive approach and the interview method in collecting data, the study attempts to explain the functioning mechanism of the site "GuinRank" and analyze its main tools. The website data analysis produces keywords that better present the organizations website and makes them clear for the search engines to give accurate results. It is a professional tool for search engine optimization, which makes the site getting a better ranking on Google, by using better keywords. Through the artificial intelligence of the site (GuinRank), it is possible to coordinate better content that Google prefers. It also searches for gaps in the text in a record time and analyzes the level of competition comparing to the biggest competitors on the virtual space.

Keywords Digital marketing · SEM · SEO · Content marketing · Web analytics · GuinRank

K. Bentameur (✉)
Department of Commercial Sciences, Faculty of Economic Sciences, Business Sciences and Management Sciences, Mohammad Al-Bashir Al-Ibrahimi University, Bordj Bou Arreridj, Algeria
e-mail: Keltoum.bentameur@univ-bba.dz

I. Belmihoub
Department of Economics, Faculty of Economic Sciences, Business Sciences and Management Sciences, Mohammad Al-Bashir Al-Ibrahimi University, Bordj Bou Arreridj, Algeria
e-mail: isma.belmihoub@univ-bba.dz

S. Sedkaoui et al. (eds.), *International Conference on Managing Business Through Web Analytics*, https://doi.org/10.1007/978-3-031-06971-0_23

JEL Classification M21, M31, M39

1 Introduction

Over the last decades, the world of digital marketing is becoming more and more complex and exciting, where small businesses are using simple digital marketing strategies to expand their online presence and acquire new customers; they become successful companies at a large scale by harnessing the power of enterprise digital marketing. In our modern era, every online entity must have digital tools for strategic analysis, in order to improve its competitiveness on Internet market. Modern tools of online data analysis are now more powerful and more structured, where artificial intelligence (AI) and machine learning algorithms can make analysis faster and more accurate by speeding up data processing. Artificial intelligence can standardize an organization's data sources, which may help in getting more value than it really is. Today's digital marketing provides many great ways where marketers can start using AI for analysis like Google Analytics.

Google as the most used search engine in the world translates the contents to other platforms as well. The optimization for Google is still being the best bet for today's organizations, when talking about SEO; we refer to search optimization "first from Google." It is about the ongoing optimization of the website content in order to appear among the top of search results, without the need to pay. In this context, a specialized team worked on the creation and the designing of the website GuinRank, which stands on artificial intelligence, an understanding of SEO rules, accurate research, and analysis. It also works according to scientific foundations and algorithms to improve digital content and develop user content online. This study tries to find out to what extent the artificial intelligence of GuinRank can contribute to the progress of SEO.

The study is related to different literature review, including a range of research papers and many interesting theoretical insights into SEO (Terrance 2018; Zilincan 2015; Spais 2010; Baye and De los Santos 2016). This research focuses on studying the relevance of search engine marketing and search engine optimization, as well as the impact of keyword analysis, among other SEO-friendly technologies that positively influence digital marketing, while, there are few empirical searches on advances in SEO by developing artificial intelligence. The focus is on the mean of web analytics by GuinRank, as one of the latest innovations and recent applications in the field of web analytics, which started in August 2020. However, there are many studies and theoretical literature and demos about search engines and web analytics. They focus on the sponsored links that appear alongside the organic results. The originality of the research stems from an attempt to combine three basic variables: SEO through continuous content improvement, using GuinRank's web analytics results to gain insight into the likelihood of a website appearing in first search results through content optimization, in order to make marketers having access to valuable data that helps them to achieve their organization's marketing goals.

2 Digital Marketing Through Search Engine Marketing

Digital Marketing digital marketing is best considered as how digital marketing tools such as web sites, CRM system, and database can be used to get closer to customers, to be able to identify, anticipate, and satisfy their needs efficiently and effectively (Dave Chaffey 2017, p. 21). It is a term used to describe the integrated marketing services used to attract, engage, and convert customers online. According to the Digital Marketing Institute (DMI), the digital marketing refers to "The use of digital technologies to create an integrated, targeted and measurable communication, which helps to acquire and retain customers while building deeper relationships with them" (Reshmi 2019, p. 429). It is also known as "how to harness the power of digital media and use it to achieve the utmost success in business, now and in the future" (Jones 2009, p. 2). Digital marketing uses multiple channels such as content marketing, influencer marketing, SEO, social media, and online advertising to help brands connect with customers.

Search Engine Marketing The SEM are practical marketing measures aiming to get more visibility in search engines either by getting more free traffic or paid traffic, whereas, the SEM incorporates search engine optimization (SEO), which adjusts or rewrites the website content and site structure in order to achieve higher page ranking in search engine result pages, and enhance pay-per-click (PPC) listings. To generate best results, both organic SEO and paid SEO practices must have shared goals and combined metrics, evaluate the data to determine future strategy, or find out right tools to get traffic for selected keywords in national and local search results (Terrance 2018). The SEO tools can help search engine marketing (SEM) through the overall improvement of landing pages by technical auditing the website pages (Terrance 2018). This helps to remove poor-quality links in order to improve content and increase performance and gain more traffic and conversions towards the site.

3 The Correlation Between SEO and Content Marketing

SEO Search engine optimization (SEO) is a series of tweaks and technologies that make it easier for search engines to crawl, index, and understand website content (Zilincan 2015, p. 506). It is also considered as one of the most dynamic newest channels for marketing online promotion, as they are a prime source for getting more and more online customers (Spais 2010, p. 9). SEO can be divided into two different groups (Zilincan 2015, p. 506): on-page (modifying the structure of a website) and off-page (techniques independent of website's structure). A good combination between both of them can give a higher position in search engine result pages (SERPs) and then a significant increase in traffic.

Marketing research is necessary to know the customer, the market, and the institution's special capabilities, before thinking about the site optimization itself, and it is impossible to make a search engine optimization for every word on a page (Zilincan 2015, p. 506); therefore, it is important to select keywords, which most accurately represent the content. In addition, optimization is always performed according to a user-defined search method. SEO includes – on-page – the elements that are in direct control of a publisher, such as contents, titles, domain name, URL structure, headings, internal links, meta tags, page speed, structured data, and site map (Zilincan 2015, p. 507).

Content Marketing Since the term is relatively new, many definitions were introduced to Content Marketing (CM) that are similar in meaning. It is defined by Puilizzi as "the marketing and business treatment to create and distribute valuable, compelling and relevant content, to attract, acquire and engage a clearly defined and understood target audience, with the aim of driving profitable business for clients" (Puilizzi 2014, p. 5).

The Content Marketing Institute (CMI) defines the content marketing as: "Content marketing is the marketing technique for creating and distributing content relevant and value to attract, acquire, and engage a specific goal, and clearly understood by an audience, with the aim of driving profitable clients' business" (group.com 2018, p. 5). In more detail, it is the creation and distribution of articles, art, blogs, videos, graphics, etc., from pieces to reach and communicate with existing and new clients, while there are those who assume the content marketing to be a creation of relevant content, with attractive and entertaining value. This content must be provided continuously, to maintain or change customer behavior. Content marketing is very important to retain existing customers and getting new ones. It also helps companies to build a strong brand (Duc 2013, p. 3). The purpose of content marketing is to provide valuable information to consumers, and then good content would create brand loyalty, to be better purchased in the future.

From that, we conclude that content marketing is the developing, implementing, and providing relevant content, which is very necessary to create the company's customer base. This can be realized by creating text, videos, images, graphics, e-books, white papers, and other content, also by sponsoring, developing, and sharing them with new content formats, in order to attract a clearly defined audience. Both search engine optimization and content marketing have altered the digital marketing world. However, SEO as a technical process raising traffic quality and attracting a huge number of visitors to a website, on the other hand, content marketing pays a particular attention to the use of a valuable and relevant content to drive more lucrative client business. Therefore, SEO strategy is inserted around content marketing since every website needs keywords, content, graphics, images, etc. They both have to go together to achieve marketing purposes. SEO's key component is to search for, find, and leverage relevant keywords in website content in order to rank higher on search engine result pages (SERPs). It is significant to use keywords in a suitable manner and take a strategic approach. Thus, SEO works with content optimization in a balanced way.

Web Analytics Web analytics in general can be known as a collection, analysis, measurement, and reporting of web data in the purpose of understanding and optimizing web use. In addition to being a process to measure web traffic but also are used to study business and market and to improve the efficiency of a website. Web Analytics permit the measuring data related to the website, including visitor behavior, traffic volume, conversion rate, web server performance (Cao Truong 2017, p. 14). In other terms, web analytics interest in user experience, and other information to understand and demonstrate the results, and continuously improve the efficiency of a website by the mean of collecting and analyzing web data, in order to maximize the web use of the site. It measures things that the webmaster takes into consideration, and this includes the time it takes to load the page, number of views, time spent by the visitor on the site, and other factors that help in improving website performance (Jennifer 2020). Web analytics help to measure the effectiveness of advertisement campaigns. They are both reporting and analyzing, as they provide the business with targeted insights into its endeavors and can be a powerful tool to raise the institution's digital marketing efforts.

To the highest level, with this insight, it can continue to improve the site's effectiveness, by using web analytics. Furthermore, the organization can track the number of visitors to its home page and the number who reaches its page. The most important goal that companies should achieve in general is to increase conversions. However, Web analytics can also quickly tell us the percentage of total daily, weekly, or monthly visitors, with in-depth marketing analytics, we can review all business's marketing efforts and identify the places that bring the highest return on investment. Whether analyzing payments per click, SEO strategy, or inbound marketing efforts, marketing analytics provides us with the next step that the organization should take next.

Building a content marketing strategy requires specific expertise that goes beyond the editorial dimension. Because if readers are the privileged target of content posted on the web, the mistake would be to neglect Google. Improving the organization's strategy by combining content marketing with SEO is an essential step for a content strategy successful, in producing its own content and measuring its performance and then rating this special SEO report in a content marketing service. Web analytics are important for a number of reasons; they help to understand the behavior of a website's visitors and use such data to optimize it more effectively. They also allow to find out things that are broken on website of organization, It helps to track the source of traffic like blogs, links, and others and to plan our marketing strategy, SEO campaigns, and design of the website.

In the world of artificial intelligence supporting technologies, web analytics would enhance the business analytics capabilities to improve the business intelligence. The increasing volume and complexity of business data is driving the commercial adoption of artificial intelligence for business analytics tools. Machine learning in business intelligence (or BI) and the mainstream use of artificial intelligence are helping businesses to pull out actionable insights from large and complex datasets; it also helps make business recommendations that can be understood by any business user. That is why the tools used to analyze results are so important – to get a better picture of customers and how they interact with content. Web analytics has changed the way companies and executives design their website. Artificial intelligence takes it to a higher level, where smart algorithms can tell us exactly what visitors are looking for on a website and what products to display and which products they do not want.

4 The Web Analytics GuinRank

GuinRank is a content optimization tool that can help content creators boost their site's ranking in Google search results, by recommending relevant phrases and keywords (klml, What is GuinRank 2020). GuinRank helps improve content results Based on search guidelines from Google. This tool identifies underutilized content in content strategies (klml, What is GuinRank 2020). In order to make Google relies on artificial intelligence in reading and understanding content, GuinRank website was designed to be an excellent toolkit for content owners. It enable novice SEO professionals to analyze their competitors' keyword content and offer ideas for writing articles that easily lead to search results. In addition to being the best in terms of analysis, thinking, and essay writing, it supports multiple languages. The site is not completely free like Google Analytics, but it has three kinds of subscribing plans. It is either for free with some tools for the first choice or payable with additional tools for the two other choices, but as a beginner organization, the free plan can be used as a starter to content optimization and analyzing competitors' results (Fig. 1).

GuinRank is a professional SEO toolkit that improves the ranking of the business content in Google. Based on Google's guidelines and LSI, GuinRank helps to improve the business content score (klml, medium.com 2020), enabling the user to generate more traffic and influence the right people into buying decisions.

Fig. 1 Site analysis indicators. (Source: www.Guinrank.com)

Fig. 2 The number of subscribers to the site until October 2020. (Source: Webmaster GuinRank)

GuinRank provides users with access to a rich collection of resources (including but not limited to keyword research tool, content optimization tool, content rewriting tool, and content translation); it is related to SEO content marketing, SEO copywriting, search engine optimization, and Google Ranking algorithm (GuinRank.com 2020). GuinRank also helps to delve into the competitors' content; this gives users an outline of what to write to be competitive. It has a built-in editor, which shows how in-depth our topic is and what are the areas of improvement. This tool not only shows us the best keywords that describe our website content but also ensures search engines understand them better (klml, What is GuinRank 2020); it is thus an SEO content optimization tool and a professional suite of tools for content creators.

An Egyptian Engineer called Ahmed Hashem, designed the site in 2017, and then several copies of the web were developed, but the upgraded version was launched early August 2020. However, in few months, the website achieved great popularity and subscriptions by site owners and bloggers, as the number of subscribers approached To 13,000 subscribers from all over the world (webmaster 2020); it also offers discounts for early subscribers by 70%, as a site promotion policy, which is a valuable opportunity for emerging institutions.

Figure 2 shows that the number of subscribers on the site has reached 12,410 (the date of preparing this paper). Within 4 months of the subscription's launching to the tool, searches for keywords have reached 72,911 keywords, as most subscribers work on blogging and take advantage of the free website analytics (500 words) scheme as a first stage. The following figure shows the geographical distribution of the subscribers (Fig. 3).

According to Alexa rank,[1] the site subscribers GuinRank are from different countries, and their number varies from country to another. The largest number of subscribers was 51% from Egypt, followed by Algeria with 10% of subscribers,

[1] A site's Alexa rank is based on estimates of traffic and visitor engagement over a period spanning the last 3 months and serves as a useful metric for judging a website's overall popularity in relation to all other currently live websites.

Fig. 3 Geographical distribution of the subscribers on the site. (Source: Webmaster GuinRank)

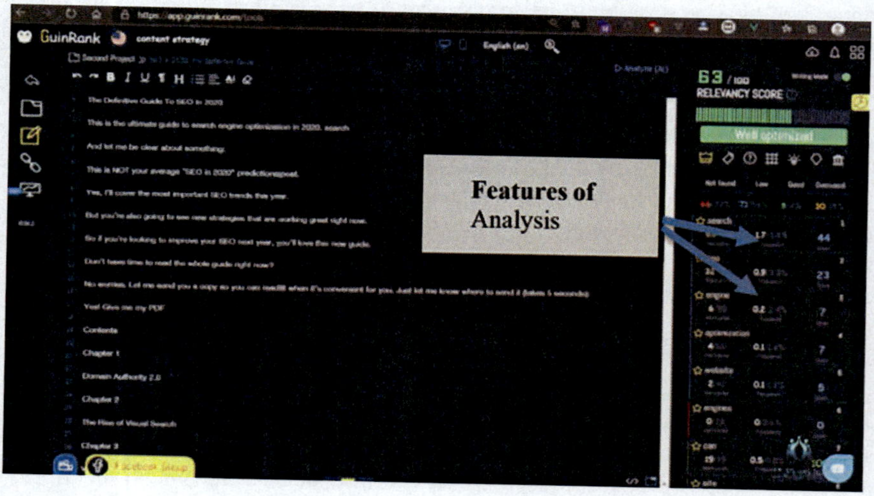

Fig. 4 GuinRank features. (Source: www.GuinRank.com)

where it beats the number of subscribers from Tunisia and Morocco. The reason is that the site supports the Arabic language along with other languages.

GuinRank Features (SEO) The SEO content optimization is a professional toolkit used by "content kings" to boost the business website ranking in Google, by only including recommended keywords and phrases. Content score is constantly being improved based on Google's guidelines (klml, medium.com 2020). The features of this tool are (Fig. 4).

Key Features of Ultimate SEO Content Optimization AI Tool GuinRank The main features of GuinRank's ultimate SEO optimization tool are the following (appsread.com 2020):

- It creates content that Google wants to rank and boost our ranking in Google by just including the recommended keywords and phrases.
- Helps identify underutilized content on the content strategy and improve the content score based on Google's guidelines and LSI.
- Its optimizer tool offers blueprint of what needs to be written to be competitive.
- Real-time text editor increases the keyword relevance score.
- Increasing content relevancy score would locate different topic difficulty, on-page SEO checker.
- Keyword analyzer (KA).

Competitor Analyzer: Here the basic tools for previously used keywords are developed. Now, the marketers need to consider the competitive environment to get an idea of the fierce competition by classifying a single keyword. It measures and evaluates top ranks management when it comes to critical SEO scores, including keywords (appsread.com 2020). The goal is to have a panoramic view of what they encounter and their opportunities; this is GuinRank's area of expertise.

Content Management: Certainly, content management is one of the most important and effective parts of this tool, allowing the website to be ranked at the top of the page and thus obtaining better results; this is how GuinRank does the analysis of the best sites on the page and compares them (appsread.com 2020).

Keyword Tracking: The growth of the company depends on the strategy. However, tracking keywords are the essential aspect in the world of SEO. It allows the website to gain value and improve its ranking and brings the website of business closer to the best results. This also allows us to advance content, release data, and prepare a flawless website. It is the human tendency to be the first. However, this trend needs to be closely monitored. It allows us to understand the ups and downs of our website's ranking. If necessary, we can make changes (appsread.com 2020).

User Management: The user management system has entrusted the administrator with the responsibility of authorizing or prohibiting access to IT resources to internal and external users. This system gives us the rights in terms of users' access to different IT resources such as storage systems (appsread.com 2020), networks, services, devices, applications, systems, etc.

5 Key Factors to Achieve Success for Using the Site

The main factors to achieve success in using the site are (Hachem 2020):

- *The first factor*: Notice the first ten words: the tool placed an asterisk in front of it and achieve a score for each word above 80.

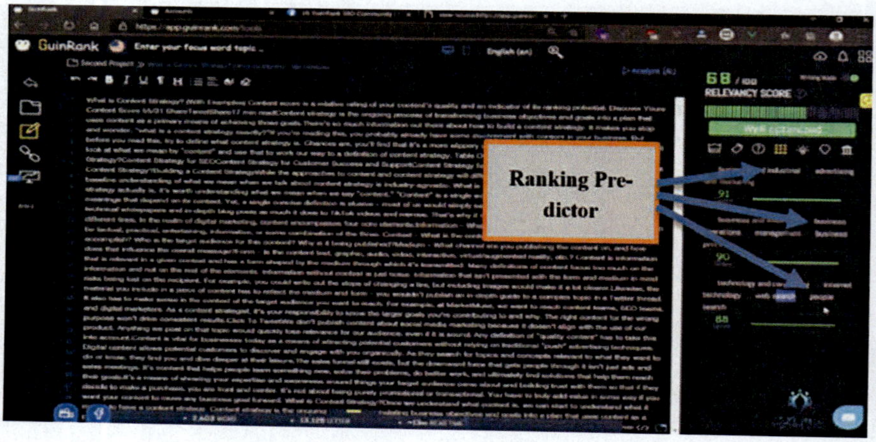

Fig. 5 Ranking predictor. (Source: www.GuinRaink.com)

- *The second factor*: An essay score is generally from 55 to 75, preferably from 70 to 74.
- *The last factor*: Ranking predictor is a website's AI analysis of the article; its goal here is that the article – after adhering to the first and second factors – can succeed in a large percentage in achieving a correct ranking predictor.

The success here is that the keyword is the first word in the chance of appearing (to Rank) with a number closer to 100%, meaning 60 is better than 50 and 70 is better than 60 (webmaster 2020). The presence of the number 1 keyword in the ranking predictor is in itself a success for an article; here we need to raise its percentage. By raising its value, we guarantee the best success of the article; if the ranking predictor reads a keyword more than the words that were targeted, this is an indication that the article has errors and must be re-reviewed (Fig. 5).

Before publishing the enhanced article, it must be ensured that the ranking predictor reader indicates the keyword correctly. Furthermore, the tool provides the most important successful factor, which is the content, but it is not the only one. Through this tool, articles are written in way to be more understandable to search engines and compatible with SEO factors. However, the site's building must be without problems. As a result of the use of website tools, some sites are leading the results soon, whereas, the other takes some time; as a practical example (webmaster 2020), we put the following figure (Fig. 6).

6 Conclusion

Measuring the effectiveness of digital marketing is one of the greatest challenges facing organizations today. Content marketing consists of creating and publishing

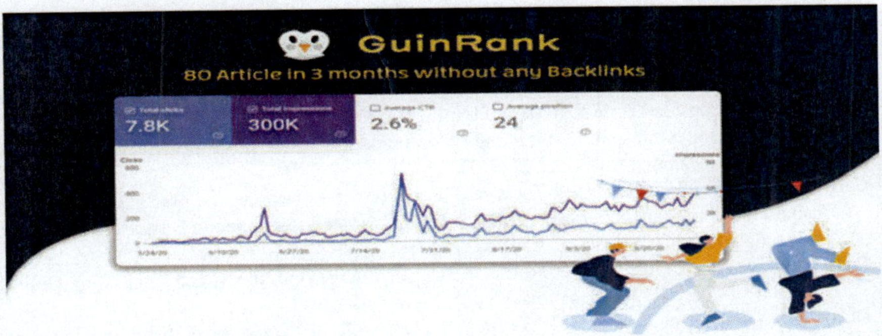

Fig. 6 Sample of work results with the tool without the need for links. (Source: webmaster GuinRank)

content that opens up opportunities in order to attract a specific audience. That content can take different forms: blog posts, which provide a useful and relevant response to the needs of Internet user to generate traffic in a sustainable way. It is necessary to inject good SEO practices into each key stage of the content marketing cycle, while SEO is more technically oriented (Google expertise, keywords, on-page optimization, technical SEO, etc.). Content marketing is developing relevant editorial mechanisms to attract readership, and content and SEO go hand in hand. Without one of them, for example, keyword research would not be of much use unless we used those keywords in our content. In addition, creating content without knowing what the organization's audience is looking for would make it without any significant traffic. Moreover, high-quality content is what earns us backlinks and is what search engines crawl when they land on the organization's site.

Keyword strategies are essential for effective search engine marketing (SEM). So searching for the best website analytics and competitive *keywords* that are closely related to the specific business or industry you work with is Jane Rank site; it uses Google *keyword* planner, a keyword analysis tool, to analyze the size of the most popular and competitive keywords (high or medium or low). Working with GuinRank site analytics tools will inevitably take strategic content marketing to a higher level, which would inevitably lead to the development of SEO. As the site knows continuous developments and updates, for example, a new tool will be added, called "question map" which is a tool for great ideas; it will also be supported with five new languages: (Finnish (fi), Dutch (nl), Danish (da), Czech (cs), and Polish (pl)). The challenge is to use the GuinRank site in the analysis of the global scientific journals' site and to influence the journal impact factor by analyzing scientific articles.

Acknowledgments We would sincerely like to express our gratitude to Mr. Engineer Hashem Ahmed for precious information about the GuinRank Website. This work has benefited tremendously from the discussions we have had with him.

References

appsread.com.: Récupéré sur GuinRank – The Ultimate SEO Content Optimization Tool (2020). https://www.appsread.com/2020/08/guinrank-the-ultimate-seo-content-optimization-tool/

Baye, M.R., De los Santos, B.: Search engine optimization: what drives organic traffic to retail sites. J. Econ. Manag. Strateg. **25**(1), 6–31 (2016)

Cao Truong, H.P.: Web Analytics Tools and Benefits for Entrepreneurs, Bachelor's Thesis in Business Information Technology. India: Sciences Degree Programme in Business Information Technology, Lahti University of Applied (2017)

Dave Chaffey, P.S.: Digital Marketing Excellence: Planning, Optimizing and Integrating Online, 5th edn. Routledge/Taylor & Francis Group, New York (2017)

Duc, L.M.: Content marketing, p. 3. Bachelor's Thesis, DP in International Business, University of Applied Sciences, Germany (2013)

group.com.: Content Marketing for Ministries 1Q1, website. Récupéré sur www.group.com (2018). https://www.agroup.com › files › uploads › ContentMar

GuinRank.com.: Récupéré sur What is guinrank services (2020). https://www.guinrank.com/site/about

Hachem, A.: GuinRank SEO Community. Récupéré sur www.facebook.com/groups/GuinRank SEO Community (2020). https://www.facebook.com/groups/GuinRankSEOCommunity/188011705808751/user/100001302195821

Jennifer.: Customer Analytics vs Web Analytics. Récupéré sur www.educba.com (2020). https://www.educba.com/customer-analytics-vs-web-analytics/Importance

Jones, D.R.: Understanding Digital Marketing: Marketing Strategies for Engaging the Digital Generation. Softback, 2 (2009)

klml, k.: What is GuinRank. Récupéré sur www.saasworthy.com (2020). https://www.saasworthy.com/product/guinrank

klml, k. medium.com. Récupéré sur Guinrank Review 2020 | #1 SEO Content Optimization Ai Tool (2020). https://medium.com/@aakarman/guinrank-review-2020-1-seo-content-optimization-ai-tool-2a8921bd14c3

Puilizzi, J.E.: Content Marketing. Mc Graw Hill Education, New York (2014)

Reshmi, V.C.: Digital Marketing- A Key to Development of Digital India, p. 429. National Seminar on Empowering India Through Digital Literacy, Annamalai University, India (2019)

Spais, G.S.: Search Engine Optimization (SEO) as a dynamic online promotion technique. The implications of activity theory for promotion managers. J. Innov. Mark. **6**(1), 9 (2010)

Terrance, A.R.: Importance of Search Engine Marketing in the Digital World. In: Proceedings of the First International Conference on Information Technology and Knowledge Management, Proceedings of ICITKM, vol. 14, p. 155. ACSIS, New Delhi (2018)

webmaster, G.: guinrink. (chercheur, Intervieweur) (2020)

Zilincan, J.: Search Engine Optimization, p. 506. CBU International Conference on Innovation (2015)

Part IV
Business Value Creation from Web and Social Media Analytics

The Impact of the Absence of E-payment on E-marketing: Case of Tourism Sector in Algeria

Mohamed Bendehiba and Nesrine Zerrouki

Abstract Although Algeria knows a noticeable development in the use of information and communication technologies (ICT) in all sectors, foremost the financial and the bank one, it is witnessing a significant delay in the application of e-payment in various fields, particularly the tourism sector. Hence, this study highlights the impact of the absence of e-payment on e-tourism in Algeria. It concluded that this absence affects directly the tourism industry in Algeria as it affects consumer satisfaction on the traditional payment method on the one hand and gives consumer the freedom to cancel his reservation on the other prompting the national tourism establishments to resort to foreign e-payment systems and affecting the national economy. Thus, this study recommends the need to accelerate the application of e-payment at the local level.

Keywords E-payment · E-marketing · Foreign tourism · Algeria

1 Introduction

With the advanced development in the information and communication technologies (ICT), new methods have emerged, among these e-tourism that highly affected the way the tourism business is executed, the decision-making process both within the tourism institution and the tourist destinations, and the way tourists interact with each other. These technological developments have a continuous impact through the introduction of continuous changes in the tourism sector which is characterized by its global nature that makes the use of ICT necessary.

In the current era, e-tourism is the main channel that most world's countries depend on to achieve tourism development due to the second generation of the Internet 2.0 through which dialogue and interaction between the various tourism dealers

M. Bendehiba (✉) · N. Zerrouki
University of Khemis Miliana, Khemis Miliana, Algeria
e-mail: mohamed.bendehiba@univ-dbkm.dz; n.zerrouki@univ-dbkm.dz

S. Sedkaoui et al. (eds.), *International Conference on Managing Business Through Web Analytics*, https://doi.org/10.1007/978-3-031-06971-0_24

and tourists has become easier. Thus, tourism marketing using ICT, especially the Internet, has become one of the main activities that tourist destinations depend on in promoting their tourism services and competing within the international tourism markets.

The e-payment system is considered as a result or part of the great development in ICT, especially the Internet, as it replaced the traditional means of payment by saving time and effort devoted to the exchange process. Hence, it encouraged the establishment of e-banking services and broadened the horizons for e-tourism and opened the way for it to access international tourism markets.

In view of the previous elements, we will try to answer, in this paper, the following question:

"To what extent does the absence of e-payment affect e-tourism?"

To answer the main question of the study, we present the following hypothesis:

"The absence of e-payment have placed a considerable financial losses that were incurred on the national economy in the context of foreign tourism due to the high rates of canceling e-reservations by tourists, and the orientation of national tourism institutions towards foreign e-payment systems".

2 Theoretical Framework for the Research

2.1 E-tourism as a Natural Partnership Between Tourism and ICT

The term e-tourism refers to the use of ICT, especially the Internet, in the travel and tourism industry by bringing about drastic changes in the behavior of tourist dealers and tourist consumers alike activating the role of tourism dealers by creating a direct interaction that allows them to identify and meet the ever-changing needs of tourist consumers (Bendehiba and Mebirouk 2015, p. 299).

E-tourism is the process of integrating ICT into tourism business by providing tourism services to e-tourists in high quality and at the lowest cost (Alipour 2011, p. 270).

2.2 Improving the Competitive Priority of Tourism Services Through E-tourism

E-tourism is defined as every form of commercial exchange in tourism services that takes place between tourism operators and tourist consumers via electronic media on the Internet that would enhance the e-tourists' service. This has an increasing impact on tourism promotion, marketing, and sales and has enabled the delivery of information in both directions between the tourism industry and tourists (Pourfakhimi 2014, p. 104).

Moreover, e-tourism marketing has several characteristics that make it a new tourist marketing method. The following are among the most important ones:

The ease of using the Internet to search for desirable tourism destinations and obtaining information about its tourism services and its various prices, as well as the ease of comparison between these tourist destinations.

The possibility of direct purchase of the tourism service via the Internet, so that the tourists do not incur the move to the place that provides the service and avoid the continuous attempt by tourist dealers to persuade them to buy the service in question and not the other.

Helping tourism operators so that their tourism services reach a large number of tourists in various global markets (Bruyère 2010, p. 40).

The ability of home shopping with continuous availability of information 24/24 that provides ease and speed of transactions and faster access for tourist consumers.

E-tourism allows the exit of tourism services from local to international markets.

E-tourism marketing allows reducing the number of tourist brokers and the direct dealing between the tourism service providers and tourists "B to C".

In this type of marketing, it is not possible to distinguish between small and large tourist institutions. This gives the opportunity for young tourist dealers to compete in international tourism markets via the Internet and enter the world of e-tourism marketing.

Reducing the costs of contacting the tourist consumer with the possibility and ease of modifying the offers and tourist publications that are included in the tourist sites on the Internet.

2.3 Using the Internet to Increase Clarity, Attract New Tourists, and Increase Profit. This Is Done Through the Following

2.3.1 Promoting the Website

Once the site is completed and started operating, it must be promoted. There are multiple means used in this field as there are companies specialized in promoting the website of the tourism establishment in a thoughtful and targeted way to its target tourism market and potential tourists which achieves the best positive impact with the least effort and time possible. The process of attracting e-tourists, enhancing their confidence, and increasing their satisfaction is a factor that helps in converting them into buyers of the tourism service.

2.3.2 Attractive Website

The use of electronic promotion is considered as one of the tools and methods that attract tourists and expand the customer base of the tourism establishment. To attract tourist to browse the tourism services provided by the website, the following must be done:

Integrated electronic content: this is through identifying the contents of the website in an integrated manner at the top and bottom of the page that enables and facilitates the e-tourist to browse it. The website must be attractive as well to push this tourist to browse it (Wen 2012, p. 19).

Indexing of electronic content: which means that there is a complete animated index that often appears next to the page of the website identifying the different contents of the website and making it easy for e-tourist to choose (Jill 2009, p. 754).

Correct technical performance of the website: by changing and updating the attractive tools used by this website so that thee-tourist does not feel bored, especially the e-tourist consumer whose loyalty has been gained by the tourism establishment (Bernardo et al. 2012, p. 343).

2.3.3 A Dynamic Website

The strategy of online marketing lies in how to convert e-tourists browsing the tourism institution's website to buyers of tourism services because one of the ways to measure the effectiveness of the website is the extent of its conversion of online tourists who brows it to tourists who buy services presented in it. This is done through the following:

Presenting a valuable and attractive offer on the main page of the website that is considered one of the steps that can drive the e-tourist to go deeper in browsing the website, hence, increasing the chance that this tourist browser will turn to buyer of the service.

Keeping up with competitors in terms of tourism services' price while providing means of comparison between the tourism services offered on the institution's website and the various competitive price offers. Unquestionably, other competitive aspects must be available. Reliability, responsiveness, and assurance are among the most important factors that gain the confidence and satisfaction of the e-tourist and drive him to make the purchase decision.

The ease of dealing with the website. The ease and speed of dealing and access to tourist information in the website are one of the encouraging and helpful factors that can transform the visiting e-tourist in the website into a buyer of the offered tourism services.

Providing the means of e-payment through the e-payment system, especially via credit card. This reflects the advanced stage reached by website. It is considered one of the basic elements that guarantee the gain of thee-tourist's loyalty.

Providing gifts and rewards to e-tourists browsing the website is considered one of the main factors that enhance their loyalty and contribute to pushing them to purchase the tourism services offered by the tourism establishment.

3 Methodology of the Research

The target study community is the tourism dealers in Algeria including travel agents and tour operators in Algeria.

In this study, we relied on the database obtained from the Ministry of Tourism and Traditional Industries, in addition to conducting several interviews with sector officials at the level of the ministry on the one hand and conducting a field study represented in distributing a questionnaire to Algerian tourist dealers (tourism agencies, travel and tour operators) on the other hand.

4 Analyzing and Discussing of the Results

4.1 E-Tourism and E-payment System in the Sector of Tourism in Algeria

E-payment is considered one of the necessary elements that must be provided in e-tourism work. This is done by linking e-payment systems to reservation and distribution systems through which credit cards and electronic checks can be used in the electronic process between tourist dealers and tourists via the Internet without the need for the tourism institutions and tourists to move to the financial and banking institutions in order to complete the tourism process. The availability of protection factors, privacy, security, brand, or commercial reputation during the online payment is among the most important factors that drive e-tourists to purchase online.

4.2 *The Impact of Absence of E-payment on Performance of Tourist Destination of Algeria*

The absence of e-payment has a major impact on the e-behavior and performance of e-tourists and tourism institutions, as follows:

4.2.1 The Absence of E-payment and the Postponement of the E-purchase Process for the E-tourist

The process of postponing e-purchase is the decision taken by the e-tourist consumer to delay the online purchase process of the tourist service offered on the tourist website of the tourism establishment. This is for several reasons that differ from a tourist consumer to another and from a tourist website to another, but the absence of e-payment is among the most important basic factors that drive a tourist to postpone the online purchase process. The absence of this technology forces the tourist to postpone the e-purchase of the tourist service and does not give him any other choice. This postponement has a direct negative impact on the tourism establishment as there is no guarantee for it that this tourist consumer who booked his tourism service electronically would not change his behavior.

4.2.2 The Absence of E-payment and the Process of Changing the Tourism Destination of the E-tourist

The process of changing the tourism destination by the e-tourist is a dynamic process through which interactive adjustments are made in order to adapt to the surrounding environment for different tourist destinations. Most of e-tourists often change their original travel plans in order to adapt to the tourist destinations to maximize their benefits and achieve their desires and commensurate with their capabilities during the preflight planning stage. The characteristics of the e-tourist affect primarily the process of changing the original tourist destination. Every tourist has the possibilities of change in order to deal with new and unexpected situations. Thus, the process of changing the tourist destination is the process by which the Internet user tourist changes his mind and cancels the process of reserving the tourist service that he performed previously for reasons that differ from one e-tourist to another. The absence of e-payment is among the main reasons and factors that push or allowed tourist to change his tourist destination. All the stages that an e-tourist performs during his online purchase of the tourist service can be controlled by the establishment except the stage between e-reservation and taking the tourist tour. The tourist establishment cannot control the behavior of the e-tourist consumer with the absence of e-payment that is the only way that enables the tourism establishment to earn the loyalty of thee-tourist consumer and ensure that it does not change his tourist destination.

5 Conclusion

The use of information and communication technology in Algeria has contributed in an efficient manner in the maximization of the potential tourism market. Thanks to this technology, a great number of tourists using the Internet may be incited to book tourism services supplied by the tourism companies, which can promote their tourism offers using this technology. This has enabled the said companies to expand their customer base, to improve the organization of their tourism industry, and to encourage the spread of e-commerce practices in the national economy. However, this efficiency remains relative because of the absence of an inter-sectoral approach between the sectors of tourism and banking in matters of e-payment in the context of a hard competition between the world states to attract new tourists. This state of things has had a negative impact on the satisfaction of the tourists using the Internet, mainly the foreign one. It has also particularly negatively affected the confirmed tourism market of the Algerian tourist destination and the national economy as a whole. This is due to the big financial losses (in hard currency) undergone by the national economy in the context of foreign tourism.

References

Alipour, M.: The impact of web marketing mix (4s) on development of tourism industry in Iran. Int. J. Bus. Soc. Sci. **02**(06) (2011)

Mohamed Bendehiba and Mohamed Bachir Mebirouk, The effect of the changed behaviour of the foreign tourist on the unorganised foreign tourism in the context of the absence of E-payment in Algeria, Int. J. Bus. Soc. Sci., Vol. 6, No. 2, February 2015

Bernardo, M., et al.: Functional quality and hedonic quality: a study of the dimensions of e-service quality in online travel agencies. J. Inf. Manag. **49**, 342–347 (2012)

Bruyère, S.: L'intelligence compétitive 2.0 pour le pilotage des projets e-marketing, Thèse de Doctorat en Sciences de l'Information et de la Communication, Université du Sud Toulon-Var, 18 octobre 2010 (2010)

Jill, C.: Marketing Communications: Interactivity, Communities and Content, 5th edn. Pearson Education (2009)

Pourfakhimi, S.: The impact of users' "online reviews" and "ratings" on consumers' behaviour toward hotel selection factors. In: Information and Communication Technologies in Tourism: Proceedings of International Conference, Dublin, Ireland, 21 January 2014 (2014)

Wen, I.: An empirical study of an online travel purchase intention model. J. Travel Tour. Mark. **29**(01), 18–39 (2012)

Google Trends Analysis Using R: Application on Algerian Tourism

Houssame Eddine Balouli and Lazhar Chine

Abstract In this work, we analyzed Google users' trends about tourism in Algeria using the programming language R. The main goal is to predict the number of tourists from outside Algeria depending on their activity using Google Search. This work is based on many packages: we started with "gtrendsR" package used in the selection of keyword (or keywords), which is in our case "Timimoun," search area, and the horizon. Then, "Tidyverse" package is used in the data cleaning phase. Finally, "prophet" package used to make predictions. The results shows us that the number of search operations about Timimoun using Google from outside Algeria will still very weak.

Keywords Google Trends · gtrendsR · Programming language R · Tourism · Algeria

1 Introduction

Algeria is distinguished by a set of natural, historical, and cultural characteristics that make it a regional and international tourist pole that attracts tourists from all countries of the world. The natural potential is a coastline of over 1200 km mountain ranges stretching from east to west and vast desert in addition to climate diversity. As for the historical possibilities, Algeria is considered the cradle of many civilizations: Phoenician, Byzantine, Romanian, and Islamic (Zaarour and Sabti 2018).

Despite all this potential, Algeria could not make the best use of it and thus the inability to meet the increase in internal and external demand. This is due to the

H. E. Balouli (✉)
University of Algiers 3, Khenchela, Algeria
e-mail: balouli.houssameeddine@univ-alger3.dz

L. Chine
Boumerdes University, Algiers, Algeria
e-mail: L.chine@univ-boumerdes.dz

S. Sedkaoui et al. (eds.), *International Conference on Managing Business Through Web Analytics*, https://doi.org/10.1007/978-3-031-06971-0_25

country's inability to develop tourist attractions such as tourist sites, transportation, accommodation, and infrastructure (Zayed 2018).

In an attempt to advance tourism in Algeria, the government has passed bills that encourage tourism investment through tax exemptions of up to 10 years. Also, it established several investment funds to accompany and finance tourism projects, including the National Agency for the Support and Employment of Young People (Beloudj 2006).

As a type of tourism, Algeria promoted desert tourism as an important source of winter tourism, as the desert in Algeria is characterized by moderate heat throughout the year, but the deterioration of the security conditions in the Sahel region contributed to the weakness of the tourism product (Bourbaba and Maachou 2018).

We try through this work to predict the future of desert tourism in Algeria by analyzing Google Trends about a popular tourist destination in Algeria (Timimoun) in the last 3 years.

2 Literature Review

With the heavy use of the Internet today, search engines have become an important source of information by analyzing trends. The Google search engine is among them. In this section, we present many similar recent studies. It is not necessary that these studies used the package "gtrendsR," but we explain how Google Trends are analyzed in the topic of tourism or any other topics.

Regarding the topic of tourism, Bokelmann and Lessmann (2019) presented a study in which they explained how we can use Google Trends data in the tourism demand forecasting. The researchers collected the data from two websites: GNTB and Tripadvisor using many keywords as an indicator (Deutschland Flüsse, Deutschland Seen, and Deutschland Berge). As a result, the study confirmed that Google Trends analysis is a good tool to elaborate significant forecasts about tourism in Germany in the short term (Bokelmann and Lessmann 2019).

In the same context, Ballatore et al. (2019) demonstrate how Google Trends data are useful in the tourism management by analyzing spatial and temporal variation in interest in places at multiple scales between 2007 and 2017 in Amsterdam metropolitan area. The comparison between the hotel visits to Amsterdam and search interest for Amsterdam shows to the researchers that there is a similarity between the trend lines (Ballatore et al. 2019).

Similarly, Irem Önder (2017) realized a comparison between Google Trends for two cities (Vienna and Barcelona) in one hand and two countries (Austria and Belgium) in the other hand. The main objective is the elaboration of tourism demand model (Önder 2017).

In the context of ecology and environment, Nghiem et al. (2016) studied the impact of media reports of news networks and scientific publications on the volume of research, by studying time series on Google Search data from 2004 to 2013.

Seven keywords in environmental topics were selected starting from January 2004 until December 2013: climate change, ecosystem service (or ecosystem services), deforestation, orangutan (or orangutan), invasive species, endangered species, and habitat loss. All the necessary tests was realized to detect trends. The next step is the elaboration of SARIMA model that showed a pattern of positive and contemporaneous correlation between the number of news articles and search volumes for climate change (Nghiem et al. 2016).

As a summary to this section, we can say that the literature review shows us how we can benefit from Google Trends as a source of information in one hand and a good tool in decision-making in the other hand.

Our added value is that we will analyze Google Trends data about desert tourism in Algeria using open-source programming language R to make forecasts about this sector in Algeria.

3 Method and Results and Discussion

The first step in our work is the installation and the loading of the necessary packages: gtrendsR (Philippe et al. 2016), Tidyverse (Wickham et al. 2019), lubridate (Garrett and Wickham 2011), prophet (Taylor and Letham 2020), and maps (Brownrigg 2018) (Fig. 1).

The second step is the selection of the keyword, the geographic region, and limits the search on specific time, using gtrends function (Fig. 2):

Fig. 1 Loaded packages

```
library (tidyverse)
library (lubridate)
library (gtrendsR)
library (prophet)
library (maps)
```

```
Timimoun <- gtrends(keyword = "Timimoun",
                    qeo = C("US", "FR", "DE", "RU", "GB" ),
                    time = "today+5 – y"
                    )
```

Fig. 2 Execution code (extraction of trends about Timimoun)

```
> names(Timimoun)
[1] "interest_over_time"  "interest_by_country" "interest_by_region"  "interest_by_dma"
[5] "interest_by_city"    "related_topics"      "related_queries"
```

Fig. 3 The organization of the extracted data

```
Timimoun_Top <- Timimoun$interest_by_region %>%
  filter(Timimoun$interest_by_region$hits >= 1 )%>%
  arrange(desc(hits))
```

Fig. 4 Google Trends by regions (execution code)

```
                     location hits  keyword geo gprop
                   California  100 Timimoun  US   web
                      England  100 Timimoun  GB   web
                 ÃŽle-de-France  100 Timimoun  FR   web
       North Rhine-Westphalia  100 Timimoun  DE   web
          Nord-Pas-de-Calais   97 Timimoun  FR   web
  Provence-Alpes-CÃ´te d'Azur   77 Timimoun  FR   web
             Midi-PyrÃ©nÃ©es   60 Timimoun  FR   web
                     Lorraine   58 Timimoun  FR   web
                  Rhone-Alpes   57 Timimoun  FR   web
       Languedoc-Roussillon   49 Timimoun  FR   web
              Franche-ComtÃ©   47 Timimoun  FR   web
              Upper Normandy   36 Timimoun  FR   web
                       Alsace   34 Timimoun  FR   web
                    Aquitaine   33 Timimoun  FR   web
         Centre-Val de Loire   26 Timimoun  FR   web
                      Picardy   21 Timimoun  FR   web
             Pays de la Loire   16 Timimoun  FR   web
```

Fig. 5 Interest by geographic location (Rstudio output)

The script code is as follow:

- *Keyword*: Timimoun
- *Geographic location*: Five countries: the USA, France, Germany, Russia , and Britain
- *Time*: last 5 years

Using "names" function we get a look on how our data is organized (Fig. 3).

The most important components are interest over time, interest by country, and interest by region.

Using filter and arrange functions, we select only top Google searches by region (Fig. 4).

Top interest by region is dominated by France region, with more than 600 search operation using Google. Regarding US, GB, and DE regions, we have only 100 operation search for each in last 5 years. Meanwhile, we note Russia's absence from the list (Fig. 5).

Using "ggplot2" package we realized the bar plot in below to confirm the previous analysis. The most important interest (HITS) in the last 5 years about Timimoun is from FR region (Figs. 6 and 7).

We transform our data using "as.tibble" function. This transformation is necessary for the next steps (visualization, forecast) (Fig. 8).

The result is as follow (Fig. 9):

```
ggplot(data=timimoun_top, aes(x=GEO, y=HITS, fill=GEO)) +
geom_bar(stat="identity", position=position_dodge())+
geom_text(aes(label=HITS), vjust=1.6, color="white",
          position = position_dodge(0.9), size=3.5)+
scale_fill_brewer(palette="Paired")+
theme_minimal()
```

Fig. 6 Bar plot of interest by region (execution code)

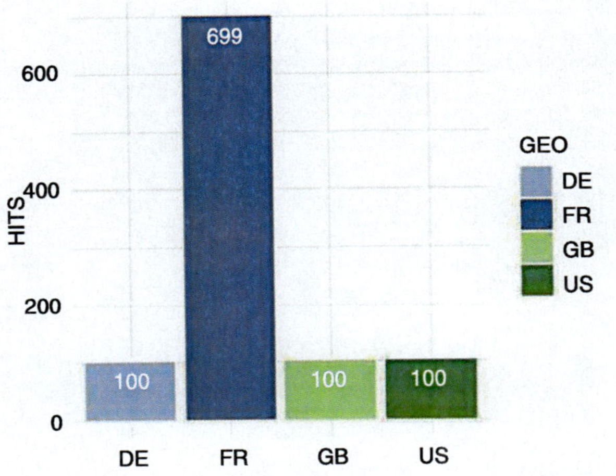

Fig. 7 Bar plot of interest by region

```
Timimoun_timeseries <- as.tibble(Timimoun$interest_over_time)%>%
                                 mutate(ymd(date))
```

Fig. 8 Necessary cleaning (execution code)

```
> Timimoun_timeseries
# A tibble: 1,305 x 8
   date                 hits keyword  geo   time       gprop category `ymd(date)`
   <dttm>              <int> <chr>    <chr> <chr>      <chr>    <int> <date>
 1 2015-07-19 00:00:00     0 Timimoun US    today+5-y  web          0 2015-07-19
 2 2015-07-26 00:00:00     3 Timimoun US    today+5-y  web          0 2015-07-26
 3 2015-08-02 00:00:00     0 Timimoun US    today+5-y  web          0 2015-08-02
 4 2015-08-09 00:00:00     0 Timimoun US    today+5-y  web          0 2015-08-09
 5 2015-08-16 00:00:00     0 Timimoun US    today+5-y  web          0 2015-08-16
 6 2015-08-23 00:00:00     0 Timimoun US    today+5-y  web          0 2015-08-23
 7 2015-08-30 00:00:00     0 Timimoun US    today+5-y  web          0 2015-08-30
 8 2015-09-06 00:00:00     0 Timimoun US    today+5-y  web          0 2015-09-06
 9 2015-09-13 00:00:00     0 Timimoun US    today+5-y  web          0 2015-09-13
10 2015-09-20 00:00:00     0 Timimoun US    today+5-y  web          0 2015-09-20
# ... with 1,295 more rows
```

Fig. 9 Interest by region for Timimoun (Rstudio output)

We have 1295 observations (1295 search operation in the last 5 years from the 5 regions). Rstudio shows us only the first ten rows. For better visualization, we plot out time series using always "ggplot2" package (Figs. 10 and 11).

```
timimoun_timeseries_plot <- Timimoun_timeseries %>%
  ggplot() +
  geom_line(aes(date, hits, color = geo), size = 0.1)
```

Fig. 10 Plot of interest by region (execution code)

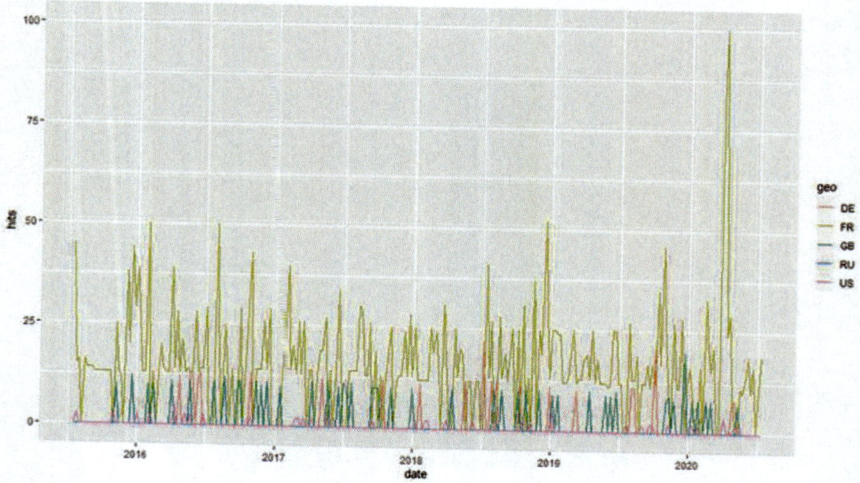

Fig. 11 Plot of interest by region time series (Rstudio output)

```
Timimoun_Fr_timeseries <- Timimoun_timeseries %>%
                          filter(Timimoun_timeseries$geo == "FR")%>%
                          select(date, hits)%>%
                          rename(ds = date, y = hits)%>%
                          arrange(ds)
```

Fig. 12 Necessary cleaning data (execution code)

From the previous graph, we note that the most important searches for Timimoun using Google were from French regions, which peaked in the year 2020.

We concentrate in the next steps on FR region. Therefore, we subset our time series using "Tidyverse" package, and we will analyze only FR data (Figs. 12 and 13).

We have 261 observations (261 search operation using Google in FR region in the last 5 years. The next chart shows us the plot of our data (Figs. 14 and 15).

Through the graph, we note the stability of searches for the city of Timimoun using Google during the past 5 years. The biggest jump in search numbers was the year 2020.

Using "prophet" package we build our model (forecast model). The first step is to use "prophet" function to make automatic statistical tests to our data and choose the best model (Fig. 16).

Taking into consideration the new model, we use "make_future_dataframe" function to create the new time series including the 365 days of predictions (Fig. 17).

```
> Timimoun_Fr_timeseries
# A tibble: 261 x 2
   ds                      y
   <dttm>              <int>
 1 2015-07-19 00:00:00    45
 2 2015-07-26 00:00:00    15
 3 2015-08-02 00:00:00    16
 4 2015-08-09 00:00:00     0
 5 2015-08-16 00:00:00    16
 6 2015-08-23 00:00:00    14
 7 2015-08-30 00:00:00    14
 8 2015-09-06 00:00:00    14
 9 2015-09-13 00:00:00    13
10 2015-09-20 00:00:00    13
# ... with 251 more rows
```

Fig. 13 Necessary cleaning output

```
> timimoun_fr_timeseries %>% ggplot() + geom_line(aes(Date, Hits), size = 1)
```

Fig. 14 Plot of interest in FR region (execution code)

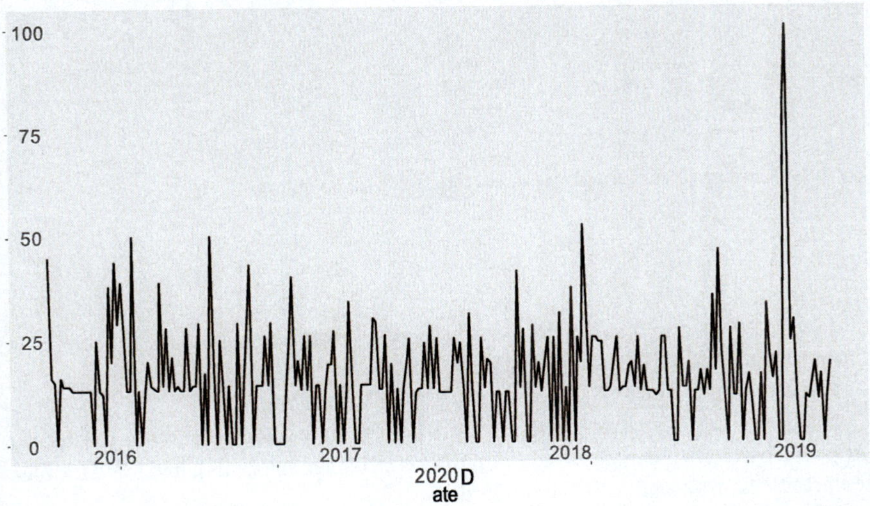

Fig. 15 Plot of interest in FR region (Rstudio output)

```
> timimoun_fr_model <- prophet(timimoun_fr_timeseries)
Disabling daily seasonality. Run prophet with daily.seasonality=TRUE to override this.
```

Fig. 16 Automatic statistical tests (execution code)

```
> timimoun_fr_present_future <- make_future_dataframe(timimoun_fr_model, periods = 365)
```

Fig. 17 Creation of the new time series including the 365 days of predictions

```
> timimoun_fr_predict <- timimoun_fr_predict(timimoun_fr_model,timimoun_fr_present_future)
```

Fig. 18 Make predictions (execution code)

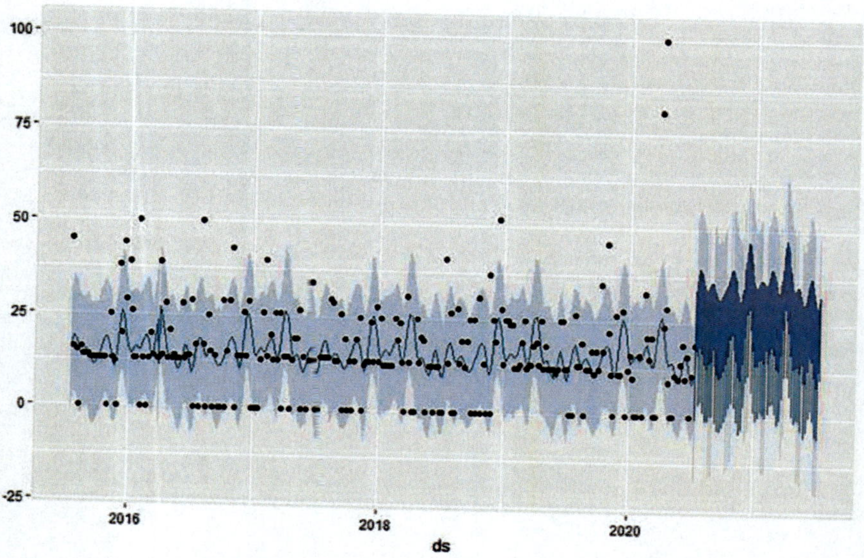

Fig. 19 Plot of original data and model and forecast (Rstudio output)

The next step is to make predictions based on the chosen model and the new time series using "predict" function (Fig. 18).

The last step is to plot the results using "plot" function (Fig. 19).

Black dots represent the original time series, light blue represents the model proposed by the Prophet package (including confidence intervals), and the dark blue represents the forecast of the next 365 days.

Forecasts show us that the interest will be stable in the next 365 days; this confirms the lack of interest of tourists from outside Algeria to visit the country in general and the desert regions in particular.

```
> prophet_plot_components (timimoun_fr_model, timimoun_fr_predict)
```

Fig. 20 The presentation of all the components: trend, weekly seasonality, and yearly seasonality (execution code)

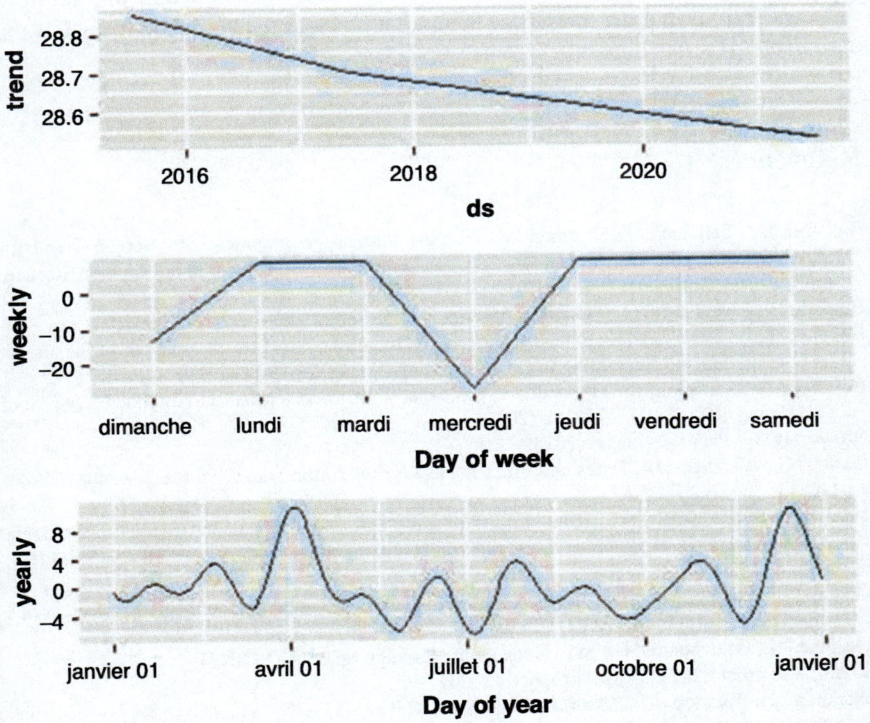

Fig. 21 The presentation of all the components: trend, weekly seasonality, and yearly seasonality

For more details about the model, we can present all its components: trend, weekly seasonality, and yearly seasonality using prophet_plot_components function (Fig. 20).

Through the general trend curve, we notice a decrease in interest in searching for Timimoun (Fig. 21).

4 Conclusion

Google Trend analysis allows researchers to study interest of millions of people around the world which the case in our study by analyzing interest of people from five regions about a famous Algerian tourist city of Timimoun. The most important interest about Timimoun is from France region, but it is still very weak in the next year, which confirms the weakness and lack of marketing and advertising of tourism in Algeria for tourists from outside the country.

In the other hand, we can say that the open-source programming language R and "gtrendsR" package is a good tool to extract information about people interest from different regions in order to analyze them to improve decision-making, which is the case in the Algerian tourism.

References

Ballatore, A., Scheider, S., Spierings, B.: Tracing tourism geographies with Google Trends: a Dutch case study. In: The Annual International Conference on Geographic Information Science, pp. 145–163. Springer, Cham (2019)

Beloudj, B.: Obstacles to investment in Algeria. N. Afr. Econ. J. **4**(1), 65–85 (2006)

Bokelmann, B., Lessmann, S.: Spurious patterns in Google Trends data – an analysis of the effects on tourism demand forecasting in Germany. Tour. Manag. **75**, 1–12 (2019)

Bourbaba, S., Maachou, L.: The reality and prospects of desert tourism in Algeria. J Legal Soc. Sci. Univ. Djelfa. **12**(1), 175–183 (2018)

Brownrigg, R.: Package 'maps' (2018)

Garrett, G., Wickham, H.: Dates and times made easy with lubridate. J. Stat. Softw. **40**(3), 1–25 (2011)

Nghiem, T.P., Papworth, S.K., Lim, F.K., Carrasco, L.R.: Analysis of the capacity of Google Trends to measure interest in conservation topics and the role of online news. PLoS One. **11**(3), e0152802 (2016)

Önder, I.: Forecasting tourism demand with Google trends: accuracy comparison of countries versus cities. Int J Tour. Res., 1–13 (2017)

Philippe, M., Eddelbuettel, D., Massicotte, M.P.: Package 'gtrendsR' (2016)

Taylor, S., Letham, B.: Package 'prophet' (2020)

Wickham, H., Averick, M., Bryan, J., Chang, W., McGowan, L.D., François, R., et al.: Welcome to the Tidyverse. J. Open Source Softw. **4**(43), 1686 (2019)

Zaarour, N., Sabti, W.: The tourist attraction in Ageria. Al-Hoggar J. Econ. Stud. **3**(1), 249–260 (2018)

Zayed, M.: Tourism investment opportunities in Algeria. Ijtihad J. Legal Econ. Stud. **7**(5), 65–89 (2018)

Deep Learning-Based Automated Learning Environment Using Smart Data to Improve Corporate Marketing, Business Strategies, Fraud Detection in Financial Services, and Financial Time Series Forecasting

Zair Bouzidi, Mourad Amad, and Abdelmalek Boudries

Abstract The automated learning environment is based on recurrent neural network trained with long short-term memory, used because of its ability to learn, in taking the concatenated function and time series data as input, while integrating encapsulations of several search results, using Smart Data (data streaming of objects connected and all the Web). We've used some of its ideas to improve corporate marketing, business strategy, and financial time series forecasting. This model is shown to be capable of learning to understand and localize different aspects of corporate marketing and business strategies. We put it through its paces on the challenging tasks of detecting fraud and forecasting financial time series, demonstrating that it is more reliable and uses less parameters and computation than other approaches.

Z. Bouzidi (✉)
LIMPAF Laboratory, Computer Science Department, Faculty of Sciences and Applied Sciences, University of Bouira, Bouïra, Algeria

LMA Laboratory, Science Commercial Department, Faculty of Economy, Bejaia University, Béjaïa, Algeria
e-mail: zair.bouzidi@univ-bejaia.dz

M. Amad
LIMPAF Laboratory, Computer Science Department, Faculty of Sciences and Applied Sciences, University of Bouira, Bouïra, Algeria
e-mail: amad.mourad@univ-bouira.dz

A. Boudries
LMA Laboratory, Science Commercial Department, Faculty of Economy, Bejaia University, Béjaïa, Algeria
e-mail: abdelmalek.boudries@univ-bejaia.dz

S. Sedkaoui et al. (eds.), *International Conference on Managing Business Through Web Analytics*, https://doi.org/10.1007/978-3-031-06971-0_26

Keywords Business · Fraud detection · Long short-term memory · Marketing · Recurrent neural network · Smart data · Times series forecasting

1 Introduction

We propose a new real-time deep learning-based automated learning environment. It is based on a recurrent neural network trained with long short-term memory (LSTM) that integrates encapsulations, using smart data (connected objects and all the Web). Thus, we provide, not only a solution to this challenge, but, also, we propose better performances.

In this paper, we're looking for information about financial time series that is relevant. As soon as this data is retrieved (distinguished from abusive information), it is used to improve marketing, business, fraud detection, and forecasting of financial time series. Mentioned below are our main contributions.

1. We develop a RNN-based model that uses low-level content learning capabilities to automatically separate relevant information from redundant or abusive.
2. We develop a RNN-based model that uses content learning capabilities of smart data (all the Web and connected objects) to automatically and efficiently capture real-time dynamics of financial data. Using a set of knowledge related to financial market, this model collects, using smart data, as per their lexical similarity, accurate prediction of volatility from financial time series.
3. We adapted some algorithms to streaming (smart data).
4. An event-independent model for filtering content on multiple sources at once is developed, taking into account the limitations of the previous work's model. Experiments on multiple financial market-related flows of contents with diverse characteristics demonstrate that our suggested model outperforms other approaches. While our approach filters content issued from all online channels from social media (all the Web) to connected objects.
5. Once this real-time RNN-based model implemented and manually annotated the first information deduced with multisource content learning capabilities to capture automatically and efficiently accurate real-time prediction of volatility from financial time series, we use a collection of financial market expertise and a set of tagged contents to obtain reliable future volatility predictions that are essential for investment planning, hedging fiscal risks, and determining government policy, etc.

The remainder of the paper is structured as follows: in Sect. 2, the background and related works are presented. We outline, in Sect. 3, our proposed model as well as details on forecasting financial time series volatility successfully. Preliminary results are followed by an examination of the findings. Finally, we conclude and make some suggestions for future researches.

2 Background and Related Works

2.1 *Marketing, Business, Fraud Detection, and Financial Time Series Forecasting*

The most important tool in retail is the organization of shelves on which items are displayed, notably with a wide variety items and shopping practices. Retailers will not only increase sales but also reduce costs, by properly managing shelf allocation and product display (Aloysius and Binu 2013). From learning models in the organization of supermarket shelves by grouping of products, generally bought together, we can extract the following relation: "the customers who buy product X at the end of the week, during the summer, in general also buy product Y."

Also, the credit institution making it possible to decide whether or not to grant credit according to the profile of the credit applicant, his request, and his past loan experiences is used in data mining. There is also overbooking (optimization of the number of seats on planes, hotels, etc.), the targeting of offers (organization of advertising campaigns, promotions), and the analysis of commercial practices, strategies, and their impact on sales in data mining. This awareness, initially unknown, may be associations, patterns, or general trends in this information. To verify the correctness of the system or the estimation of some parameters difficult for mathematical modeling, experimental data is needed. Data mining is a field that has emerged with the explosion in the amount of information stored, with major advances, particularly in processing speeds and storage media. The purpose of data mining is to uncover valuable information in large amounts of data that can help understand the data or predict the behavior of future data. Since its inception, data mining has used several statistical and artificial intelligence tools to achieve its goals. It is an essential component of big data technologies, big data analysis techniques, and recently intelligent data streaming. It is also defined as the process of discovering new knowledge.

Element set extraction is the model where certain sequence exploration problems lend themselves to the discovery of frequent sets of elements and their order. For example, we look for rules of the type "if a customer buys a car, he is likely to take out insurance within a week," or, in the context of stock prices, "If A and B are up, it is possible that C and D will be up within a week," according to market analysis. Traditionally, the extraction of a set of elements is used in marketing applications to detect patterns among competing elements, in large transactions, for example, by analyzing the shopping cart transactions of customers in a supermarket (Bouzidi et al. 2021a, c).

Sequential model extraction has many real-world applications because data in many fields such as bioinformatics like genomics and proteomics are encoded as sequences (Bouzidi et al. 2021a). We also see the development of the analysis of the consumption basket which consists in studying sales (Analysis of receipts) (Boghey & Singh 2013).

Table 1 Comparative table of all economic tasks used

Economic Tasks	References	Our new approach (ONA)
Detect novel frauds	Pumsirirat and Yan (2018), Schreyer et al. (2017), Dong et al. (2018), Gomez et al. (2018) and Fiore et al. (2019)	Our New Approach
Trading performance	Sermpinis et al. (2019)	
Exchange rate prediction	Kodogiannis and Lolis (2002)	
Stock prediction	Fischer and Krauss (2018)	
Trade on the stock market	Calvez and Cliff (2018)	
Company stock prices	dos Pinheiro and Dras (2017)	
Forecasting of financial time series	Bodyanskiy and Popov (2006), Lai (2006), Ghazali (2009), Pradeepkumar and Ravi (2017), Tk and Verner (2016), Lasfer (2013), Gudelek et al. (2017), Bao (2017) and Mohammad (2018)	

Web mining is the model where the World Wide Web is fertile ground for mining data research, due to the large amount of knowledge available online. Web mining research is at the crossroads of research carried out by many research communities, such as databases, data recovery, and artificial intelligence (AI), especially the sub-domains of learning and natural language processing (Kosala and Blockeel 2000).

Text mining is a data mining branch that specializes in enterprise word processing to analyze content and extract information. The most important tasks are to recognize and interpret the information in the document. These are language technologies that make it possible to switch from text to a digital vector (full text) (presence-absence or frequency).

String mining is the pattern for which channel mining usually deals with a limited alphabet of elements appearing in a sequence, usually very long. Those of the ASCII character set used in natural language text, nucleotide bases, may be examples of the alphabets “A,” “G,” “C,” and “T” in DNA sequences or amino acids for protein sequences and examining sequenced genes and proteins to determine their properties (Bouzidi et al. 2021a).

The latest sequential pattern mining pattern is all about discovering unexpected and useful patterns in datasets. Subsequences can be interesting based on a variety of factors, including how often they appear, how long they last, and how much money they make, among others (Table 1).

Detection of fraud is a significant subject in that it can be formulated in supervised and/or unsupervised category. While in the unsupervised learning group, class labels are either unknown or assumed to be unknown and clustering techniques are used to find out distinct clusters containing bogus samples or far away bogus samples not belonging to any cluster, it is treated as an outlier detection problem when all clusters contain genuine samples. Class labels are recognized in the group of supervised learning, and a binary classifier is designed in order to identify fraudulent samples. Detection of fraud (involving cyber fraud) is becoming increasingly threatening, and fraudsters often tend to be a few notches ahead of organizations in terms of discovering and easily circumventing new loopholes in the system. On the other hand, in order to anticipate fraud in near real time, if not real-time, organizations make immense investments in capital, time, and resources and try to minimize the effects of fraud. Financial fraud exists in numerous sectors, including finance, insurance, and investment (stock markets). It can be both offline and online as well. Online fraud involves credit/debit card fraud, purchase fraud, and security-related cyber fraud, while offline fraud involves accounting fraud, falsification, etc.

There has been an increase in intelligent automated trading and decision-support systems in financial markets, particularly stock markets, due to technological advancements and breakthroughs in deep learning models.

However, time series issues are difficult to predict, especially financial time series (Cavalcante et al. 2016). Despite the contrary study of the efficient market hypothesis, neural and deep learning models have shown great success in forecasting financial time series. The stock market follows a random walk (RW), and any profit made is the result of chance (Fama 1995). This is due to the ability of neural networks to adapt to any nonlinear dataset without making static assumptions or without prior knowledge of the dataset (Lu et al. 2009). Deep learning uses both basic and technical analytical data, the two most widely used techniques for forecasting financial time series, to train and develop models (Cavalcante et al. 2016).

Fundamental analysis is the process of forecasting market movements by using or exploiting textual information such as financial news, company financial results, and other economic factors such as government policies.

2.2 *Automated Learning*

It's a set of processes that lead to the acquisition of knowledge and skills. Although machine learning is a branch of artificial intelligence (AI), it deals with creating algorithms to perform complex tasks without programming it, making extensive use of AI tools and principles, mathematics, other cognitive sciences, etc. It can be based on statistical approaches to give the possibility of "learning" from data in two phases. The first, namely, model design phase (training) consists in estimating a model from observations (data). Knowledge and skills are acquired through a series of actions. For example, in handwriting recognition, two comparable characters are never exactly the same. In order to learn to identify characters by observing "examples," i.e., identified characters, a machine learning model can be created (Table 2).

Table 2 Comparative table of all AI concepts used including our approach

AI concepts	References	Our new approach (ONA)
Elman ANN (EANN)	Adhikari and Agrawal (2014) and Bouzidi et al. (2019)	ONA
Multilayer feedforward	Adhikari and Agrawal (2014) and Bouzidi et al. (2020)	
ConvNets/Autoencoder	Lai (2006), Lasfer (2013), dos Pinheiro and Dras (2017), Bao (2017) and Gudelek et al. (2017)	
RNN	Adhikari and Agrawal (2014), Mohammad (2018) and Sermpinis et al. (2019)	
LSTM	Adhikari and Agrawal (2014), Bouzidi et al. (2021b, e), Cavalcante (2016), Tk and Verner (2016), Pradeepkumar and Ravi (2017) and Gudelek et al. (2017)	
Memory Networks	Adhikari and Agrawal (2014), Cavalcante (2016), Tk and Verner (2016), Pradeepkumar and Ravi (2017) and Gudelek et al. (2017)	
Social Media	Adhikari and Agrawal (2014), Bouzidi et al. (2019, 2020, 2021b, e), Cavalcante (2016), Tk and Verner (2016), Pradeepkumar and Ravi (2017) and Gudelek et al. (2017)	
Analysis	dos Pinheiro and Dras (2017), Bao (2017), Fischer and Krauss (2018) and Mohammad (2018)	

Automated learning is used for a variety of purposes, such as aiding diagnosis (Bouzidi et al. 2021a), to identifying outliers, detecting missing data, and extracting relevant knowledge from various sources (social media) (Bouzidi et al. 2019, 2020), fraud detection and financial market analysis (Boghey & Singh 2013; Bouzidi et al. 2021b, e) etc. In order to avoid overusing networks of neurons and hidden layers, it's important to use an adequate number of layers (Bouzidi et al. 2021a, b, c, d, e), to detect and so avoid overfitting. Data are divided into two subsets (Bouzidi et al. 2021a, c, d): the learning set making it possible to change the weight of the neural network. The validation set makes it possible to check the relevance of the network, avoiding overfitting.

2.3 Social Networks and Smart Data

Networking is a substantial component of website users' online activities. Recent developments in using of networking reflect the fact that networking has not only

Table 3 Comparative table of all techniques and methods

References	Identification methods	Used OSN and smart data
Zaini et al. (2020)	Flood disaster game-based learning	Twitter
Vivakaran and Neelamalar (2018)	Educational purposes among the faculty of higher education with special reference	
He et al. (2016)	Summarization with social-temporal context	
Dussart et al. (2020)	Capitalizing on a TREC track to build a tweet summarization dataset	
Lamsal and Kumar (2020)	Semiautomated artificial intelligence-based classifier for disaster response	
Rudra et al. (2019)	Summarizing situational tweets in crisis scenarios: An extractive-abstractive approach	
Bouzidi et al. (2019)	Based on artificial neural network (ANN)	Twitter and Facebook
Bouzidi et al. (2020)	Based on feedforward neural network (FFNN)	All the Web
Bouzidi et al. (2021b, e)	Based on long short-term memory (LSTM)	All the Web & **Smart Data**
Our New approach	Marketing and business can benefit greatly from deep learning from social media and big data. Fraud detection and financial time series forecasting	

a growing number of users but also a substantial rise in the number of those applications (Table 3).

In a short time, social media have invaded the daily lives of users of Internet and Web professionals. The giants Twitter and Facebook were seen establishing, growing, and evolving. They have been followed by a number of other networks more specific: LinkedIn, Instagram, etc. The list is long. Among the existing studies, a group of studies identifies useful networking information, using neural learning, to successfully extract structured information from unstructured textual networking contents. People use networking to post situational updates in various forms such as text messages, images, and videos.

While the predominant function of networking remains social interaction, networking sites are also considered the fourth most popular source of information. The sharing of data can be facilitated by networking before, during and after an eventual event. With the proliferation of social media, knowledge is transformed from expert knowledge to everyday knowledge co-produced by various stakeholders thanks to Web 2.0. Studies on the use of social networking data to gain insight into human activity, including disease epidemic detection and stock market forecasting, have been increasing in recent years. However, understanding these voluminous and high velocity data is a difficult task.

Information retrieval models in general, and those from multiple sources, are the focus of this subsection.

Contents were collected from all online channels tracked automatically by the online listening tool, namely, Radian6 (Imran et al. 2020) or any its competitor, such as Awario, Brand24.com, Brandwatch, Mention, Keyhole, Socialert.net, SocialPilot.co, Simplify360, etc., from websites to social media, such as Twitter, Facebook, LinkedIn, Instagram, Google+, YouTube, and so on. According to the API (application programming interface) of many networking platforms, you can access their data (Imran et al. 2020). Using online listening tools, you can get a reasonable representation of the fundamentals, such as the gathering of information (such as that found in conversations on social media, the news, or anywhere else on the Internet), and removing information that has been dubbed, such as a retweet, and any information that is harmful or redundant from the data; thanks to neural learning and the tagged messages, relevant information can be obtained from the learning corpus. Volunteers create the tagged messages, which are then analyzed to make sure the data is accurate.

Our listening and monitoring strategy for online methodology reflection consists of the following components: cleaning up the data to remove duplicates and replications, enabling relevance through neural learning, and utilizing the learning corpus generated by tagging of messages in social media and elsewhere on the Web. For disaster managers, the emergency management model verifies and analyzes information in order to ensure that it is accurate.

Some of the advantages include finding out what the final discussion will be about, as well as foretelling and stopping an impending disaster from occurring. There is no commercial value in making decisions that could save lives, but there is a moral value in doing so.

Our proposed model's functional architecture can be seen in Fig. 5. When a message is properly cleaned, the artificial neural network-based automated learning environment extracts it from social media, based on predefined keywords, so that it is not redundant and does not re-record (insult). The message must be checked against the content manually annotated by volunteers in our laboratory to ensure it is free of duplicates and replications. These checks have found the message to be relevant. It will be sent to disaster managers as soon as possible, so that they can make quick and efficient decisions that could save or relieve lives.

Smart data, a different concept from big data, is based primarily on real-time data analysis. This term refers to a method of data analysis that analyzes data at the source directly, without the need to send it to a centralized system. Video streaming, however, requires connection to an Internet server.

2.4 Related Works

Much research exists for financial time series forecasts (Adhikari and Agrawal 2014; Bao et al. 2017; Bodyanskiy and Popov 2006; Bouzidi et al. 2021b, e; Fischer and Krauss 2018; Ghazali et al. 2009; Gudelek et al. 2017; Kodogiannis and Lolis 2002; Lai et al. 2006; Lasfer et al. 2013; Lu et al. 2009; Mohammad et al. 2018; dos Pinheiro and Dras 2017; Pradeepkumar and Ravi 2017). Research for fraud detection is common (Bouzidi et al. 2021b; Dong et al. 2018; Fiore et al. 2019; Gomez et al. 2018; Pumsirirat and Yan 2018; Schreyer et al. 2017; Wu et al. 2018). On the other hand, there are only two studies for the market price (Fama 1995; Calvez and Cliff 2018) and only one for placing products in a supermarket (Aloysius and Binu 2013).

3 Automated Learning Environment for Fraud Detection and Financial Time Series Prediction

3.1 Automated Learning Environment

Artificial intelligence confirms the combination of reinforcing of learning and deep learning (Abiodun et al. 2019), mathematically represented, as:

$$\mathrm{AI} = \mathrm{RL} + \mathrm{DL} \tag{1}$$

where

AI represents artificial intelligence, RL represents reinforcement of learning, and. DL represents deep learning.

A feedforward NN (FFNN) is an automated learning classification algorithm which is made up of organized in layers, similarly to human neuron processing units. Unlike FFNN, the feed-backward NN (FBNN) can use internal state "memory" to process sequence of data inputs, such as RNN.

3.2 Recurrent Neural Network (RNN)

As a class of feedforward NNs, recurrent neural networks (RNNs) are augmented by the inclusion of recurrent edges connecting adjacent time steps. Figure 1 shows the well-known RNN (Wu et al. 2018).

Definition
To describe this form of RNN, we can use two equations; all calculations required for computation at each phase of the moment on the forward pass are:

$$h_t = \alpha\ (\sigma_{hx}.x_t + \sigma_{hh}.h_{t-1} + b_h) \tag{2}$$

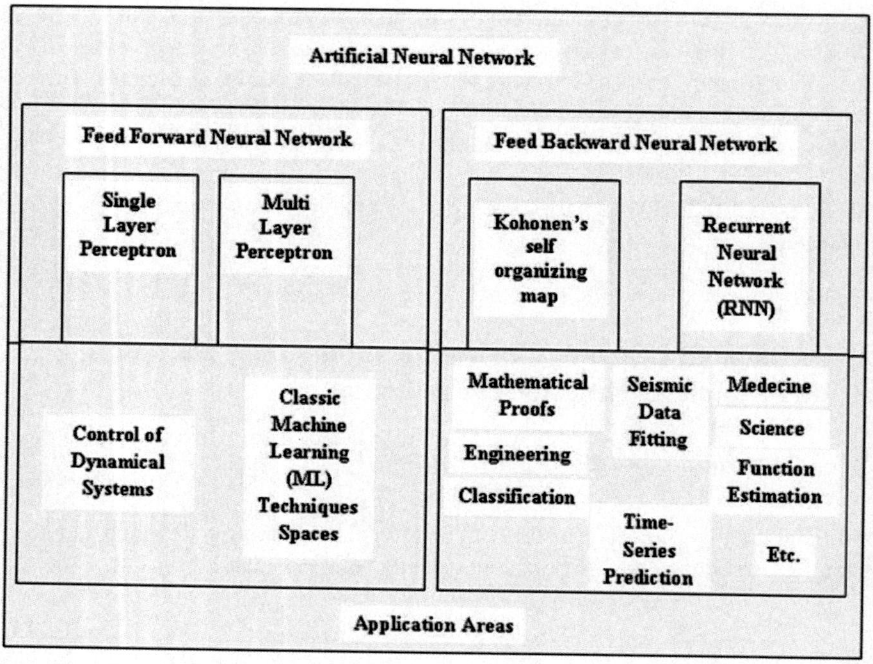

Fig. 1 Artificial NN classification with their application areas

$$y_t = \beta\left(\Phi_{yh}.h_t + b_y\right) \tag{3}$$

where

the σ terms denote weight matrices (e.g., σ_{hx} is the weight matrix between input and hidden layers).

The terms b denote vectors of bias (e.g., b_h is hidden bias vector) which allow each node to learn an offset. α denotes the hidden layer function at level t.

As a general rule, the output layer function is an element-wise application of the sigmoid function. Recurrent neural networks (RNNs) are neural networks that can be extended over time and have edges that feed into the next time step rather than the next layer of neurons. For example, an RNN can be used to identify text or speech sequences. It has periods during which the presence of short memory in the net indicates. RNN is like a hierarchical network where the input in the form of a tree needs to be hierarchically processed since there is no time to input sequence. RNNs are powerful sequence learning tools that can be adapted. To date, RNNs have proven to be excellent pattern recognition and prediction engines, especially in tasks involving machine learning of sequences, such as text or speech recognition. The recurring layer of RNNs contains feedback loops. This enables them to retain information for a longer period of time in their "memory" (Fig. 2).

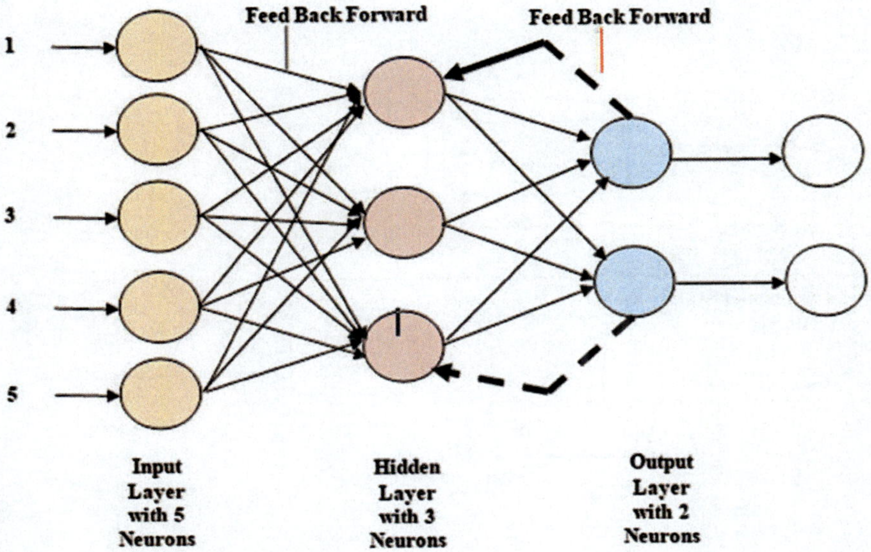

Fig. 2 Recurrent Neural Network (RNN) classification. (Abiodun et al. 2019)

Since each hidden state depends on the previous hidden states, RNN is intrinsically deep in time even though it is not deep in space (Wu et al. 2018). When propagating errors over several time stages, the problems of disappearing and bursting gradients can occur. A standard RNN, on the other hand, may not be able to solve problems that require long-term learning. To put it another way, this is because the loss function gradient degrades exponentially over time (this is called the gradient vanishing problem). During this time, they only have hidden states and the hidden states work like RNN memory. An RNN structure can be shown in Fig. 3. The arrows in Fig. 2 illustrate the relationships between FFNN components. RNN has achieved remarkable success in sequential learning problems. The success achieved in terms of application to public relations includes the accuracy of speech recognition and timing.

3.3 *Long Short-Term Memory (LSTM)*

Long short-term memory, one of the most promising RNN architectures for sequence learning, was suggested to address the lack of RNN. It introduces the memory cell, a computing unit replacing traditional artificial neurons in the secret layer, compared to the RNN. It is a kind of RNN that uses specific units with standard units. A memory cell is a part of LSTM units capable of keeping data in memory for a long time. It is often referred to as sophisticated RNN.

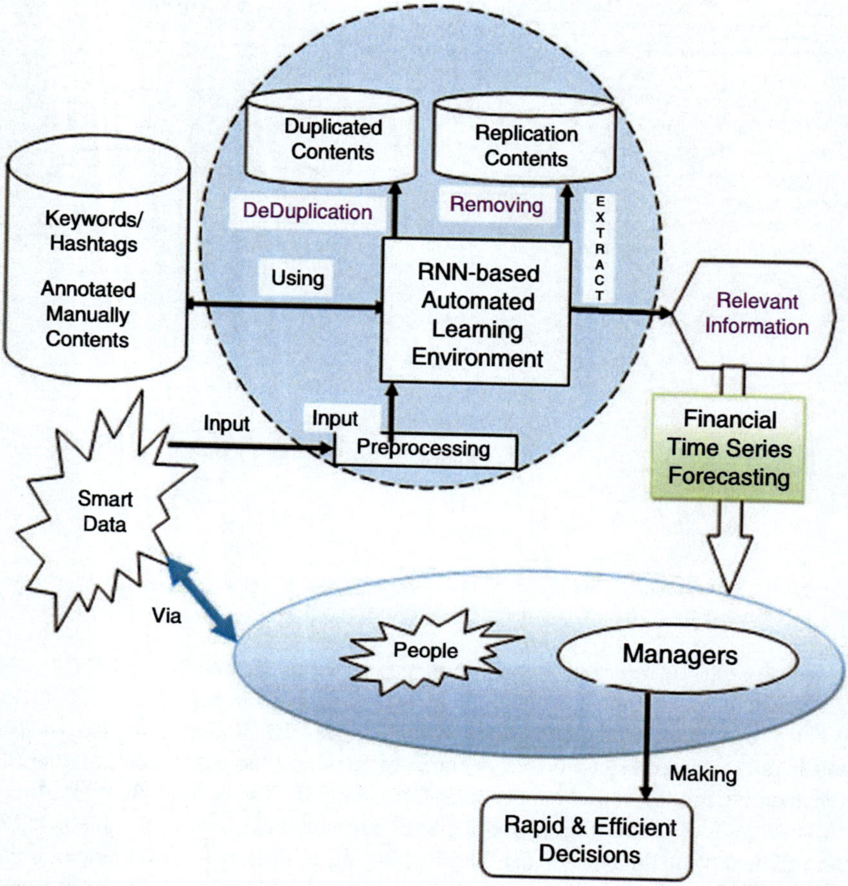

Fig. 3 RNN-based automated learning environment to retrieve relevant information

Definition

Mathematically, it is possible to formulate the LSTM model computation as follows:

$$i_t = \Omega\ (\Phi_{ix}.x_t + \Phi_{ih}.h_{t-1} + b_i) \tag{4}$$

$$f_t = \Omega\ \left(\Phi_{fx}.x_t + \Phi_{fh}.h_{t-1} + b_f\right) \tag{5}$$

$$c_t = f_t \odot\ c_{t-1} + i_t \odot \tanh(\Phi_{cx}\ .\ x_t + \Phi_{ch}\ .\ h_{t-1} + b_c\ ) \tag{6}$$

$$\beta_t = \Omega\ \Big(\Phi_{\beta x}.x_t + \Phi_{\beta h}.h_{t-1} + b_\beta \tag{7}$$

$$h_t = \beta_t \Theta \tanh(c_t) \quad (8)$$

where

Θ denotes element-wise multiplication.

Φ is the logistic sigmoid function.

i, f, β, and c are, respectively, the input gate, forget gate, output gate, and cell activation vectors, all of which are in the same size as the hidden vector h at level t.

3.4 RNN-Based Automated Learning Environment

Artificial NNs (ANN) are inspired by the human brain and biological neural networks. These systems learn to perform tasks by considering examples of task-specific rules, usually without programming. Neural network is based on collection of connected units/nodes called neurons, which model biological brain neurons. Each connection, like the synapses of a biological brain, can transmit a signal from artificial neuron to another.

In order to get the best possible outcome, it's not just about a set of algorithms but also the following steps:

- Where the success of a project is based on collecting relevant and sufficient data, the term "data acquisition" is used to describe the process.
- The data preparation and cleaning.
- Model creation.
- Model evaluation: it is composed of evaluating trained model on the other (second) set of data.
- Validation/deployment: For example, if new input data is available, it could be used to further refine and improve the model in the field.

  To avoid overlearning, it is important to use a sufficient number of neurons and hidden layers. Consequently, there are two subsets of the data:
- Learning set: allows changing the weight of the NN and.
- Validation set: to verify the relevance of the network.

We can reinitiate learning by shrinking the network. A method of regularization is used to prevent overlearning. To prevent a neural network from overlearning, regularization techniques like the weight decay method are employed.

Researchers have developed neural networks to model the behavior of the human brain. But if you really think about it, normal neural networks don't quite do their original intention justice. The reason for this claim is that feedforward neural networks cannot remember the things they learn. Every iteration you train the network it starts over; it doesn't remember what it saw in the previous iteration when you process the current dataset. This is a big drawback when identifying correlations and data patterns. To solve this problem, we turned to recurrent neural networks (RNNs). Since they can store data in short-term units of memory (called "hidden state"), RNNs are uniquely equipped to model short-term dependencies. For this reason, RNNs are widely used in time series forecasts to identify correlations and data patterns.

3.4.1 RNN: Modeling

An investigation of the entire ANN modeling process, including the collection of learning data, pre- and post-processing of data, activation functions, initialization of weights, algorithms of learning, and error functions, has been undertaken to implement systematic methods that often lead to efficient ANN models. While all of these factors influence the efficiency of ANN, greater attention has been paid to finding the best architecture. There is no theoretical basis for how this architecture will be discovered or how it will appear. A repeated trial and error procedure is the most common approach adopted, in which a large number of different architectures are tested and compared to each other. This method is also time-consuming and relies primarily on the human expert's expertise and intuition, which implies a high degree of uncertainty. As a matter of fact, we list a number of different approaches, including empirical or statistical methods, hybrid methods such as fuzzy inference, constructive and/or pruning algorithms, and evolutionary strategies, which all have their own advantages and disadvantages. The training data comes from the Algiers Floods of November 10, 2001, and the Boumerdes earthquake of May 21, 2003. Volunteers manually annotate this data, which can be gleaned from neural networks.

We use a neural network with a hidden layer, taking as input and gives as output. The input to the network is a content e, as:

$$e = (w_1, \ldots, w_i, \ldots, w_n) \tag{9}$$

containing words w each coming from a finite vocabulary; $\mathbf{Y}$ is the set of contents issued from the social media.

Let

$$e_i \in \check{C}_{n=E}, \text{ with } i \in [1, N] \text{ and } e_i = (w_{i1}, w_{i2}, \ldots, w_{in}) \tag{10}$$

containing words each coming from the set of words $\hat{W}$ where each word comes from a finite vocabulary $\mathbf{Y}$, the incorporation of a content of the source message i relevant for, at least, a keyword or a hashtag such as:

$$\exists j \in [1, M] / \quad h_j \in \mathbf{H} \tag{11}$$

We want the learning of a generic space with the neural network, as:

$$\widetilde{E} = \max\{\, e_k \,\} \tag{12}$$

$$\boldsymbol{k} \in [\mathbf{1}, \boldsymbol{K}]$$

normalizing the differences:

$$\widetilde{E} = [E - \mathbf{RDF}] \text{ where } \mathbf{RDF} = [\boldsymbol{R} + \boldsymbol{D} + \boldsymbol{F}] \tag{13}$$

Thanks to the RNN-based model, the transformation of $\{e_i\}$ into $\{e_k\}$ can be explained by:

$$\exists \boldsymbol{j}, \boldsymbol{l} \in [\mathbf{1}, \boldsymbol{M}] / \boldsymbol{Hj} \in \boldsymbol{H}, \boldsymbol{wj} \in \mathbf{W}$$

$$\max\left\{e_i \rightarrow e_k = \left\{e_i/e_i \text{ is relevant for } h_j \text{ and } w_i\right\} \text{with } i \in [1, N] \right\} \tag{14}$$

$$\boldsymbol{k} \in [\mathbf{1}, \boldsymbol{K}]$$

with

$$e_i \notin [\boldsymbol{R} + \boldsymbol{D} + \boldsymbol{F}]$$

where

$\boldsymbol{D}$, $\boldsymbol{R}$, and $\boldsymbol{F}$ denote the number of retweets, tweets, and alerts that were sent out more than once.

The goal is to maximize the size K of the set $\boldsymbol{E_K}$. Figure 4 shows the RNN-based emergency management architecture. Table 4 presents the set of features for the content extraction task. Using keywords, Algorithm 1 identifies information that must be manually annotated to enhance neural network learning. Following passages, Algorithm 2 demonstrates how this emergency management model can learn the relevant information to inform public opinion and in particular disaster managers so that they can make quick and effective decisions which could save lives.

Functioning of recurrent neural learning: Fig. 4 shows the functioning the RNN-based automated learning environment to retrieve relevant information.

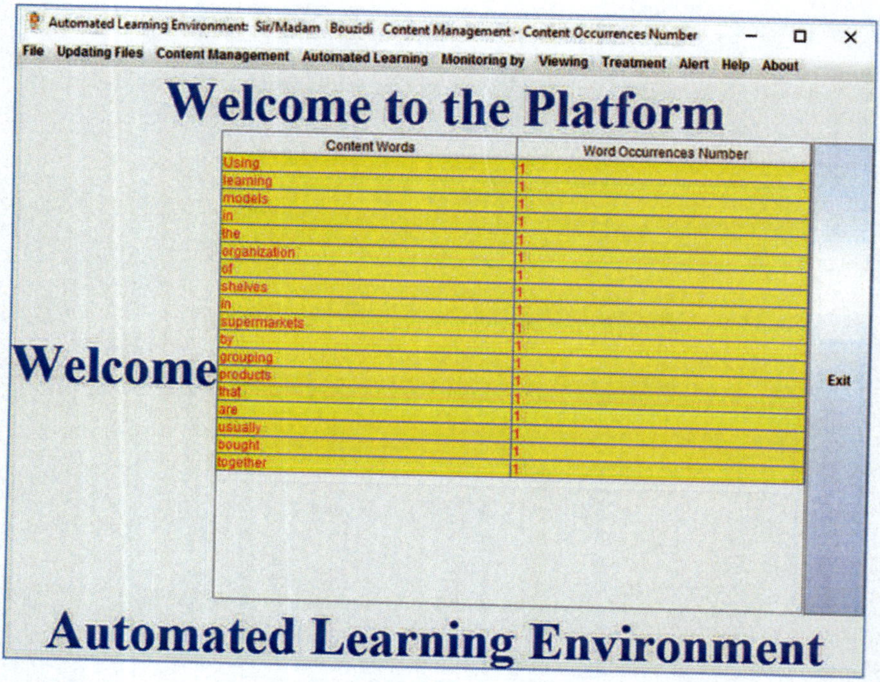

Fig. 4 Example of content occurrences number in the automated learning environment

Table 4 Comparative table of all techniques and methods with big data used in models including ours

Models	Identification usage of big data
Ofli et al. (2016)	Human and machine intelligence to understand big data
Horita et al. (2017)	Decision-making and emerging big data
Immomen et al. (2015)	Quality of big data's social media data
Smith et al. (2012)	In public social media, big data privacy is at risk
Our Previous approaches Bouzidi et al. (2021b, e)	Smart data privacy in public social media
Our new approach	Automation and big data can help in many areas, including marketing, business strategy, fraud detection, and financial time series forecasting, to name a few

3.4.2 Sequential Pattern Mining: Implementation of Algorithms

The medium tests the frequency of a pattern: the higher it is, the more regular the pattern. Frequent patterns from infrequent patterns can be distinguished using a threshold (Boghey & Singh 2013). Streaming is a method of playing content commonly used on the Internet. It is opposed to file downloading which requires retrieving all data from a file. Video streaming, however, requires connection to an Internet server.

3.4.3 Dynamic Counting Algorithm

The DIC algorithm (Bouzidi et al. 2020), to reduce the number of database executions, is suitable for streaming. In order to process the content blocks, the DIC is therefore suitable for partitioning the database into transaction blocks.

As the number of scans in the database decreases, a single pass of material occurs, while DIC simultaneously considers candidate item sets of different sizes. This poses the problem of storing candidate item sets processed simultaneously and the cost of measuring candidate media which is higher than for the Apriori algorithm (Agrawal et al. 1996).

The Apriori algorithm, requiring N passages on the database, is a possible optimization consists in generating in memory, during the first passage, the identifiers (TID) of the transactions. For streaming, this algorithm is also suitable. The TID lists corresponding to the K content of each k-item package are kept. After determining the frequent 1-item sets, we generate a list of TIDs, which is more efficient in streaming. This removes infrequent items when playing the first material and thus reduces the TID lists in memory. Generate TID lists in parallel memory as soon as the first content becomes more efficient.

3.4.4 The AprioriTID Algorithm

The streaming-friendly AprioriTID algorithm (Agrawal et al. 1996) is shown in Algorithm 1. One of the goals of optimizations in this algorithm is to facilitate streaming with many calculations and storage in a single read. The threshold is set by the analyst. This can be accompanied by an iterative method by setting a threshold at the start and adjusting the value of the threshold according to the result. The algorithm proposed by Savasere (Bouzidi et al. 2020) solves the memory space problem of the previous algorithm. As an advantage of the algorithm, it only requires one reading at most (Fig. 5).

3.5 Discussion About Sequential Pattern Mining and Dynamic Counting Algorithm

Figures 6 and Algorithm 1 show examples of content occurrences number in the automated learning environment. Algorithm 2 shows an example of content size in the automated learning environment (Algorithm 3).

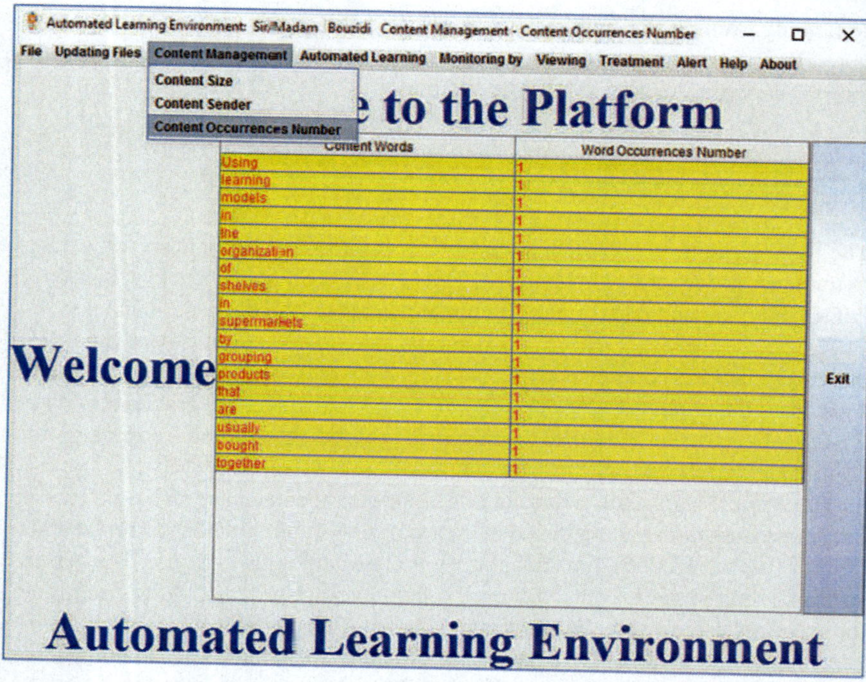

Fig. 5 Example of content occurrences number in the automated learning environment

3.6 *Experimental Process*

In this study, we chose a pool of 6900 messages from thousands of pieces of data. The 6900 messages were chosen on the basis of consumption. Some consumers were used to frequenting this place already, while others had only limited data available. We selected data from June (January 01, 2020) until the end of June (June 30, 2020), when all data was available. We excluded the seasonal impact from our analysis. We divided the data into three categories: training, validation, and testing, with each category receiving 60% of the total.

The developed model was tested with different configurations of various analysis time steps. In addition, the model was evaluated for different look ahead/forward time steps.

3.7 *Experimental Results*

In this study, two different models, including this model were established and compared based on the input sample. The details of input sample and models are shown in Table 5. The input sample is captured by RNN-based automated learning

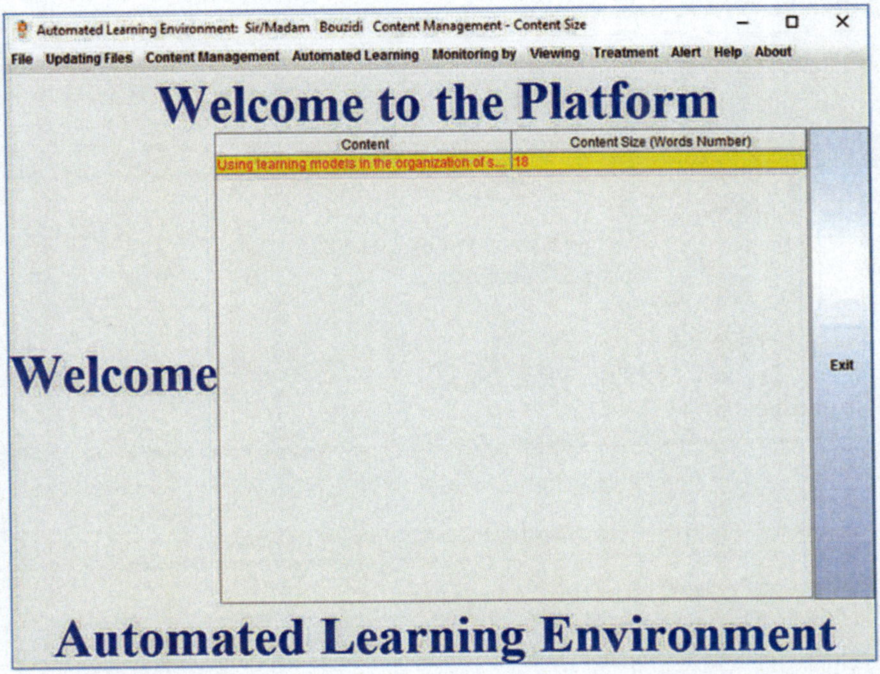

Fig. 6 Example of content size in the automated learning environment

Algorithm 1 Determining relevant information with annotated information

begin
Require: V: Content of Social Networks; σ : Minimum Support Threshold;
Ensure: Set of frequent items;
input (a content $V(w_1, w_2, \ldots, w_i, \ldots, w_n)$) and σ
initialization $i \leftarrow 1$
initialization $C_i \leftarrow$ 1 set of size 1 patterns (one item)
while ($C_i \neq \emptyset$) **do**
    Calculate the Support of each pattern $m \in C_i$ in the content set
    $F_i \leftarrow \{m \in C_i\}$ support $(m) \geq \sigma\}$
    $C_{i+1} \leftarrow$ {all possible combinations of F_i patterns of size $i + 1$}
    incrementation $i \leftarrow i + 1$
endwhile
output $\bigcup (i \geq 1) F_i$
end

Algorithm 2 Pseudo codes for one-step performance prediction

```
Require: Input feature series S = {S_t}, t = 1, 2,..., Ts
function OneStepPrediction(S)
        Initialize h_0 = 0, f = 0, (m_1, m_2) = (1, 0)
        Initialize} S_0 = (f, m_1, m_2), t = 0
        while t < Ts do
                Generate h_{t+1} and y_{t+1} by h_t and S^t
                Sample S^{t+1} using y_{t+1}
                t = t + 1
        end while
        return S = {S^t}, t = 1, 2,..., T_s
end function
```

Algorithm 3 Pseudo codes for present-to-the-end performance prediction

```
functionPresentToTheEndPrediction(S)
Initialize h_0 = 0, f = 0, (m_1, m_2) = (1, 0)
Initialize S0 = (f, m_1, m_2), t = 0
        while (m_1, m_2) ≠ (0, 1) do
                Generate h_{t+1} and y_{t+1} by h_t and St
                Sample S^{t+1} using y_{t+1}
                t = t + 1
                if t > tp then
                        S_t = S^t
                end if
        end while
Ts = t
return S = {S^t}, t = 1, 2,..., Ts
end function
```

environment. The results are evaluated with the RMSE, MSE, and R2 calculated with the formulas (15, 16, and 17). Final results are shown in Table 5.

This section presents the experiments carried out to compare the performance of models, including our proposed LSTM model, tested with the dataset, introduced in the following subsection, which have been preprocessed. The mean squared error (RMSE), the mean absolute error (MAE), and the (R-Square) were the measures used to assess model performance across all experiments.

Table 5 Examples of relevant content for a hashtags and keywords set from social media and smart data

Model	RMSE	MAE	R^2
Neural Network (Bouzidi et al. 2019)	17,088.3797	18,471	0.3284
Feedforward NN (Bouzidi et al. 2020)	16,100.9272	19,461.5	0.4645
Our new approach	13,359.4722	19,962.5	0.4805

$$\mathbf{RMSE} = \frac{\mathbf{1}}{\mathbf{N}} * \sum_{\mathbf{i=1}}^{\mathbf{N}} \sqrt{(\mathbf{y_i} - \mathbf{y_i^*})^2} \tag{15}$$

$$\mathbf{MSE} = \frac{\mathbf{1}}{N} * \sum_{\mathbf{i=1}}^{\mathbf{N}} | \mathbf{y_i} - \mathbf{y_i^*} | \tag{16}$$

$$R^2 = \frac{1}{N} * \frac{\sum_{i=1}^{N} (y_i - y_i^*)^2}{\sum_{i=1}^{N} \left(y_i - \hat{y_i}\right)^2} \tag{17}$$

where

N represents the number of content flow, y_i is the real content in flow i, and y_i^* is the relevant content flow. $\hat{y_i}$ is the mean value of the relevant content number.

3.7.1 Data Description

We partitioned the data (6900 messages) into training, validation, and test data at 60%, 20%, and 20%, respectively. We have divided the datasets into a training set and a verification set. The learning set is applied to train different deep learning models, while updating the weights and bias of the neural cell. The validation set is applied to validate the models. The verification set checks the skill of these models.

3.7.2 Results

RNN is an important part our framework and provides vector characteristics based on historical information. The final experimental results are presented in Table 5.

In this section, we have checked the effectiveness of the proposed RNN model against the benchmarks: the feedforward prediction method is the widely used deep learning model. In the experiment, the deep learning model must learn (finding best hyper-parameters), including find the number of neurons, the number of layers of neural networks, and the activation function of the neural network. After a complete experiment, we obtained the final configuration results of this model through the evaluation of the verification set.

To be fair, the number of relevant contents is taken as the historical information for NN, FFNN, and our new approach. Further, from the RMSE and MAE, it is obviously that our new approach is more accurate than other models since combining the advantages of both. This result indicates that our model is more suitable to retrieve relevant content than the Neural network and the feedforward neural network models.

4 Conclusion and Perspectives

A new ad hoc real-time deep learning-based automated learning environment with accurate forecasting of volatility from financial time series is presented here. As a result of using smart data, it utilizes a brand new multi-view recovery model. Financial decision-making can be greatly aided by using this type of strategy, as well as strategic decision-making. The following are some of the limitations of this work:

1. Social media content has been described as informal, nonsensical, and chaotic, with possible spelling mistakes, abbreviations, and other errors.
2. As a result, domain-specific biases may exist in the dataset.
3. Side by side, data of connected objects may contain different types of reasons in relation to social media content.
4. When analyzing specific content, the features of the automated learning environment have been developed.

There are a slew of ways this research could be used in the future. In order to complete the model, additional research questions and viewpoints should be considered. The following are good places to start when making pure improvements:

1. Before releasing this new information, the information should be better validated to prevent errors in accurate volatility forecasting from financial time series, which can be caused by information that is being abused.
2. Using multiple sources (connected objects and social media) can have other improvement ideas.

References

Abiodun, O., Jantan, A., Omolara, O., Dada, K., Umar, A., Linus, O., Arshad, H., Aminu Kazaure, A., Gana, U., Kiru, M.: Comprehensive review of artificial neural network applications to pattern recognition. IEEE Access. (2019). https://doi.org/10.1109/ACCESS.2019.2945545

Adhikari, R., Agrawal, R.K.: A combination of artificial neural network and random walk models for financial time series forecasting. Neural Comput. & Applic. **24**, 1441–1449 (2014). https://doi.org/10.1007/s00521-013-1386-y

Agrawal, R., Mehta, M., Shafer, J., Srikant, R., Arning, A., Bollinger, T.: The quest data mining system. In: Proceedings of the Second International Conference on Knowledge Discovery and Data Mining KDD'96, pp. 244–249 (1996). https://doi.org/10.5555/3001460.3001511

Aloysius, G., Binu, D.: An approach to products placement in supermarkets using Prefix Span algorithm. J. King Saud Univ. Comput. Inf. Sci. **25**(1), 77–87 (2013). https://doi.org/10.1016/j.jksuci.2012.07.001

Bao, W., Yue, J., Rao, Y.: A deep learning framework for financial time series using stacked autoencoders and long-short term memory. PLoS One. **12**(7), e0180944 (2017)

Boghey, R., Singh, S.: Sequential pattern mining: a survey on approaches. Proceedings - 2013 International Conference on Communication Systems and Network Technologies, CSNT 2013, 670–674 (2013a). https://doi.org/10.1109/CSNT.2013.142

Bodyanskiy, Y., Popov, S.: Neural network approach to forecasting of quasiperiodic financial time series. Eur. J. Oper. Res. **175**(3), 1357–1366 (2006). https://doi.org/10.1016/j.ejor.2005.02.012

Bouzidi, Z., Amad, M., Boudries, A.: Intelligent and real-time alert model for disaster management based on information retrieval from multiple sources. Int. J. Adv. Media Commun. **1-26** (2019). https://doi.org/10.1145/253260.253325

Bouzidi, Z., Boudries, A., Amad, M.: Towards a smart Interface-based automated learning environment through social Media for Disaster Management and Smart Disaster Education. In: Advances in Intelligent Systems and Computing. SAI 2020, vol. 1228, pp. 443–468. Springer, Cham (2020). https://doi.org/10.1007/978-3-030-52249-0_31

Bouzidi, Z., Boudries, A., Amad, M.: Chapter 2: Deep learning and social media for managing disaster: Survey. In: Intelligent Systems Conference 2–3 September, 2021, (IntelliSys 2021), Amsterdam, The Netherlands, Volume 294, Intelligent Systems and Applications, IntelliSys 2021 LNNS 294, vol. 1. Springer Nature (2021a). https://doi.org/10.1007/978-3-030-82193-7_2

Bouzidi, Z., Boudries, A., Amad, M.: LSTM-based automated learning with smart data to improve marketing fraud detection and financial forecasting. In: 5th International Scientific Conference on Economics and Management, (EMAN 2021), Serbia (2021b). https://doi.org/10.31410/EMAN.2021.191

Bouzidi, Z., Boudries, A., Amad, M.: Enhancing crisis management because of deep learning, big data and parallel computing environment: survey, ID:443. In: Proceedings of the 3rd International Conference on Electrical, Communication and Computer Engineering (ICECCE), 12–13 June 2021, Kuala Lumpur, Malaysia (2021c). https://doi.org/10.1109/ICECCE52056.2021.9514189

Bouzidi, Z., Amad, M., Boudries, A.: A Survey on Deep Learning in Big Data and its Applications, International Conference on Innovations in Energy Engineering & Cleaner Production IEECP'21, ID:124. Silicon Valley (2021d). https://doi.org/10.6084/m9.figshare.14737953

Bouzidi, Z., Boudries, A., Amad, M.: Deep LSTM-based model using smart data to improve awareness and education in marketing, business strategy and financial forecasting. MC Medical Sciences (MCMS). **1**(5) December 2021 (2021e). https://doi.org/10.55162/MCMS.2021.01.034

Calvez, A., Cliff, D.: Deep learning can replicate adaptive traders in a limit-order-book financial market. In: 2018 IEEE Symposium Series on Computational Intelligence (SSCI), pp. 1876–1883 (2018)

Cavalcante, R.C., Brasileiro, R.C., Souza, V.L., Nobrega, J.P., Oliveira, A.L.: Computational intelligence and financial markets: a survey and future directions. Expert Syst. Appl. **55**, 194–211 (2016)

Dong, M., Yao, L., Wang, X., Benatallah, B., Huang, C., Ning, X.: Opinion fraud detection via neural autoencoder decision forest. Pattern Recogn. Lett. (2018) http://www.sciencedirect.com/science/article/pii/S0167865518303039

Dussart, A., Pinel-Sauvagnat, K., Hubert, G.: Capitalizing on a TREC track to build a tweet summarization dataset. In: Text REtrieval Conference (2020) (TREC' 2020)

dos Pinheiro, L.S., Dras, M.: Stock market prediction with deep learning: a character-based neural language model for event based trading. In: Proceedings of the Australasian Language Technology Association Workshop 2017, pp. 6–15 (2017)
Fama, E.F.: Random walks in stock market prices. Financ. Anal. J. **51**(1), 75–80 (1995)
Fiore, U., Santis, A.D., Perla, F., Zanetti, P., Palmieri, F.: Using generative adversarial networks for improving classification effectiveness in credit card fraud detection. Inf. Sci. **479**, 448–455 (2019) http://www.sciencedirect.com/science/article/pii/S0020025517311519
Fischer, T., Krauss, C.: Deep learning with long short-term memory networks for financial market predictions. Eur. J. Oper. Res. **270**(2), 654–669 (2018)
Ghazali, R., Hussain, A.J., Nawi, N.M., Mohamad, B.: Non-stationary and stationary prediction of financial time series using dynamic ridge polynomial neural network. Neurocomputing. **72**(10–12), 2359–2367 (2009). https://doi.org/10.1016/j.neucom.2008.12.005
Gomez, J.A., Arvalo, J., Paredes, R., Nin, J.: End-to-end neural network architecture for fraud scoring in card payments. Pattern Recogn. Lett. **105**, 175–181 (2018). Machine Learning and Applications in Artificial Intelligence. http://www.sciencedirect.com/science/article/pii/S016786551730291X
Gudelek, M.U., Boluk, S.A., Ozbayoglu, A.M.: A deep learning based stock trading model with 2-d cnn trend detection. In: Computational Intelligence (SSCI), 2017 IEEE Symposium Series on, pp. 1–8. IEEE (2017)
He, R., Liu, Y., Yu, G., Tang, J., Hu, Q. & Dang, J. (2016). Twitter summarization with social-temporal context. In World Wide Web, 20(2), pp. 267–290, DOI: https://doi.org/10.1007/s11280-016-0386-0
Horita, F.E.A., De Albuquerque, J.P., Marchezini, V., Mendiondo, E.M.: Bridging the gap between decision-making and emerging big data sources: an application of a model-based framework to disaster management in Brazil. Decis. Support. Syst. **97**, 2–22 (2017). https://doi.org/10.1016/j.dss.2017.03.001
Immomen, A., Paakkonen, P., Ovaska, E.: Evaluating the quality of social media data in big data architecture. IEEE Access. **3**, 2028–2043 (2015). https://doi.org/10.1109/ACCESS.2015.2490723
Imran, M., Ofli, F., Caragea, D., Torralba, A.: Using AI and social media multimodal content for disaster response and management: opportunities, challenges, and future directions. Inf. Process. Manag. **57**(5), 1–9 (2020). https://doi.org/10.1016/j.ipm.2020.102261
Kodogiannis, V., Lolis, A.: Forecasting financial time series using neural network and fuzzy system-based techniques. Neural Comput. & Applic. **11**, 90–102 (2002). https://doi.org/10.1007/s005210200021
Kosala, R., Blockeel, H.: Web mining research: a survey. ACM SIGKDD explorations newsletter 2, 1 (2000). Proc. ACM Meas. Anal. Comput. Syst. **37**(4), Article 111 (2000)
Lai, K.K., Yu, L., Wang, S., Huang, W.: Hybridizing exponential smoothing and neural network for financial time series predication. In: Proceedings of Computational Science – ICCS 2006, Lecture Notes in Computer Science, ICCS 2006, vol. 3994, (2006). https://doi.org/10.1007/11758549_69
Lamsal, R., Kumar, T.V.V.: Classifying emergency tweets for disaster response. In*International Journal of Disaster Response and Emergency Management (IJDREM)*. **3**(1), 14–29 (2020). https://doi.org/10.4018/IJDREM.2020010102
Lasfer, A., El-Baz, H., Zualkernan, I.: Neural network design parameters for forecasting financial time series. In: Modeling, Simulation and Applied Optimization (ICMSAO), 5th International Conference on, pp. 1–4. IEEE (2013)
Lu, C.-J., Lee, T.-S., Chiu, C.-C.: Financial time series forecasting using independent component analysis and support vector regression. Decis. Support. Syst. **47**(2), 115–125 (2009) http://www.sciencedirect.com/science/article/pii/S0167923609000323
Mohammad, A.H., Rezaul, K., Ruppa, T., Neil, D.B.B., Yang, W.: Hybrid deep learning model for stock price prediction. In: IEEE Symposium Symposium Series on Computational Intelligence SSCI, pp. 1837–1844. IEEE (2018)

Ofli, F., Meier, P., Imran, M., Castillo, C., Tuia, D., Rey, N., Briant, J., Millet, P., Reinhard, F., Parkan, M., Joost, S.: Combining human computing and machine learning to make sense of big (aerial) data for disaster response. Big Data. **4**(1) (2016)

Pradeepkumar, D., Ravi, V.: Forecasting financial time series volatility using particle swarm optimization trained quantile regression neural network. Appl. Soft Comput. **58**, 35–52 (2017). https://doi.org/10.1016/j.asoc.2017.04.014

Pumsirirat, A., Yan, L.: Credit card fraud detection using deep learning based on auto-encoder and restricted boltzmann machine. Int. J. Adv. Comput. Sci. Appl. **9**(1), 8–25 (2018)

Rudra, K., Goyal, P., Ganguly, N., Imran, M. & Mitra, P. (2019). Summarizing situational tweets in crisis scenarios : an extractive-abstractive approach. In IEEE Transactions on Computational Social Systems, 6(5), pp. 981–993, DOI: https://doi.org/10.1109/tcss.2019.2937899

Sermpinis, G., Karathanasopoulos, A., Rosillo, R. and Fuente, D.(de la), (2019), Neural networks in financial trading, Ann. Oper. Res., doi: https://doi.org/10.1007/s10479-019-03144-y

Smith, M., Henne, B., Szongott, C. and Voigt, G. von, (2012), Big Data Privacy Issues in Public Social Media,6th IEEE International Conference on Digital Ecosystems Technologies, pp. 1–6,Doi:https://doi.org/10.1109/DEST.2012.6227909

Schreyer, M., Sattarov, T., Borth, D., Dengel, A., Reimer, B.: Detection of anomalies in large scale accounting data using deep autoencoder networks. CoRR, abs/1709.05254 (2017). http://arxiv.org/abs/1709.05254

Tk, M., Verner, R.: Artificial neural networks in business: two decades of research. Appl. Soft Comput. **38**, 788–804 (2016) http://www.sciencedirect.com/science/article/pii/S1568494615006122

Vivakaran, M. V. & Neelamalar, M. (2018). Utilization of social media platforms for educational purposes among the Faculty of Higher Education with special reference to Tamil Nadu. In Higher Education for the Future, 5(1), pp. 4–19, DOI: https://doi.org/10.1177/2347631117738638

Wu, Q., Ding, K., Huang, B.: Approach for fault prognosis using recurrent neural network. J. Intell. Manuf. **31**, 1–13 (2018). https://doi.org/10.1007/s10845-018-1428-5

Zaini, N.A., Noor, S.F.M. & Zailani, S.Z.M. (2020). Design and development of flood disaster game-based learning based on learning domain. In International Journal of Engineering and Advanced Technology (IJEAT), 9(4), pp. 679–685, DOI:https://doi.org/10.35940/ijeat.C6216.049420

The Role of Web Analytics in Supporting the Effectiveness of Electronic Customer Relationship Management at the Jumia Store in Algeria

Miloud Ferhoul, Youssouf Boukedroune, and Nawel Chicha

Abstract Many researchers have been interested in talking about the importance of big data analytics and how it can help companies improve their customer relationship management, especially in light of recent changes to the business environment, the opening up of markets to one another, and the rapid advancements in information and communications technology, because the customer is one of the most profitable assets for institutions that are good and valuable in managing their relationship with him, especially its ability to reach the largest possible number of his preferences and consumer needs, and thus techniques have evolved to manage the relationship with the customer, especially with the use of web analytics, which facilitates this task. E-CRM (electronic customer relationship management) is required for the company's customer relationship management. We tried to include a very important case study in this study, embodied in the "Jumia" store in Algeria. Web analytics have a direct and positive impact on E-CRM, according to the findings of the study. When it comes to keeping its current customers happy, "Jumia" store in Algeria is relying solely on word-of-mouth advertising.

Keywords Web analytics · CRM · E-CRM

1 Introduction

There has been an increase in business environment complexity, regardless of size or type of business activity, as a result of the technological developments that have occurred at the end or beginning of this century, including wiki economics, e-marketing, and e-government, among others. This means that any organization must be flexible in order to keep up with the rapid changes in the technological field, especially as it relates to web analytics and other data processing and research

M. Ferhoul (✉) · Y. Boukedroune · N. Chicha
Université of Khemis Miliana, Rout Thiniet El Had, Khemis Miliana, Algeria
e-mail: miloud.ferhoul@univ-dbkm.dz; y.boukedroune@univ-dbkm.dz; n.chicha@univ-dbkm.dz

S. Sedkaoui et al. (eds.), *International Conference on Managing Business Through Web Analytics*, https://doi.org/10.1007/978-3-031-06971-0_27

tools, which could provide a wealth of new opportunities for the organization to both better serve its customers and achieve its strategic goals.

However, traditional institutions are more concerned with hearing from machines and employees than customers, making it impossible for them to deliver competitive performance at the level of effectiveness. In contrast, modern institutions that adhere to a marketing philosophy and strategic management outlook are more concerned with raising the level of customer satisfaction. Internet technologies have revolutionized customer relationship management and given rise to E-CRM, a modern form of the practice that requires a high degree of technical expertise. Web analytics, for example, is among the most important tools that contribute to improving customer relations electronically, as it opens avenues for the institution and its customers to communicate more effectively.

The "Jumia" e-commerce store in Algeria is one of the most important institutions interested in e-commerce, particularly at the international level. In the past, this online store has stated its desire to improve its electronic relationship with customers through the use of web analytics, one of the most important of these tools. Because it wants to keep its current customers and attract as many new ones as possible, it offers a wide range of products.

Using the previous statement, it is possible to ask the following key question:

> In the Algerian "Jumia" store, how does web analytics support the efficiency of electronic relationship management with customers?

As a follow-up to this main question, here are a few sub-questions:

What exactly is web analytics, and how does it work?
What do organizations hope to achieve by automating the management of customer relationships?
When it comes to customer relationship management, how does web analytics play a role for organizations?

The study's significance can be summed up as follows:

If an institution is serious about sustaining its long-term viability in today's competitive marketplace, it must be aware of the importance of web analytics.
The challenge of managing customer relationships electronically is also a complex one, illustrating the importance of utilizing digital applications and web analytic tools to keep in touch with current and prospective customers.

2 Theoretical Background of Web Analytics

When it comes to business management and administration, web analytics is a critical component, especially with today's widespread manifestations of technological globalization necessitating businesses to rely on the Internet and its accompanying applications. Web analytics is well understood in the company, as are its significance, role, and most critical processes.

2.1 *What Is Web Analytics?*

Despite numerous attempts, researchers were unable to come up with a single consensus definition of web analytics. Rather, the definitions of this application tool differed depending on the media and communication technology used in the various fields of use. Web analytics can be described as a technology and method for the collection, measurement, analysis, and reporting of data from websites and web applications (Zheng and Peltsverger 2015, p. 13). Since the advent of the World Wide Web, the field of web analytics has grown tremendously. It has evolved from a simple HTTP (Hypertext Transfer Protocol) traffic logging function to a more comprehensive set of usage data tracking, analysis, and reporting. In addition, the web analytics market and industry are booming, thanks to the proliferation of tools, platforms, positions, and businesses in the field.

2.2 *The Historical Background of Web Analytics*

Even though web analytics has its origins in recent decades of technological advancement and application in various fields, most studies on the subject confirm that the introduction of the World Wide Web's first browser in 1989 had a direct impact on its narrow concept of web analytics' emergence.

Website analytics pioneered by Webtrends, a Portland, Oregon-based company, was based on data gathered from web server logs. During the same year, Webtrends developed the first commercial website analytics application. Analog, the first free log file analysis program, was developed by Dr. Stephen Turner in 1995. This service was offered in 1996 by WebSideStory for websites that displayed banners. The types of data that can be recorded in web server logs are restricted. When it comes to things like screen sizes, user interactions, and mouse events like clicking and hovering, they were unable to provide this information. Page tagging, a relatively new method for circumventing this restriction, has been gaining traction recently.

Because of this, web analyses are still the focus of current research, especially in light of what the International Internet Society ISOC publishes, which has generated a lot of interest in keeping up with technological advancements.

2.3 The Importance of Web Analytics

It is the collection and analysis of user data that forms the foundation of web analytics. There are a wide range of uses for web analytics in various industries, including traffic monitoring, e-commerce optimization, marketing/advertising, and more. Web analytics can also be used to improve a website's performance. Some common purposes for web analytics data collection are as follows:

An **E-CRM** is a strategy that focuses on enhancing customer orientation, acquisition, and retention (Cresss and Veytsel 2000, p. 374). In order to better understand their customers' needs, more and more companies are analyzing website usage data.

Diagnose and improve the performance of web applications for enterprise organizations. Users are more likely to convert when a page takes longer to load, according to a study by Tag Man (2012).

It is possible to track and measure the success of commercial campaigns using web analytics. It is essential that web analytics distinguish between different types of traffic, marketing channels, and visitors if they are to be of any use.

2.4 Uses of Enterprise Web Analytics

Using web analytics, you can track the number of people who come to your site via various search engines.

You can find out what search terms people use to find your business using web analytics (Google Analytics).

Web analytics can demonstrate the effectiveness and impact of your content. It will also let you know if the keywords you've chosen are bringing in more people to your site.

With the help of web analytics, you can learn which keywords to bid on, how to tailor your ads for better results, and how to calculate the bid amounts for the topics that are the most cost-effective for you.

This is a good thing because it means that you will be able to see whether or not your fixes and modifications have raised your conversion rate from, say, 1% to, say, 5%.

You can use web analytics to figure out how much money you should spend on advertising to ensure that you get a good return on your investment.

Your investment will pay off more than it would have without analytics, as it will show which areas of your business are performing the best.

2.5 *Web Analytics Process*

There are many reasons why an organization might use web analytics, including improving their ability to communicate with the outside world as well as enhancing the user experience on their website. Here, we'll take a look at some of the most critical aspects of web analytics.

As a result of the business issue being raised, it is necessary to identify the analysis's goals.

Prioritize the analysis of web-based data by establishing KPIs that measure your progress toward your goals (Waisberg and Kaushik 2009, p. 02).

Continuously and accurately collect data from the surrounding environment, taking into account time, effort, and cost.

Enabling decision-makers to make better decisions by providing them with data analysis, data sorting, and data storage in information management systems.

We try to define the reading and its precise meaning in order to test the alternatives after obtaining the exact information (Bekavac and Praničević 2015, p. 376).

Insights gleaned from either data analysis or website testing should be acted upon (Fig. 1).

For organizations that want to stay on top of their ever-changing environment, web analytics is a must-have in order to avoid threats and take advantage of opportunities, especially in the area of website management and customer loyalty. We can see this through this figure, which shows that the web analytics process is periodic and continuous.

2.6 *Web Analytics Tools*

Finding just one tool to answer all of your organization's questions, according to "Avinash Kaushik," author of *Web Analytics 2.0* and *Web Analytics: An Hour a Day*, will land your company in a trench. This means that you need to look for multiple web analytics tools to keep up with changing customer behavior and preferences.

2.6.1 Google Analytics

Free service that generates detailed statistics about visitors to an enterprise's website is the most powerful web analytics tool. According to the site's usage statistics, more than half of the world's top 20,000 websites are currently using it. By doing this, the organization can learn about the origins of its visitors, what they do while on the site under surveillance, and how often they return.

Because Google Analytics collects so much information from the site, it is able to provide the organization with a greater level of detail in its reports.

Fig. 1 Web analytics process. (Source: Based on the following reference: Zheng and Peltsverger (2015), Web Analytics Overview, Southern Polytechnic State University USA)

2.6.2 Yahoo Web Analytics

An enterprise environment survey can be better understood with the help of a similar Yahoo view. Better access control options and a simpler approach to multi-site analyses, timely data collection, and the availability of all visitor behavior data, demographic reports, and custom options are also provided. In terms of profiling, filtering, and customizing, Google Analytics is a great and free option for those who want to dig a little deeper.

2.6.3 Crazy Egg

You can build heat maps and track your visitors' every click using Crazy Egg, which is a long way of saying that you're looking into how your website is used by your customers. You can also see and click on the parts of the Foundation's website that website visitors find more interesting using the Crazy Egg service.

Improved website design and conversion rates are two additional benefits. However, this service is typically paid for by the institution.

The effectiveness of your organization's activity depends largely on your ability to facilitate and monitor the organization's website, and among the most important of these, we find Adobe Analytics, Web Site Optimizer, Optimizely.com, Qualaroo, and Facebook Insights all fall under the category of web analytics tools. Most of them don't come for free.

3 The Theoretical Background of Electronic Customer Relationship Management E-CRM

There can be no success in an organization's strategic vision without a clear understanding of its surrounding environment, especially in relation to customers, who are the foundation of its activity and its continuation. Customers are clearly the main driver of success in any organization.

It doesn't matter what industry you work in or how big or small your company is; you always want to build strong relationships with your current and potential customers. The success or failure of an organization and the people who work for it depends on the quality of their relationships. Organizations strive to meet the needs and expectations of their customers in order to maximize profits and gain a competitive advantage through positive customer relations. Companies that provide the best value to their customers retain them, attract more customers, and consolidate their relationships with them are the most successful. Direct mail, public relations, exhibition spaces, press releases, the spoken word, personal selling, and other traditional methods of interaction and communication have been used by organizations in the past, as well as progress. When it comes to customer service in the knowledge society, organizations have shifted to the Internet in an effort to save money while still delivering high-quality service at a rapid pace.

Customers' behavior has changed significantly as a result of the introduction of the Internet and its applications, and the widespread use of smart phones among consumer circles has contributed to altering consumer tendencies. As a result, organizations have been forced to keep up with technological advancements in order to facilitate communication and build relationships with their current and potential customers, and it has provided opportunities for organizations to shift from customer relationship management to enterprise resource planning.

3.1 Conceptual Foundation of CRM and E-CRM

To put it another way, CRM helps companies to tailor their offerings around the needs of customers. In their 2003 article, Croteau and Li explain that CRM (customer relationship management) refers to a business strategy focused on improving customer satisfaction and loyalty by providing more responsive and customized service to each individual customer (Croteau and Li 2003, p. 23). CRM has improved an organization's ability to connect with customers and suppliers via the internet since the advent of the Internet. E-CRM stands for electronic customer relationship management. E-CRM is being improved by the Internet, which has features that are appealing to both customers and businesses. It is the underlying technology and how it interacts with users and other systems that distinguish CRM from E-CRM. Customers using a self-service browser-based window in E-CRM can perform a slew of tasks, including placing orders, checking order status, viewing purchase history, requesting more details about products, sending emails, and more. In terms of location and time, these capabilities give customers a lot of flexibility.

3.2 E-customer Relationship Management (E-CRM) Implementation Stages

Modern business weapons and strategic tools, such as E-CRM, give organizations a clear picture of their progress in customer management, how their efforts compare to those of their competitors, and how they can make improvements. It is easier for customers to use and more cost-effective for businesses to use web-based CRM or E-CRM. The implementation of an enterprise-wide E-CRM strategy is fraught with complications. In order to implement an enterprise-wide E-CRM strategy, the following are the most important steps: (1) understanding customers, (2) developing infrastructure in information and technology, (3) understanding people and organization, (4) management commitment, and (5) process management (Kennedy 2006, p. 61). As a result, organizations will perform better and retain customers in the long run if this is implemented (Fig. 2).

3.3 The Significance of Implementing an Electronic System for Managing Customer Relationships

As a result of the incorporation of new technology, such as web analytics, into the company's marketing strategy, the level of marketing within the organization has been elevated (Usman et al. 2012, p. 501). Most businesses that have implemented E-CRM properly see more benefits than drawbacks, and they also maintain a high level of competition in the strategic business environment.

Effective use of web analytics enables the company to communicate with its customers in a more flexible manner, which in turn fosters greater customer loyalty. You can learn more about what customers expect in terms of quality, price, and delivery time (the mainstays of competition).

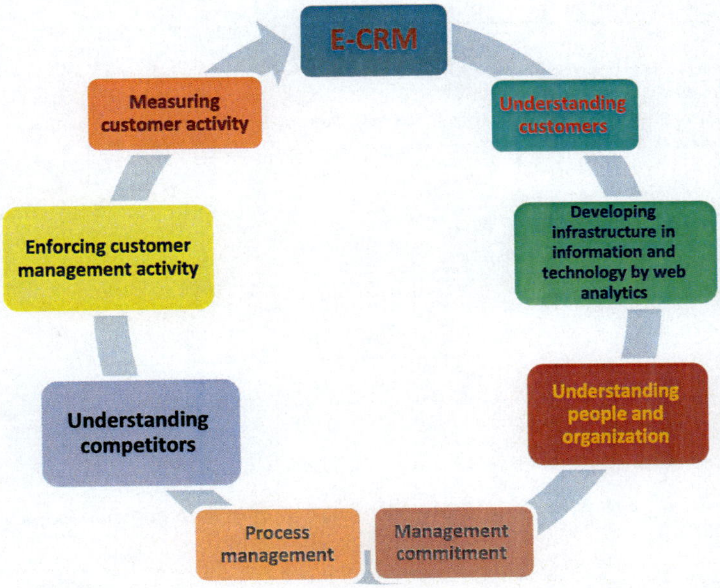

Fig. 2 Stages of establishing E-CRM. (Source: based on the following reference: Dawn and Chowdhury (2011), Electronic Customer Relationship Management (E-CRM): Conceptual Framework and Developing a Model, Center for Management Studies, JIS College of Engineering (Under West Bengal University of Technology))

Organizations must understand and realize the technological level of their customers, suppliers, and competitors in order to identify strengths and exploit available opportunities and diagnose weaknesses to avoid threats, in light of the current business environment's rapid transformations. A company's ability to effectively activate the elements of its marketing mix is perhaps its most important point of strength, which it can achieve through the careful application of web analytics (Fig. 3).

3.4 *Customer Relationship Management (CRM) Through the Use of Web Analytics*

Using web analytics for electronic customer relationship management is a goal for many companies, and we've attempted to summarize the most significant advantages here.

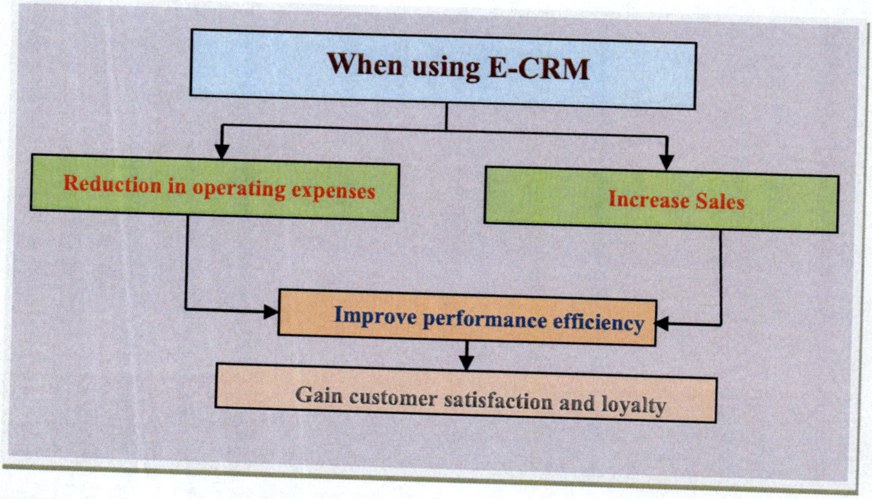

Fig. 3 The importance of E-CRM. (Source: Prepared by researchers with access to Aileen Kennedy 2006)

3.4.1 Web Analytics, E-CRM, and Profitability

Using web analytics, companies can activate their electronic customer relationship management system, giving their internal managers better control over customer segments and the ability to increase profits while reducing costs. A McKinsey consultant office study found that a 10% increase in repeat customers can increase profits by 10%. But a 10% reduction in total marketing expenditures required to bring in new customers only adds 0.7% to the company's profit margins. Even if a company is able to reduce the cost of attracting new customers, keeping current customers happy is more profitable than chasing new customers. Delivering value on their own terms is the best way to keep these customers happy.

It was discovered by Anderson Consulting that a typical $1 investment in equipping management with web analytics systems can earn up to $13 in profit by improving their ability to manage customer relationships (Scullin et al. 2010, p. 04). E-CRM performance, according to Anderson Consulting, accounts for as much as 64% of the difference in return on sales between average and high-performing companies. Customers and businesses alike stand to benefit from a well-executed E-CRM implementation, as shown by this data. Increased customer satisfaction leads to increased profitability because customers are happier when the overall customer experience is improved.

3.4.2 Web Analytics and E-CRM the Basis for Increasing Customer Loyalty

It is possible for a company to speak to its customers in a single, consistent voice with an effective E-CRM system. Everyone in an organization has access to the same transaction history and customer information when using web analytics software. Individual customer acquisition and retention costs can be quantified using an E-CRM system's data. Because of this information, the company can better allocate its time and resources to serving its most valuable clients and prospects (Jukić et al. 2003, p. 22). An organization can better manage its "best" customers by classifying them as a premium group and recognizing that treating every customer the same way is neither necessary nor advisable.

Personalization is one strategy a business can use to win over new customers and keep them coming back (Waltner 2001). Data from various sources, including customer databases, click stream data, and transaction systems, is used to create real-time profiles for each customer. Using what it "knows" about a customer's shopping habits, the tool picks out the best deal each time that customer visits the company's website. Whenever a customer accepts or rejects an offer, the personalization engine builds this information into the customer's profile, making it available for more informed future offers (Greenberg and Paul 2001, p. 68).

Personalization works just as well on B2B sites as it does on B2C ones. In sites with a wide variety of products, services, and information, many people see it as a helpful tool for locating things quickly and easily. It is a time saving mechanism that drives the advertising and content displayed on a website based on the interests of the customers.

3.4.3 Customer Service Can Benefit from the Use of Web Analytics

Customers' data is centrally stored in the web analytics system. To avoid the customer's frustrating and time-consuming "hunt" for assistance, a company can serve customer needs at all possible contact points (epiphany.com 2001a). Among the features of web analytics are search engines, live assistance, email management, news feeds/content management, and multi-language support. Electronic customer relationship management is built on the foundation of all of these tools.

4 How the "Jumia" Store in Algeria Used Web Analytics to Activate Its E-CRM System

4.1 *Who Is the Jumia Store?*

Jeremy Hodara and Sasha two former McKinsey consultants, founded an African e-commerce company for electronic goods in 2012 in Lagos, Nigeria. The company now serves more than five million customers annually and has rapidly expanded to 5 African countries, including Egypt, Morocco, the Ivory Coast, Kenya, and South Africa. It was announced in 2014 that Jumia would open offices in 14 African countries by 2018, beginning with the establishment of its first offices in Uganda, Tanzania, Ghana, Cameroon, and Algeria.

Among Jumia's many services are Jumia Travel, an online hotel reservation platform, and the Jumia Food app, which allows users to order and have food delivered, both of which were launched in June 2013 and the Jumia Deals app, which was launched in April 2015, both of which were launched by Jumia before the end of 2017, and in the same year, Jumia launched the Jumia One app, which allows users to pay bills and other payments like airtime. Additionally, Jumia and Amadeus collaborated on a platform for purchasing airline tickets.

In 2015, Jumia is expected to grow by up to 265% over 2014, bringing in revenues of up to $234 million. With a market capitalization of over $1 billion, Jumia became Africa's first unicorn company in 2016, and in late November 2018, it partnered with the crypto currency company Telco to facilitate payment services across operating regions. That same month, Jumia and Carrefour signed an agreement to sell their products electronically at the African level in partnership.

4.2 *The Jumia Store in Algeria*

As a pioneer in online shopping, it has been operating for more than 6 years, and it employs more than 290 people at 3 organizational levels. For domestic and international orders, they are able to fulfill customer requests through their network of national branches, which includes all of the United States. Quality and after-sales services are closely monitored at all of its locations throughout Canada, which are viewed as both delivery and promotion points in implementing total quality management.

Those in charge of Jumia's strategic planning state that this online retailer is always looking for ways to better manage its relationships with current and potential customers in order to gain their satisfaction and loyalty, as it attempts to arrive at an analysis of the preferences and tastes of the consumers targeted by it, especially on the level of digital media.

4.3 *Benefits of Web Analytics on E-CRM at Jumia Algeria*

According to an annual report from Algeria's Ministry of Post, Media Technology, and Communication, Internet use in Algeria will reach 15 million users by the end of 2019, and the ideal spread of smart phone use that relies on web applications has led the Jumia store to conclude that the Algerian market is a popular one. With a growth rate of 32.4% in 2017 and a projected increase to 49.2% in 2019, the electronic market in Algeria, which has over 17 million devices and over 15 million users, is now ranked 37th globally by the "Mobile Market Report" and "Newzoo Global." In Algeria, 61% of Jumia's customers use their phones to access the basic system for shopping, and the rest may be distributed on laptops or even orders placed via simple phones via direct wireless communication.

4.3.1 Uses of Web Analytics for Market Research by Jumia Algeria

Jumia Algeria uses web analytics, particularly Google Analytics and Thermal Analysis, to help customers research the market and gain a better understanding of their customers. Jumia receives over 1.7 million monthly visits, with 34% of those visitors being young adults (18–24 years old). Sixteen percent of those aged 35–44 years old, 13% of those aged 45–54 years old, and 9% of those aged over 54 years old were found to be in this age group. If we were talking about Jumia Travel or Jumia Food, the items most in demand through the Jumia website would be furniture and electrical household items (such as smart phones), toys, games, luxury products, home maintenance, clothes, and other services (Table 1).

To summarize, we see that, over time, the Jumia store has shifted its focus away from training employees in web analytics in favor of sales and communication skills, as shown in the table. However, Jumia's market share did not exceed 43.11% in 2014. Jumia has a 5.61% share of its annual expenditures preparing and using web analytics, while the rate of spending for training employees in sales and communication techniques increased by 4.81% to record 10.42% in

Table 1 The relationship of web analytics to an E-CRM

The year	Web analytics rate spending	The amount of money spent on sales and communication training for employees	Algeria's market share growth rate
2014	5.61%	10.42%	43.11%
2015	7.02%	9.96%	49.32%
2016	10.21%	9.30%	51.34%
2017	13.83%	8.77%	51.21%
2018	15.78%	8.24%	49.27%
2019	18.21%	7.03%	57.91%

Source: based on the annual report between 2014 and 2019 of Jumia

2014. Faced with fierce competition from other Algerian online marketplaces like Amazon, OuedKniss, Batolis.com foorshop.dz., Dzshop, Dzboom, and others, using web analytics allowed the company to reduce its annual costs for customer communication and employee training in this area, so in 2019, the company reduced the volume of its spending on employee training.

4.3.2 Web Analytics and Shopping Carts

For example, the Jumia store used web analytics to increase sales, develop e-marketing, find new ways to accept electronic payments, and make use of mobile applications and technologies such as shopping carts. The organizers are excited to continue using web analytics to accomplish these goals. Customer relationship management is an important part of this website's design, and it aims to identify the best motivating factors to take full advantage of these sites and present them in a unique way to customers. To keep them coming back to the site and keeping them loyal, the company uses web analytics to keep track of what he adds to his shopping carts, so they can send him offers based on the quality, price, or even the method by which he acquired the product. This helps to keep them coming back to the site and keeping them active.

4.3.3 "Tag Icon" and "Crazy Egg" Services Contribute a Lot to Studying Purchasing Decisions

In this context, some web analytics tools work to ensure the improvement of electronic marketing channels and assist customers in completing the purchase process without any difficulties. As a result of web analytics, electronic shopping carts can now include a tag icon to track a customer's purchase journey, from the time he selects the items for purchase to when he receives the goods. Verifying customer performance and purchase progress is an important factor in the success of a site and increasing user loyalty, for example, for customers. For example, customers who were unable to complete their purchases as a result of site technical difficulties should be noted. As well as highlighting the site's products page, search engines and social networks can also be used to make it easier for visitors to find it. Product names are listed on each page, as well as relevant keyword titles. There are a number of things to keep in mind when it comes to optimizing a website, such as working to discover the preferences and desires of site visitors in general, all of which can be done by using services like Crazy Eggs or even AdSense to measure the temperature distribution of the site's visitors. Some services available on some devices, such as location based services, allow you to track mobile device visits to the site in order to provide specific offers based on the geographic location of the device being used. As an example, separate ads can be placed directly on the main product page so that customers can complete the purchase with one click and avoid early exits from the site.

5 Conclusion

Analysis software for e-commerce sites can be used to gain insights into the habits and patterns of customers, the way they move around on the site, and what motivates them to buy. Web analytics and electronic customer relationship management, in our opinion, offer businesses a great opportunity to expand their customer base, as well as a necessity that must be taken care of to remain and compete in the world of online shopping. Customers and consumers are not just numbers in this web analytics; they are people who need to be protected and cared for. It is possible for smaller e-commerce businesses to compete with the likes of Jumia and Amazon, which are known for their emphasis on data security and electronic transactions, by implementing E-CRM.

According to the findings, the following are the most significant:

Customers' satisfaction and costs can be improved and reduced through the use of current technology and websites, which are key components of electronic customer relationship management.

Personalization and speedy response from web analytics allow any organization to provide personalized and long-term services and offers to its customers, which increases customer satisfaction and makes him or her an evangelist for your company to the rest of your potential customers.

Because of this freedom to use data exchange capabilities and build and maintain relationships with customers, web analytics gives businesses a great opportunity to benefit from personalization and tailor their products and services to each individual customer's personality and temperament.

5.1 Suggestions

To maximize customer value of maximizing value and benefit for customers, and gaining their loyalty within the company's strategic vision, web analytics should be used to support the use of electronic customer relationship management systems, creating a separate department to handle customer electronic relationships in light of the rapid growth in the number of people around the world using technology and its applications.

Teach employees how to effectively manage electronic customer relationships by utilizing web analytics.

References

Bekavac, I., Praničević, D.G.: Web analytics tools and web metrics tools: an overview and comparative analysis. Croat. Oper. Res. Rev. **2**, 373–386 (2015)

Cresss, & Veytsel.: Web Analytics: Translating Clicks into Business. Croat. Operat. Res. Rev. (2000)

Croteau, A.M., Li, P.: Critical success factor at CRM technological initiative. Can. J. Adm. Serv. **1**, 92 (2003)

Dawn, S.K., Chowdhury, R.: Electronic Customer Relationship Management (E-CRM): Conceptual Framework and Developing a Model. Centre for Management Studies, JIS College of Engineering (Under West Bengal University of Technology) (2011)

Greenberg and Paul: Capturing and Keeping Customers in Internet Real Time. McGraw-Hill (2001)

Jukić, N., Jukić, B., Meamber, A.L.: Implementing Polyinstantiation as a strategy for electronic commerce customer relationship management. Int. J. Electron. Commer. **7**, 9–30 (2003)

Kennedy, A.: Electronic Customer Relationship Management (E-CRM): Opportunities and Challenges in a Digital World. Irish Marketing, Technological University Dublin (2006)

Scullin, S., Allora, J., Lloyd, G.O., Fjermestad, J.: Electronic Customer Relationship Management: Benefits, Considerations. Pitfalls and Trends. New Jersey Institute of Technology (2010)

Tag Man.: Just One Second Delay In Page-Load Can Cause 7% Loss In Customer Conversions (2012). Retrieved from https://buzzquake.com/just-one-second-delay-in-page-load-can-cause-7-decrease-in-conversion/

Usman, U.M., Jalal, A.N., Musa, M.A.: The impact of electronic customer relationship. Int. J. Adv. Eng. Technol. (2012)

Waisberg, D., Kaushik, A.: Web Analytics: Empowering Customer Centricity. SEMJ-org (2009)

Waltner, C.: CRM Makes On-Line Shopping Personal, InformationWeek (2001). Retrieved from https://searchsqlserver.techtarget.com/definition/data-mining/

Zheng, G., Peltsverger, S.: Web Analytics Overview, Southern Polytechnic State University USA, 3rd edn. Published in Encyclopedia of Information Science and Technology (2015)

Sentiment Analysis on COVID-19 Tweets

Soraya Sedkaoui, Mounia Khelfaoui, and Ouakli Keltoum

Abstract This study aims to identify Twitter users' feelings and beliefs regarding the measures promulgated during the global health crisis related to COVID-19. It presents an opinion analysis regarding tweets generated during COVID-19. The purpose is to discover patterns in tweets using natural language processing techniques sentiment analysis and obtain results that allow an analysis of what people think. The data analyzed were unstructured (text) carried out during COVID-19 measures between March and August 2020. The study found a link between tweets about the new coronavirus and its spread around the world.

Keywords COVID-19 · Tweets · Sentiment analysis · Machine learning · NLP

1 Introduction

COVID-19 disease marked a turning point in a global society (Crokidakis 2020). Contagion spreads, and infection rates are a source of concern for people. As a result, governments worldwide have enacted stringent public health measures to halt the virus's spread and avoid the oversaturation of healthcare systems (Steffen et al. 2020). The mandatory use of masks, border blocking, the closure of educational institutions, and social isolation are among the decisions to contain the pandemic (Weible et al. 2020).

In addition, in light of the current pandemic, quarantine periods have been established in several countries. During this time, many people around the world have stayed at home. Consequently, there has been an increase in communication channels via social networks and digital services. People use social media to express how they are dealing with the pandemic's effects and how the rhythms of their lives

S. Sedkaoui (✉) · M. Khelfaoui · O. Keltoum
University of Khemis, Miliana, Algeria
e-mail: s.sedkaoui@univ-dbkm.dz; m.khelfaoui@univ-dbkm.dz; m.khelfaoui@univ-dbkm.dz

S. Sedkaoui et al. (eds.), *International Conference on Managing Business Through Web Analytics*, https://doi.org/10.1007/978-3-031-06971-0_28

and daily activities have changed. In such a situation, where information generates a large amount of data on a global scale, it is critical to store the said information.

Nowadays, social networks present an essential data source, and managing this type of information generates valuable insights. People share their experiences and opinions, creating a rich source of textual data (Sedkaoui and Khelfaoui 2020). The data transmitted provide descriptions and opinions that can be useful for decision-making. This abundance of information, primarily obtained through social Networks, can explain individuals' perceptions.

Currently, businesses are using social networks extensively to promote their products and services. These unstructured textual data sets can be used to generate insights into mass behavior, thoughts, and emotions on a wide range of topics, including product reviews, opinions, political trends, and market sentiment values.

Twitter, founded by Jack Dorsey, is a social networking service continuously fed by millions of people's opinions from different parts of the world. The social network Twitter has many users, and during the COVID-19 pandemic, many tweets were generated. There is a need for Twitter opinion mining because it allows quantifying user interest and opinion on a specific topic or case study.

Many studies applied sentiment analysis to analyze comments on Twitter. Several studies emphasized the need to analyze the material available in these social media to know people's personalities, interpersonal relationships, and social situations. Therefore, analyzing the emotions that people worldwide profess about quarantine generates valuable sentiment analysis contributions.

This study carries out an analysis of feelings and networks social media to Twitter comments. As a part of natural language processing (NLP), this technique can extract information from websites automatically and identify the feelings or emotions that texts contain. In this context, the purpose of this study is to identify the underlying feelings of the comments on social networks related to COVID-19 effects and to analyze the most frequent words or topics that users have noticed.

It presents a case study on the relationship of topic modeling with sentiment analysis in the context of the COVID-19 pandemic. For this, a Twitter message dataset was used to train sentiment modeling. Machine learning (ML) techniques are applied using Python software packages.

This study is organized as follows. Section 2 presents a literature review related to this topic. Section 3 details the applied methodology. Section 3 discusses the results of the study. The most relevant conclusions, limitations, and opportunities for future research are presented in Sect. 4.

2 Literature Review

2020 has brought a whole new disease that has hit all the world's countries: COVID-19. Reported in 2019 in Wuhan, China, COVID-19 has spread rapidly worldwide. As a result, it was declared a pandemic in March 2020 due to its high transmission rate.

In addition, the contagion rates were so high that they had plunged health organizations around the world into tremendous concern. For this reason, governments have decreed stringent measures to contain the virus, which seek to mitigate its transmission and keep health systems running (Steffen et al. 2020).

Of all the measures put in place, social isolation, known as quarantine, has sparked the most debate, owing to the benefits and drawbacks it entails (Meier et al. 2020). The popular opinion reflects the debate over the benefits and drawbacks of quarantine. Several voices have emerged in countries where the government's position has been focused on complying with the recommendations of health agencies that ask to maintain mandatory isolation.

Of course, the debate around the relevance of different measures has reached the social networks scene, which offers great opportunities for qualitative and quantitative research in social sciences (Reyes-Menendez et al. 2020). Faced with the quarantine, millions of publications have been disseminated on social networks such as Facebook, Twitter, Instagram, etc.

It indicates that social networks have become an essential thermometer for measuring Internet users' perceptions of applying these measures. However, this situation has resulted in the appearance of comments with questionable content information, which elicits a range of emotions. Particularly, tweets about COVID-19 spark debate among users, and the opinions expressed elicit various responses on the social media platform.

Currently, one of the methods used to determine perception based on people's opinions is the use of sentiment analysis techniques. This technique has been disseminated in social networks to determine trends in user opinions or study their behavior over time. It corresponds to a branch of NLP that aims to identify and evaluate the emotional value through its structure so that the text's polarity classification is obtained in three possible categories: positive, neutral, and negative.

Sentiment analysis is defined as the set of computational techniques for the extraction, classification, understanding, and evaluation of opinions expressed in sources published on the Internet, comments on web portals, and in other content generated by users (Pang and Lee 2008; Cambria et al. 2012; Saura et al. 2019). This type of technique has been widely used in research to understand the perceptions and feelings of social network users about various issues.

Numerous studies have been developed to help formulate public health policies, especially during epidemiological outbreaks. Table 1 presents a list of studies developed that used data from social media, especially Twitter, to monitor trends, feelings, content, and information accessed and shared by users.

In addition, during 2020, many studies have been published on the same topic but concerning the new coronavirus using sentiment analysis technique.

Concerning COVID-19, Baker et al. (2020) used sentiment analysis techniques to detect influenza using machine learning techniques in tweets from Arab countries. The classification models used were decision trees, support vectors machine (SVM), Naive Bayes, and k-nearest neighbors (K-NN).

To determine the emotions Indonesians expressed regarding the 2019 presidential campaign, Wongkar and Angdresey (2019) applied this technique. For their

Table 1 Social networks data, sentiment analysis, and infectious diseases in the literature

Disease	Author(s)	Year	Data source	Aim/results
H1N1	Scanfeld et al.	2010	Twitter	Examining the content of tweets containing the "antibiotic(s)" terms to identify categories and investigate evidence of antibiotic misunderstandings or misuse
	Signorini et al.	2011	Twitter	Analyzing the content of tweets to track population information and feelings about the rapid evolution of H1N1 and track and measure disease activity
Ebola	Odlum and Yoon	2015	Twitter	Examining twitter content to provide an instant picture of virus-related tweets by monitoring trends in information dissemination, studying the possibility of early detection of epidemics, and understanding knowledge and attitudes of the population
	Towers et al.	2015	Twitter and search tools	From the analysis of daily Ebola-related internet and twitter research data in the United States, the study adjusted a mathematical model of contagion to see if news coverage was a significant factor in the temporal patterns of Ebola-related internet and twitter data
	Househ	2016	Twitter and Google news trend	The study indicated a relationship between publications in electronic media and twitter activity around significant events, such as Ebola
	Wong et al.	2017	Twitter	Examining the characteristics of tweets posted by representatives of local health departments about Ebola. Through a temporal analysis of Ebola tweets, the authors found the presence of five distinct waves, each corresponding to the main Ebola news events
MERS-CoV	Shin et al.	2016	Twitter and Google	Evaluating the possibility of using a digital surveillance system based on Google and twitter to monitor an outbreak of MERS-CoV (Middle East respiratory syndrome coronavirus) in Korea

(continued)

Table 1 (continued)

Disease	Author(s)	Year	Data source	Aim/results
Flu	Chorianopoulos and Talvis	2016	Twitter	Built and made available an open-source database that can detect flu-related symptoms and share the data in real time with the population
Zika	Stefanidis et al.	2017	Twitter	Analyzing the content of tweets about Zika from the perspective of three aspects: Location, actors, and concepts, to understand how a public health emergency of international interest occurs in social media
	Daughton and Paul	2019	Twitter	From twitter posts about the 2015–2016 Zika virus outbreaks, the study identified and described a relevant change in user behavior, specifically from the spread of travel cancellations in South America

Source: Authors' elaboration

classification, they used the Naive Bayes algorithm. Roy et al. (2020) analyzed the attributions of responsibility for the disease's spread to specific social actors, such as the government, migrants, the media, residents of territory with infected people, or global health authorities. The study identified guidelines that shape network users' interpretive frameworks in the face of the epidemic.

G7 political leaders' use of Twitter during the pandemic is examined by Rufai and Bunce (2020). The content revealed three types of use: the informative, the predominant one, the moral reinforcement of the population, and the political discussion that seeks to raise points of debate, which occurs mainly in the case of the president of the United States.

For Kullar et al. (2020), Twitter has a primary value in disseminating health messages for medical professionals, both in the current situation and past pandemics of avian flu in 2009 and Ebola in 2014.

However, Han et al. (2020) performed a content and sentiment classification of the messages produced on the Sina Weibo microblogging network during the first stage of the pandemic. The study classified the most frequently referred topics and subtopics by social network users, such as recommendations for social isolation, blessings and prayers, objective comments about the disease, and protection measures.

All these studies recognized the role of Twitter in the dissemination of relevant public information in times of health crisis. The previous studies demonstrated the importance of sentiment analysis in analyzing users' opinions of various services in various application contexts.

In this sense, the sentiment analysis technique can analyze COVID-19 data. The current study aims to contribute in that direction, understanding the platform primarily as a space for the dissemination of public messages but also as a space where ordinary users can build meanings around the coronavirus pandemic.

3 Data and Methodology

Sentiment analysis, better known as opinion mining, determines the emotional tone behind a series of words. This type of analysis is commonly performed with the information generated from social networks because of the vast information collected, which corresponds to opinions.

It creates a significant challenge that involves storing and processing the large volume of information generated from Twitter in the case of this study. This social network deserves special mention, as it is the data source for this study.

The knowledge discovery in database (KDD) steps were used as a reference to carry out the opinion analysis work, which proposes the following stages: data selection, data preprocessing, data transformation, data mining, and data interpretation. The phases illustrated in Fig. 1 were considered for the development of this study.

Python libraries were chosen as programming languages to complete the data mining and interpretation.

Data Collection The selection criteria were established to obtain data in this stage. A free Twitter API was used through Python from March to August 2020.

For many countries around the world, the quarantine began in March 2020, and it is convenient to analyze the data that is as close as possible to the months where the highest volume was evidenced. On the other hand, according to Fig. 2, the dataset created shows the volume of tweets created, especially between March and May 2021.

On average, more than 1500 tweets related to COVID-19 were observed daily, with a minimum of 1200 and a 5100 maximum by day, indicating considerable variance.

The words and hashtags, such as "COVID-19," "coronavirus," "social isolation," "confinement," "quarantine," "curfew," etc., were chosen because they returned publications related to the research target. People's feelings due to social isolation are mainly tiredness, fear, sadness, being hungry, and more.

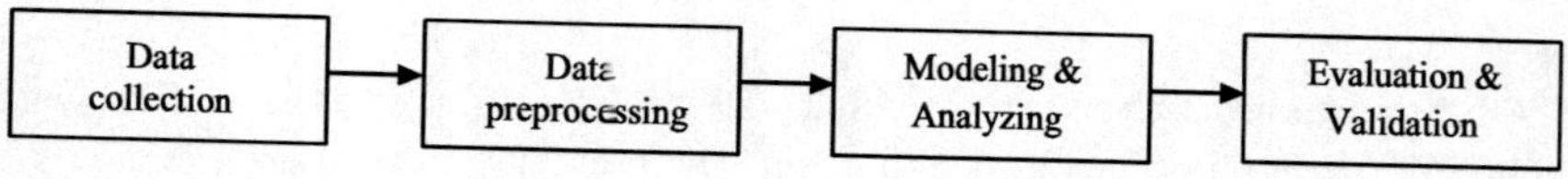

Fig. 1 Methodology phases. (Source: Authors' elaboration)

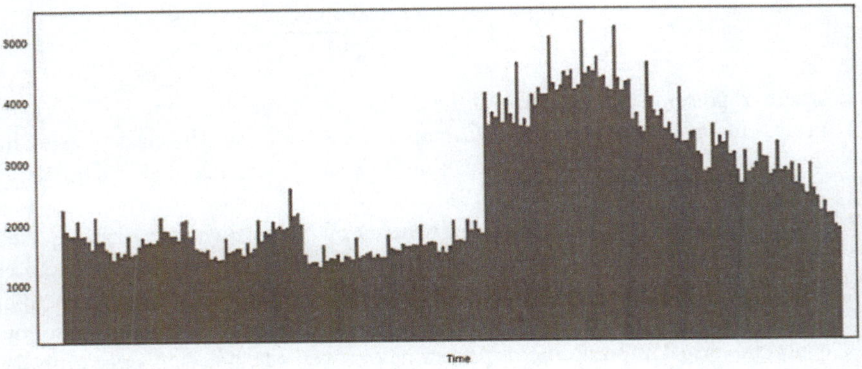

Fig. 2 Worldwide Tweets frequency by hour. (Source: Python outputs)

The chosen time and amount of data allowed an important analysis of the opinions and feelings regarding the global health crisis caused by COVID-19.

In *data preprocessing*, publications are subjected to a series of techniques that allow cleaning and reducing the characteristics of the texts, facilitating sentiment analysis using machine learning. In this stage, a data transformation was applied to eliminate symbols, numbers, and patterns that are not significant.

It is important to mention that after cleaning the dataset, the work was divided into several columns to use the function that allows grouping words according to their frequency.

Modeling To build a model for classifying comments based on feelings, there are two ways documented in the literature.

- Predefined dictionaries, or lexicons: classify the words of the texts according to the sentiment they represent (Sauter et al. 2010).

Build a machine learning algorithm that learns, from training and test data sets, to recognize feelings and emotions in the texts.

Evaluation The sentiment analysis model was built with training and test data corpus. Additionally, various machine learning algorithms were tested to classify texts according to emotions. Three algorithms software were tested: Naive Bayes, Naive Bayes Kernel, and Deep Learning (Kotu and Deshpande 2014), as follows:

1. **Naive Bayes:** This algorithm is widely used in sentiment analysis models. It is a supervised classification method based on Bayes' theorem that specifies the independence of attributes in learning. First, each word's probability is defined, and then the classifier is built to classify the comments based on the category of sentiment:

$$P(Y/X) = \frac{P(X/Y)(Y)}{(X)} \tag{1}$$

where Y presents the attributes of the document, X the class, and P (X | Y) the probability of occurrence of the attribute in the given class. The class selected by the classifier will be the one that maximizes P (X | Y) (Baker et al. 2020; Han et al. 2011).

2. **Naıve Bayes Kernel:** It retains the benefits of Naive Bayes and can be used in situations where the data does not have a normal distribution (Sedkaoui et al. 2021). In nonparametric estimation techniques, a kernel is a weight function. Because they use parametric techniques when the data's behavior does not follow a normal distribution, kernel-type estimators were designed to overcome the difficulties. They are the most commonly used in nonparametric estimation, and their most important feature is that they are unaffected by the kernel function used:

$$f(X/Y) = \frac{1}{n}\sum_{i=1}^{n} K_n\left(X - X^i\right) \tag{2}$$

K is assumed to be a d-dimensional density function. Some commonly used kernels are uniform, triangular, and Gaussian (Pérez et al. 2009).

3. **Deep Learning** is a machine learning algorithm that uses neural network architectures (Sedkaoui and Khelfaoui 2020). The complexity and extraction power of deep learning algorithms increase with the depth of the layer beneath them. Nonlinear transformation is applied to each algorithm's input and used to generate a statistical model as the output. Iterations are repeated until the output reaches an acceptable level of precision.

Validation A critical component is determining which algorithm is the best and determining its performance. It necessitates the development of methods that provide useful information about the efficiency of the algorithms. Different methods could be used to evaluate the results obtained in the sentiment analysis. The most common is the confusion matrix, which is a technique for summarizing the performance of a classification algorithm. It divides the test results into the following categories:

- True positive (TP): Values predicted correctly as belonging to class A.
- False positive (FP): Values incorrectly predicted as belonging to class A.
- True negative (TN): Values rejected correctly as not belonging to class A.
- False negative (FN): Values incorrectly rejected as not belonging to class A.

From this information, a series of measurements can be made to classify the efficiency of the algorithms:

- *Precision* is defined as the ratio of positive predictions to the number of actually positive observations. Its value increases as the number of false positives decreases:

$$\text{Precision} = \frac{\text{TP}}{(\text{TP} + \text{FP})} \tag{3}$$

- *Recall:* it is the relationship between the documents correctly classified as class A and the sum of all documents of class A. It can also be viewed as the model's ability to construct classes correctly. The closer to 1, the better the various existing classes because their value increases as false negatives decrease:

$$\text{Recall} = \frac{\text{TP}}{(\text{TP} + \text{FN})} \tag{4}$$

- *F-Score:* to measure the efficiency of a classification model, the coverage and exhaustiveness values are used. For this, the F-value is presented as the harmonic mean between both measurements and is often used to compare the performance between various models. The F-value formula combines the two previous measures in a weighted way through a parameter β, which allows giving greater importance:

$$F_\beta = \left(1 + \beta^2\right) . \frac{\text{Precision.Recall}}{(\beta^2.\text{Precision}) + \text{Recall}} \tag{5}$$

- *Accuracy:* is the simplest and most intuitive performance measure and represents the ratio of the correct predictions. It is the number of elements classified correctly among the total number of classifications carried out:

$$\text{Accuracy} = \frac{(\text{TP} + \text{TN})}{(\text{TP} + \text{TN} + \text{FP} + \text{FN})} \tag{6}$$

Once the machine learning algorithms were established, the appearance of terms related to the emotions of interest was identified (Teso et al. 2018). This weighing gives greater relevance to the terms with greater frequency, evaluating the number of times as indicated by:

$$P_i\,(t_i) = F_{ik} \tag{7}$$

F_{ik}: is the frequency of term k in documents i.

Then a sentiment analysis was performed to determine the most frequent terms in the dataset. This analysis identified the topics with the highest frequency, which is presented as nodes in the graphs, and the connections between them were determined through modularity, which is reflected in the graph by different colors (Bastian et al. 2009). This is to say that terms in the same color are closely related.

As a final step, the algorithm was validated by comparing similar tools and validated by confusion matrix.

4 Results and Discussion

Prior to March 2020, data shows a small percentage of messages related to COVID-19. However, tweets increased as the disease progressed and gained more media proportions in the following months. The data presented in Fig. 2 demonstrate similar upward behavior.

To assess the proposed models, an experiment was performed to attest to the quality of the sorting model of feelings and selection of the topic modeling algorithm. Table 2 illustrates the performance of the three used algorithms.

The results in Table 1 show that the deep learning algorithm remains the best among the three applied methods, with an almost 92% accuracy score. The Naive Bayes model has reached 43.81% of accuracy. Therefore, this algorithm is a suitable algorithm to analyze, classify, and extract the feelings and sentiments of people during the COVID-19 health crisis from Twitter comments.

As shown in Table 3, the precision obtained in all categories was greater than 60%, indicating that the model correctly classifies the elements that belong to the corresponding category. Similarly, values greater than 55% were obtained in the recall test, indicating that the model provides an adequate adjustment.

Table 3 shows the result of the best model in the test set. The result produced was similar to that obtained in the validation set. To analyze the feelings of tweets related to the COVID-19 pandemic, Fig. 3 (left side) shows the proportions between messages classified as positive, negative, and neutral for the period analyzed.

Table 2 Algorithms' performance

Algorithm	Accuracy (%)
Naive Bayes	43.81
Naive Bayes kernel	82.56
Deep learning	91.78

Source: Authors' elaboration based on Python outputs

Table 3 Precision and recall results by class

	Tests		
Sentiments	Precision	Recall	F1-score
Positive	81%	61%	70%
Negative	71%	58%	58%
Neutral	62%	82%	57%

Source: Authors' elaboration based on Python outputs

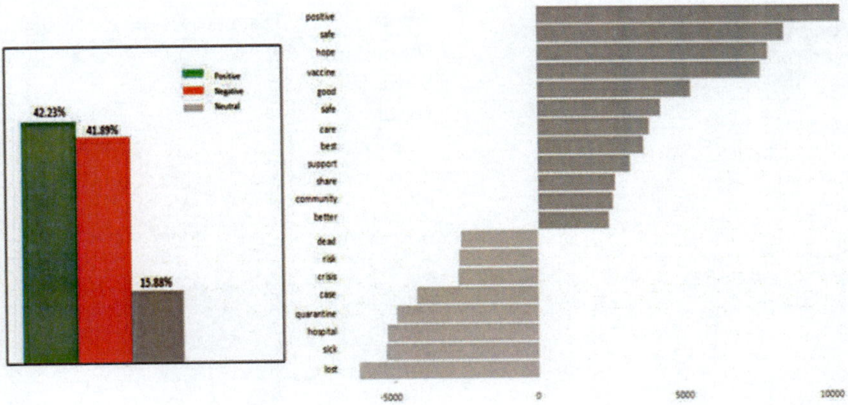

Fig. 3 Distribution of feelings classified in tweets. (Source: Python outputs)

The analysis conducted with the deep learning model shows that the predominant part of the tweets is positive cases, with 42.23% of the overall distribution. Negative tweets present approximately 41.89%, followed by neutral tweets with just 15.88%.

It is possible to observe that most users sent messages with both positive and negative feelings (right side) about the pandemic. With the growing volume of tweets in the period (Fig. 2), such data demonstrate the users' concern with the experienced context.

This is consistent with recent studies such as Pastor (2020), Xue et al. (2020). These studies also emphasized that these findings make it critical to implement public policies and strategies to address and maintain mental and psychological health.

It should be noticed that the specificities of the countries under consideration must be taken into account in comparative analyses because some studies discovered a population predominance of positive feelings (Delizo et al. 2020; Alhajji et al. 2020).

However, words such as "death," "cases," "hospital," and "quarantine," among others, can be identified as negative sentiments. Such findings denote the concern of Twitter users regarding the current situation and reflect the consequences that the spread of the virus has brought to the population. In addition, the hashtag "vaccination" is also present in messages with positive feelings about the pandemic.

This type of observation confirms that the corpus deals with the theme of quarantine and social isolation and the concerns of people here, represented by words in their isolated meanings.

Words vary across the problem, including economics, medicine, prevention, and other high-demand issues, especially during the first and second waves. In addition, there is a relationship between the occurrence of the words and the discussion of the COVID-19 pandemic. It demonstrates the modeling significance when applied to words that are discussed all over the world.

Table 4 Relationships between words found by sentiment in tweets

Words	Positive	Negative	Neutral
Quarantine	X	X	X
Stay at home	X	X	
Lockdown			X
Vaccination	X	X	
Work	X		X
Prices	X	X	
Hospital (bed)		X	
Curfew	X	X	X
Food	X	X	
Travel			X
Health	X	X	
Measures	X	X	X
Purchase		X	X

Source: Authors' elaboration based on Python outputs

Table 4 shows the relationship between the three classes of feelings and the words found applying the sentiment analysis technique. Table 4 shows ten distinct negative tweets. Among them, some tweets had exclusively negative repercussions.

The word "price" indicates the perturbation in some goods and services prices, consequently influencing consumption. The number of online purchases increased significantly with the quarantine "stay at home" and the fear of traveling or going outside. This increased the panic caused in some people and generated the need for some to stock up on food, causing shortages in supermarkets.

In addition, the number of beds in hospitals with the decreasing number of beds and the government's health measures were of great concern.

Regarding the positive topics, some words like "measures" and "vaccination" help decrease the spread of the virus and how different important measures are to raise people's awareness of the disease's retention. For "quarantine" and "stay at home," they refer to the awareness of parts of the population about social isolation; and with "food," people were commenting on price changes in various sectors.

It is important to indicate the words modeled with both negative and positive feelings, such as quarantine, food, measures, and health. For quarantine, negative comments are related to changes in their lifestyle, and staying at home was difficult for some people. However, other people reacted positively since they can purchase online. Because of this, a separate analysis was carried out.

For "health," the negative comments were primarily about the danger that health professionals face daily fighting the virus, while the positive comments were explicitly about their efforts and protection against COVID-19.

Globally, people were more positive when expressing themselves when the topic was quarantine or social isolation. So, despite the difficulties we faced during the first and second waves, there is a sense that things will improve.

5 Conclusion

First, with the appearance of the Internet and later with the development of social networks, the possibilities of interpersonal communication have multiplied exponentially. Currently, our social environment is not limited to family members with whom we live, colleagues with whom we work, and friends with whom we share our leisure time. Especially among young people, communicative activity can no longer be understood without social networks.

New digital technologies allow access to a new and valuable source of information. New technologies make it possible to identify crucial data to understand people; companies could understand their clients' behavior in all its dimensions. Recently, there has been an increase in the interest of academic professionals and researchers in the use of social media data for the most diverse purposes.

In situations such as that caused by the COVID-19 pandemic, monitoring aspects related to people's feelings and sentiments becomes important. In this sense, Twitter is a valuable tool to understand people's feelings in real time, as users widely use it to communicate their views, concerns, and perspectives. This study investigates and understands which topics are discussed and Twitter users' feelings concerning the COVID-19 pandemic. This study used the content posted by Twitter users regarding the COVID-19 pandemic to analyze individuals' feelings.

Applying sentiment analysis to people's opinions presents two important contributions to the academic literature: (i) although this type of sentiment analysis, aimed at extracting predominant emotions from social media comments, can be found in various studies; (ii) understanding the emotions that people around the world profess in the face of compulsory isolation measures can help to understand potentially transcendental psychological and social aspects for future implementation of practical actions in similar contexts.

After applying the deep learning model, it was noticed that the feelings of the positive and negative classes were in similar amounts. However, the positive feeling was slightly higher. The deep learning model allows understanding the valuable opinions and issues that are part of the context of the COVID-19 pandemic, bringing a tool that can serve to understand the events discussed in daily life.

The analysis of topics will be carried out over various times to understand how the feelings and topics addressed by users changed over time. Using other techniques in future studies can be if great interest for categorizing feelings to achieve better results. It is also intended to broaden the base of tweets about the COVID-19 for different periods.

Appendix

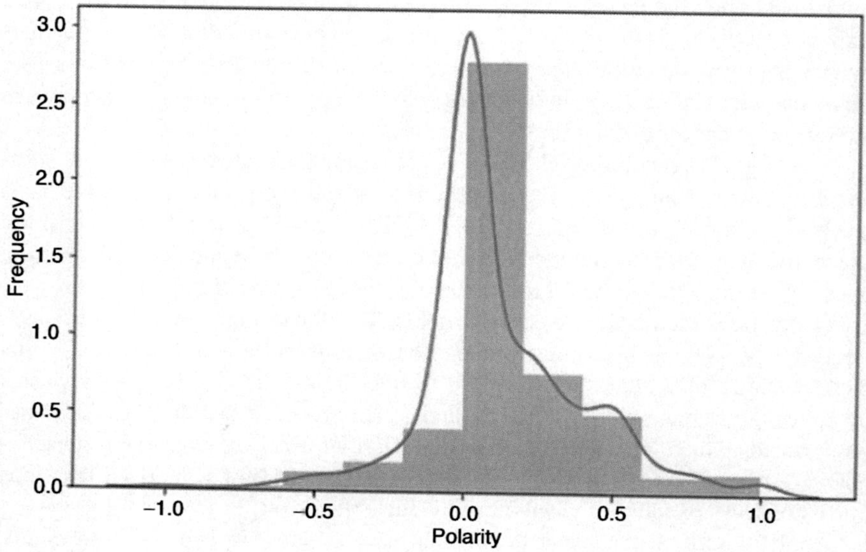

Sentiment distribution. (Source: Python outputs)

References

Alhajji, M., Al Khalifah, A., Aljubran, M., Alkhalifah, M.: Sentiment Analysis of Tweets in Saudi Arabia Regarding Governmental Preventive Measures to Contain COVID-19. Preprints (2020)

Baker, Q., Shatnawi, F., Rawashdeh, S., Al-Smadi, M., Jararweh, Y.: Detecting epidemic diseases using sentiment analysis of Arabic tweets. J. Univ. Comput. Sci. **26**(1), 50–70 (2020)

Bastian, M., Heymann, S., Jacomy, M.: Gephi: An open source software for exploring and manipulating networks. In: International AAAI Conference on Weblogs and Social Media, pp. 361–362 (2009)

Cambria, E., Grassi, M., Hussain, A., Havasi, C.: Sentic computing for social media marketing. Multimed. Tools Appl. **59**, 557–577 (2012)

Chorianopoulos, K., Talvis, K.: Flutrack.org: open-source and linked data for epidemiology. Health Informatics J. **22**(4), 962–974 (2016)

Crokidakis, N.: COVID-19 spreading in Rio de Janeiro, Brazil: do the policies of social isolation really work? Chaos, Solitons Fractals. **136**, 1–6 (2020)

Daughton, A.R., Paul, M.J.: Identifying protective health behaviors on twitter: observational study of travel advisories and Zika virus. J. Med. Internet Res. **21**(5) (2019)

Delizo, J.D., et al.: Philippine twitter sentiments during Covid-19 pandemic using multinomial Naïve-Bayes. Int. J. Adv. Trends Comp. Sci. Eng. **9**(1.3), 408–412 (2020)

Han, X., Wang, J., Zhang, M., Wang, X.: Using social media to mine and analyze public opinion related to COVID-19 in China. Int. J. Environ. Res. Public Health. **17**(8) (2020)

Han, J., Pei, J., Kamber, M.: Data Mining: Concepts and Techniques. Elsevier (2011)

Househ, M.: Communicating Ebola through social media and electronic news media outlets: a cross-sectional study. Health Informatics J. **22**(3), 470–478 (2016)

Kotu, V., Deshpande, B.: Predictive Analytics and Data Mining: Concepts and Practice with Rapidminer. Morgan Kaufmann (2014)

Kullar, R., Goff, D.A., Gauthier, T.P., Smith, T.C.: To tweet or not to tweet - a review of the viral power of Twitter for infectious diseases. Curr. Infect. Dis. Rep. **22**(6), 14 (2020)

Meier, K., Glatz, T., Guijt, M., Piccininni, M., van der Meulen, M., Atmar, K., Jolink, A., Kurth, T., Rohmann, J., Zamanipoor, A.: Public perspectives on protective measures during the COVID-19 pandemic in the Netherlands, Germany and Italy: a survey study. PLoS One. **15**(8), 1–17 (2020)

Odlum, M., Yoon, S.: What can we learn about the Ebola outbreak from tweets? Am. J. Infect. Control. **43**(6), 563–571 (2015)

Pang, B., Lee, L.: Opinion mining and sentiment analysis. Found. Trends Inf. Retr. **2**(1–2), 1–135 (2008)

Pastor, C. K. (2020). Sentiment Analysis of Filipinos and Effects of Extreme Community Quarantine Due to Coronavirus (COVID-19) Pandemic. Available at: https://ssrn.com/abstract=3574385

Pérez, A., Larrañaga, P., Inza, I.: Bayesian classifiers based on kernel density estimation: flexible classifiers. Int. J. Approx. Reason. **50**(2), 341–362 (2009)

Reyes-Menendez, A., Saura, J.R., Thomas, S.B.: Exploring key indicators of social identity in the #MeToo era: using discourse analysis in UGC. Int. J. Inf. Manag. **54**, 102129 (2020)

Roy, M., Moreau, N., Rousseau, C., Mercier, A., Wilson, A., Atlani-Duault, L.: Ebola and localized blame on social media: analysis of Twitter and Facebook conversations during the 2014–2015 Ebola epidemic. Cult. Med. Psychiatry. **44**(1), 56–79 (2020)

Rufai, S.R., Bunce, C.: World leaders' usage of twitter in response to the COVID-19 pandemic: a content analysis. J. Public Health. **42**(3), 510–516 (2020)

Saura, J.R., Reyes-Menendez, A., Palos-Sanchez, P.: Are black Friday deals worth it? Mining twitter users' sentiment and behavior response. J. Open Innov. Technol. Market Complex. **5**(3), 58 (2019)

Sauter, D., Eisner, F., Ekman, P., Scott, S.: Cross-cultural recognition of basic emotions through nonverbal emotional vocalizations. Proc. Natl. Acad. Sci. U. S. A. **107**(6), 2408–2412 (2010)

Scanfeld, D., Scanfeld, V., Larson, E.L.: Dissemination of health information through social networks: twitter and antibiotics. Am. J. Infect. Control. **38**(3), 182–188 (2010)

Sedkaoui, S., Khelfaoui, M., Kadi, N.: Does technological context support academic entrepreneurship activities in Algeria? In: Eilu, E., Baguma, R., Pettersson, J.S., Bhutkar, G.D. (eds.) Digital Literacy and Socio-Cultural Acceptance of ICT in Developing Countries. Springer, Cham (2021)

Sedkaoui, S., Khelfaoui, M.: Sharing Economy and Big Data Analytics. ISTE-Wiley, London (2020)

Signorini, A., Segre, A.M., Polgreen, P.M.: The use of Twitter to track levels of disease activity and public concern in the U.S. during the influenza A H1N1 pandemic. PLoS One. **6**(5) (2011)

Shin, S., Seo, D., An, J.: High correlation of Middle East respiratory syndrome spread with Google search and Twitter trends in Korea. Sci. Rep. **6** (2016)

Stefanidis, A., Vraga, E., Lamprianidis, G., Radzikowski, J., Delamater, P.L.: Zika in twitter: temporal variations of locations, actors, and concepts. JMIR Public Health Surveill. **3**(2), 22 (2017)

Steffen, B., Egli, F., Pahle, M., Schmidt, T.: Navigating the clean energy transition in the COVID-19 crisis. Joule. **4**(6), 1137–1141 (2020)

Teso, E., Olmedilla, M., Martínez, M., Toral, S.: Application of text mining techniques to the analysis of discourse in eWOM communications from a gender perspective. Technol. Forecast. Soc. Chang. **129**, 131–142 (2018)

Towers, S., Afzal, S., Bernal, G., Bliss, N., Brown, S., Espinoza, B., Jackson, J., Judson-Garcia, J., Khan, M., Lin, M., Mamada, R., Moreno, V.M., Nazari, F., Okuneye, K., Ross, M.L.,

Rodriguez, C., Medlock, J., Ebert, D., Castillo-Chavez, C.: Mass media and the contagion of fear: the case of Ebola in America. PLoS One. **10**(6), e0129179 (2015)
Weible, C., Nohrstedt, D., Cairney, P., Carter, D., Crow, D., Durnová, A., Heikkila, T., Ingold, K., McConnell, A., Stone, D.: COVID-19 and the policy sciences: initial reactions and perspectives. Policy. Sci. **53**(2), 225–241 (2020)
Wong, R., Harris, J.K., Staub, M., Bernhardt, J.M.: Local health departments tweeting about Ebola: characteristics and messaging. J. Public Health Manag. Pract. **23**(2), 16–24 (2017)
Wongkar, M., Angdresey, A.: Sentiment analysis using naive bayes algorithm of the data crawler: Twitter. In: Fourth International Conference on Informatics and Computing (ICIC), pp. 1–5 (2019)
Xue, J., et al.: Public Discourse and Sentiment during the COVID-19 Pandemic: Using Latent Dirichlet Allocation for Topic Modeling on Twitter. Social and Information Networks (2020) Available at https://arxiv.org/abs/2005.08817

The Role of Web Analytics in Online Marketing

Cherifi Mahfoudh and Berki Othmane

Abstract The study discusses the problematic impact of web analytics in the process of on line marketing; the purpose of this article is to identify the web analytics that is accumulating due to storage techniques and new technologies for computers and mobile phones, to devices connected to the internet and social media, in addition to the concept of electronic marketing and strategies and showing the impact of analyzing the web analytics on e-marketing by presenting the most important models for the leading international companies in this field. The study concluded with a set of results, the most important of which is the need for companies to use the web analytics to contribute to the providing services and products that fit the needs of customers more accurately and to build marketing strategies to ensure continuity.

Keywords Web analytics · Online marketing · Web analytics tools · Leaders of on line marketing companies

1 Introduction

Not long ago, there were very few website visitors, as the analytics capabilities were limited to large companies that could afford to spend a lot of money a month on software to track and report web activity.

A wide variety of web metrics measuring and tracking applications are now available, making analytics a hot topic both online and offline. For the most part, these tools are still expensive, but free alternatives exist that perform just as well.

Analytics tools have become more accessible and affordable than ever before, leading to an increase in web-based applications. Thus, web analytics has become the talk of the hour by most national, regional, and international bodies and institutions and a wide field for researching database management, methodologies,

C. Mahfoudh (✉) · B. Othmane
University of Djilali Bounaama, Miliana, Algeria

S. Sedkaoui et al. (eds.), *International Conference on Managing Business Through Web Analytics*, https://doi.org/10.1007/978-3-031-06971-0_29

and procedures that can be adopted in the exploitation of big data in all fields, and perhaps the most important of which is online marketing, whose modern guidance depends on the use of interactions between companies and communication with customers by achieving speed, shortening effort, time and rounding distances.

Based on the above, we can raise the following key issue:

What is the impact of web analytics in online marketing?

Research Structure
In order to answer the previous question, we divided this paper into three main sections:

- Conceptual Framework for Web Analytics
- Online Marketing Concept
- The Role of Web Analytics in online Marketing

2 Conceptual Framework for Web Analytics

We live now in the information age, and most of what we do is greatly influenced by our ability to analyze and exploit this information, whether it is via the Internet, our computers, or our mobile phones, and the expressive word that describes this analysis is "web analytics."

2.1 Web Analytics Definitions

To better understand and optimize the use of the Internet, web analytics measures, collects, analyzes, and reports web data. Aside from counting visitors, web analytics can be used to learn about a company's business and market, as well as boost a website's productivity (Web Analytics 2020a). This is done by measuring data related to the Internet site, including visitor behavior, traffic volume, conversion rate, web server performance, user experience, and other information to understand and demonstrate the results and continuously improve the efficiency of a website (Web Analytics Terms 2020).Web analytics applications can also help companies gauge the results of campaigns by estimating traffic to a website that changes after launching an advertising campaign. In addition, web analytics provides information on the number of visitors to the site and page views. It helps assess popular access trends that are very useful for market research.

2.2 *Types of Web Analytics*

After we have addressed the general concepts related to social networking, we will briefly address the analysis of social networks, their concept, and importance.

2.2.1 Off-site Analytics

Are analytics that refer to measurement and analysis regardless of whether you own or invest in a website. The measure includes the potential audience for accessing the site (opportunity), the potential share of voice (exposure), and the degree to which the chatter is prevalent in relation to what is happening on the Internet as a whole.

2.2.2 Website Analytics

That measure a visitor's journey on your website. This includes on-site drivers and conversion programs. For example, these programs analyze the properties of pages that convince the browser to make a purchase. It also performs analytics that measure your site's commercial performance. Usually, comparisons of aggregated data are made with key performance indicators. The results are used to improve a website or define the marketing campaigns required to attract the audience.

2.3 *Web Analytics Features and Capabilities*

The web analytics features and capabilities listed in Table 1 (TrustRadius 2020).

Table 1 Web analytics features and capabilities

Individual-level tracking	On-the-fly segmentation	In-page analytics (session recording, click tracking, mouse tracking, heat maps)
Real-time analytics	E-commerce tracking	Goal conversion tracking
A/B testing	Funnel analysis	Event tracking
Mobile analytics	Cohort analysis	Privacy compliance
Attribution modeling	Cross-device tracking	On-premise option
Benchmarking		

Source: Web analytics tools overview | TrustRadius

2.4 *Web Analytics Data Sources*

Web analytics is primarily concerned with gathering and analyzing information about how people use a website. Most of our information is derived from four sources: (Web Analytics 2020b)

- It is possible to obtain direct HTTP request data: (HTTP request headers).
- IP addresses of users, for example, are required for successful transmission of HTTP requests, but they are not part of an actual HTTP request itself.
- Session and referral data generated and processed by application level programs (such as JavaScript, PHP, and ASP.Net) are transmitted with HTTP requests. Internal logs, rather than public web analytics services, are more commonly used to track these metrics.
- External data can be used in conjunction with on-site data to help augment and interpret the website behavior data described above. As an example, geographic regions and Internet service providers, e-mail open and click through rates, direct mail campaign data, sales and lead history, or other data types may be linked to IP addresses.

2.5 *Best Web Analytics Tools*

Web analytics is a major event produced by the continuous developments in the world of technology, and it is a solution to the questions raised, as relying on it has become the task of technology companies due to their specialization and great potential to drive innovation and advancement.

Online businesses and market researchers use web analytics to collect and measure the amount of data needed to understand and improve a web site's performance. The most commonly used metrics in web analytics reports are unique visitors, visits, time on-site, bounce rate, geographic location of visitors, bounce rate, and conversion rate (Popular web analytics tools for online businesses 2020), and we will address in the following the most important of these tools.

2.5.1 Google Analytics

Google Analytics is the simplest and most comprehensive web analytics service available, and it's completely free to use. Over half of the world's top 10,000 most popular websites, according to the site's statistics, use Google Analytics to find out where their visitors come from, what they do on your site, and how often they return. While more detailed reports are available as you get more involved in site analytics, the ease of use is what makes it one of the most popular services out there (LouDuboi 2020).

"There's really only one tool for small businesses need and that's Google Analytics," notes Penn. "It's so incredibly robust in terms of what it offers and if someone tells you that Google Analytics isn't enough for a small business, then frankly they have no idea how to use it properly.".

The web analytics experts we spoke to all agreed that Google Analytics was the best option (suggested by every expert).

2.5.2 Woopra Analytics

When it comes to taking automated processes to the next level, Woopra is hands down the best new online integration tool. With Google Analytics, you'd know that they prohibit the sending of PII (personal identifiable information), which has always been the most important issue in identifying individual customer activity.

This often makes it difficult for many businesses without login systems to "join" data together if you want to gain a deeper understanding of how your individual customers are behaving on your site (e.g., lead generation sites before converting or contacting). This is where Woopra can help!

Where Woopra really excels is in its integration with other SaaS and PaaS systems which can be further extended into triggers (IF/THEN do some action) based on individual or group visitors activity.

Woopra works very much like Google Analytics – a tracking code is placed on your website which gathers information on every visitor which is then reported back to your Woopra Dashboard. For more advanced users, events can be created and triggered to identify users when they either login to your website or submit a contact form. That information you can then segment, group, report, and merge into an array of triggers (e-mail, SMS, live chat, and the list goes on!).

2.5.3 Yahoo Analytics

Google Analytics is a great place to start, but Yahoo's similar service provides a little more depth in your surveys. You can import cost of goods data (unlike Google) and get reports on visitor demographics and behavior, as well as a more straightforward approach to multi-site analytics. Customization options are also available. In terms of profiling, filtering, and customization, Yahoo Analytics is a great alternative to Google for those who want to delve a little deeper (https://policies.yahoo.com/ie/en/yahoo/privacy/topics/webanalytics/index.htm 2020).

2.5.4 Compete Analytics

Clickstream analytics tools like Compete can be used in conjunction with them. When you use Compete, you have access to creative intelligence about what your rivals are doing and how users first found you (what their clicks were both before and after). Free data on traffic volume is included as part of the service. Compete, on the other hand, has a paid search analytics service that allows you to see which keywords are driving traffic to both your site and your competitors' sites (https://en.wikipedia.org/wiki/Compete.com 2020).

"The deeper digital insights you have, the better understanding you have of your customer," says Aaron Smolick, senior director of marketing at Compete. "By using Compete products, you will have all of the information that you need to make educated decisions to optimize your online campaign, increase market share and dominate the competition."

3 Online Marketing Concept

One of the most important outcomes of globalization is the rapid decline of the industrial economy in favor of the digital economy, as the latter has made many leaps as a result of the rapid and successive development of information and communication technology, and a reflection of this strategic shift has emerged electronic marketing, which is a true application of globalization and has become an effective tool to achieve growth from on the one hand and the continuous improvement of efficient and effective customer service on the other hand.

3.1 Definition of Online Marketing

Marketing researcher Coviello from the University of Oakland knows that online marketing "is the use of the Internet and other technological and interactive means to create and create a dialogue between the organization and specific consumers (Musa 2007)".

Some also define online marketing as "the use of electronic means in conducting reciprocal commercial operations between the designated parties instead of direct contact."

From the above, we conclude that the distinctive feature in the application of e-marketing is the unification of the changing needs of customers and modern renewable technologies, which leads to a revolution in the way in which business is built, and this is why e-marketing is considered a technology of change.

3.2 Online Marketing Strategies

Companies adopt different marketing strategies to achieve their goals, by adapting them to multiple electronic means on the one hand and targeting parts of the market by relying on relationships with exchange parties on the other hand, and among the most popular electronic marketing strategies currently adopted, we mention the following:

3.2.1 Content Marketing

It is an effective way to increase sales without the need for direct sales. Content containing information about the product is published on various website platforms to target potential customers. Social media platforms are the most common usage of them to build brand awareness and increase sales.

3.2.2 Digital Marketing

It is marketing products using digital channels to reach customers; it includes marketing using social networking sites and search engines and improving content accessibility and many digital media. Digital marketing helps promote the brand and products.

3.2.3 Experiential Marketing

It mainly relies on social media to build a relationship between customers and the brand in a fun and memorable way. Customers express their opinions about the products they buy and evaluate the method used by companies to market their products, as consumers' experiences with various products determine the success or failure of each product during marketing campaigns and within experimental marketing the features of the product are promoted by testing it within a trial period, while representing the brand associated with the product through look and feel. All this leads to a personal memory for each customer about understanding brand value as well as understanding the product associated with it.

3.2.4 Social Media Marketing

It is an important way to increase communication with customers through various sites such as (Face book, Twitter, YouTube). Correct marketing strategies through social media may have an effective impact and great success, whether with the aim of promoting the brand and products or to increase sales.

3.3 *Advantages of Online Marketing*

Online marketing has created a sophisticated marketing environment that brings customers more luxury and pleasure in searching for their needs and desires and satisfying them in a way that makes this customer feel completely satisfied with these products.

At present, most organizations direct their marketing activities toward online marketing in order to benefit from the advantages and high potential of online marketing, which we mention the following:

3.3.1 Access to Global Markets

Several studies confirmed that electronic marketing leads to expanding markets and increasing the market share of organizations due to the global spread (globalization), and that electronic marketing enables customers, with different geographical locations, to obtain their needs and make comparisons between the products of different and multiple organizations, as online marketing does not recognize the limits geography.

3.3.2 Providing Products According to the Needs of Customers

Online marketing provides marketers with opportunities to adapt their products in a way that meets customers' needs electronically (*e-customization*), as the communication and interactive energies of electronic marketing have made a quantum leap in the means and methods of satisfying the needs and desires of customers and gaining their satisfaction with these products provided to them.

3.3.3 Feedback

There are great opportunities provided by electronic marketing for organizations to respond to the changes that occur in the markets as well as the changes that occur in the technical environment, which achieves the integration of customers' needs with technological developments through the so-called flexible process of product development, and this process depends on a study and a sensing of the market (sensing the market) through the interactive mechanisms of online marketing.

3.3.4 Reducing Costs and Using Flexible Pricing

Under online marketing, pricing strategies are not only a quick response to the market situation but rather take into account internal and external factors and variables, as the concept of flexible pricing is its applications through electronic marketing mechanisms, as it provides techniques that enable the buyer to search and find the best available prices.

3.3.5 Create New Forms and Channels of Distribution

Online marketing introduces a new philosophy for an electronic market in which interaction is between the parties to the exchange process without the need for intermediaries, which led to the emergence of the term "disintermediation." E-marketing also offers intermediaries in a new form and quality called "cybermediares," which are organizations whose aim is to facilitate exchanges between producers and customers, and their work is the work of commercial service providers.

3.3.6 Use Interactive Promotions with Customers

One of the most effective means of electronic promotion is electronic advertising, and electronic advertising is more attractive and widespread in light of the trend toward electronic marketing, as direct ads via the Internet are increasing at a rate of 12% annually, and their value is more than $9 billion, and electronic marketing provides a new form and concept for advertising, as companies provide marketing messages of their own and their products to targeted environments through specific websites they expect their audiences to be able to distinguish and perceive.

3.3.7 Support and Activate Customer Relations Management

The process of competition in electronic markets is going through a transitional phase due to the shift from traditional competition to competition that depends on the capabilities and capabilities of organizations, and there is no doubt that the great developments in information technology related to electronic marketing have created qualitative opportunities in supporting distinctive strategies and improving the competitive position of organizations (Sabra 2010).

4 The Role of Web Analytics in Online Marketing

4.1 The Uses of Web Analytics in Online Marketing

The uses of web analytics are multiple as it has proven its effectiveness in all fields, especially in the field of online marketing, as it helps in analyzing consumer behavior, allowing companies to serve their customers and gain a competitive advantage, by making use of the services of technical companies that help in employing big data that led to the emergence of global leading companies in the field of online marketing.

Today, companies and institutions of all kinds are able to analyze customers' movements from buying, selling, and the like more precisely so that they can accordingly know the most demanded or stagnant goods and suggest to their customers specific goods according to the purchases that are made and they have the ability to understand customers' behavior more accurately, and identify the distinguished among them and those in need of assistance or to determine their orientations or monitor their performance. This is not only for traditional sales centers but now also includes electronic stores on the Internet and on a larger scale.

The user of social networks or e-mail is often surprised by the emergence of commercial advertisements for goods that he previously searched for in other applications, but more than that there are some algorithms that use location data on the phone to suggest advertisements. Hence, we find that this occurs as a result of web analyzes resulting from these sites and their use in marketing by using every small part of the available data about users to find out their preferences and preferences in order to display the goods in the best possible way that brings e-marketing companies the greatest possible profit and does not analyze the data that gets. It should not only allow the user to browse the Internet but may also track his visit to real markets through the GPS device installed in his device. Some people may think that this is a breach of privacy, but in fact the companies that do so have immunized themselves through the terms agreement that the user signs when logging into social media applications Facebook and Twitter.

In recent years, the electronic marketing process has entered a new level of competition. In light of the vast amount of user data provided by social networks and the use of the Internet, the first concern of the e-marketing giants has become how to stay in competition at various levels; on the one hand, companies need to promote their goods, and this requires knowledge of customers 'needs, and on the other hand, they need to offer their goods at competitive prices. At the same time, it guarantees a large profit margin.

4.2 Leading Companies in Using Web Analytics in Online Marketing

4.2.1 Amazon Corporation

Amazon Corporation is an American multinational company specializing in online marketing and cloud computing, founded in 1994 by Jeff Bezos.

Amazon provides retail marketing services on the Internet to four groups: consumers, sellers, institutions, and content creators; the company provides marketing and promotion services such as advertisements on the Internet, and the company focuses on good choice of price and convenience, and websites are designed on this basis in order to achieve convenience for users so that they are the best choice (Amazon 2020).

Amazon.com handles millions of background processes every day, as well as inquiries from more than half a million sellers.

The Linux system is mainly used to be able to handle this huge amount of data, and Amazon has the three largest Linux databases in the world, which have a capacity of 24.7 terabytes.

4.2.2 eBay Corporation

eBay is an online marketing website based in California, USA, founded in 1995 by Pierre Omidyar; the company's mission is to provide a global platform for trading, and eBay is considered a leading online auction site; the company is considered one of the largest online markets in the world, as the number of website users exceeded 100 million worldwide in 2011 (Ebay 2020).

eBay has managed to achieve a tremendous impact on e-marketing, as the total value of goods sold in eBay in 2011 was $ 68.6 billion, or more than $2100 per second.

4.2.3 Walmart Corporation

Walmart is an American company specialized in retail trade, founded in 1962 by Walmart, that improves product coordination management with a better customer experience by taking advantage of predictive and descriptive analytics results by following the product at the time of entry, and this activity improves customer experience and helps retention. By classifying demand patterns in the local community, this means that the assortment varies from one store to another based on customer needs and target classifications. It also uses warehouse data and commercial operations to provide satisfactory delivery and payment options for customers, and the customer can even buy (scan and go) and receive Invoices by e-mail and payable using *Walmart Pay* card2011 (Walmart 2020).

Chain store (*Walmart*) processes more than a million commercial transactions every hour, which are imported into databases estimated to contain more than 2.5 petabytes (2560 terabytes) of data – 167 times the data in all books in the Library of Congress in United State.

4.2.4 Alibaba Corporation

Alibaba is a cloud computing and online marketing services company headquartered in Hangzhou, China, founded in 1999 by Jack Ma, and Alibaba's Tawpo site is among the 20 most visited sites in the world, and the company is a global trading platform, such as eBay, and the number of products on the site is estimated at nearly 1 billion products from 7 million users (ALIBABA 2020).

The company's sales, according to experts' estimates, exceed the sales of the social networking site "Facebook," which is estimated at $15 billion, which indicates the great potential available to "Alibaba,"and experts indicate that the company is heading to become the first e-marketing company in history. Hundreds of millions of Chinese citizens have not yet tried e-marketing, and these are a great opportunity to attract them to this service, and analysts expect that the e-commerce market in China will be greater than Current markets in the United States, Japan, Germany and France combined by 2020.

In the year 2013, transactions worth $240 billion took place in two Alibaba companies, and this number is twice the trading volume of Amazon and three times the volume of eBay.

5 Conclusion

The digital convergence in communications and media has led to the employment of its tools for analyzing and processing web data and using it in developing the online marketing process by creating and standardizing many new products and services, as well as creativity in methods of promotion and speed in implementation.

In this study we tried to stand on the concept of web analyzes and focus in particular on highlighting the role of its uses in e-marketing. The most important results reached by the study can be presented in the following points:

- Although the term web analytics has become a lot in circulation, making the most of this data remains a challenge and an obstacle for many international companies and organizations.
- The advancement of digital technology has had a great impact on marketing practices, as technology has provided innovative opportunities to capture better data for customers well and increase focus with them. As a result of these changes and the huge amount of accumulated data, companies in general and marketing in particular have to improve their working methods.
- The use of web analytics in the field of e-marketing has received increasing interest over the past few years because of its importance in increasing sales numbers and improving the corporate image of customers.
- The main goal of using web analytics by the leading companies in the field of online marketing is to provide services and products that most accurately fit and suit the needs of customers and to build marketing strategies that ensure their continuity.
- The provision of data by the technology companies of the leading companies in the field of web analytics is not without its drawbacks. It raises an important issue, which is privacy and its implications for users of various websites.

References

Web Analytics. En.wikipedia.org. Available at https://a.wekipedia.org/wiki//web analytics (2020a), 11 Aug 2020

Web Analytics Terms: Statisticalconcepts.blogspot.com. Available at http://statisticalconcepts.blogspot.com/2010/12/web-analytics-terms.html (2020), 11 Aug 2020

TrustRadius: Web analytics tools overview | TrustRadius. Available at https://www.trustradius.com/web-analytics (2020), 12 Aug 2020

Web Analytics: En.wikipedia.org. Available at https://a.wekipedia.org/wiki//web analytics (2020b), 11 Aug 2020

Popular web analytics tools for online businesses: Big data made simple – One source. Many perspectives. Available at http://bigdata-madesimple.com/popular-web-analytics-tools-for-online-businesses (2020), 12 Aug 2020

LouDuboi: 11 best web analytics tools. Available at https://www.inc.com/guides/12/2010/11-best-web-analytics-tools.html (2020), 13 Aug 2020

https://policies.yahoo.com/ie/en/yahoo/privacy/topics/webanalytics/index.htm, 14 Aug 2020

https://en.wikipedia.org/wiki/Compete.com, 14 Aug 2020

Musa, A.F.A.: Information Technology and its Role in Traditional and Electronic Marketing, 1st edn, p. 128. ATRAC Publishing and Distribution (2007)

Sabra, S.T.: Electronic Marketing, 1st edn, p. 46. Dar Al-Asyar Al-Alami for Publishing and Distribution (2010)

Amazon: https://en.wikipedia.org/wiki/Amazon_(company) (2020), 11 Aug 2020

Ebay: https://en.wikipedia.org/wiki/EBay (2020), 10 Aug 2020

Walmart.: https://corporate.walmart.com (2020), 14 Aug 2020

ALIBABA: https://en.wikipedia.org/wiki/Alibaba_Group (2020), 14 Aug 2020

Improvement of Recommender Systems with Item Link Prediction

Sahraoui Kharroubi, Youcef Dahmani, and Omar Nouali

Abstract Social networks and other web services (e.g., e-commerce) necessitate the use of recommendation techniques in order to satisfy the needs of their users. A new recommendation algorithm faces a major challenge when it comes to providing high-quality recommendations with a minimal amount of common feedback. A weighted bipartite network centered on the item entity is used in this paper for more informative modeling. Item-user connectivity can be used to uncover hidden information. Double projection forward and backward is needed to predict links that don't exist. When real datasets are used, the results look even better.

Keywords Collaborative filtering · Weighted bipartite network · Ranking · Sparsity

1 Introduction

It is common practice for companies to use recommendation systems to recommend relevant products to customers using a similarity algorithm. Large companies and websites like Netflix, Amazon, Facebook, YouTube, Twitter, and so on use recommendation techniques in their servers. The three main approaches (Adomavicius and Tuzhilinand 2005; Breese et al. 1998) to recommendation systems are the content-based approach, the collaborative approach, and the hybrid approach.

Bipartite network modeling is currently the subject of a number of in-depth studies (Musto et al. 2017; Yin et al. 2019). Bipartite networks are an excellent theoretical and practical model for a wide range of real-world systems. When it comes to online shopping or education, a bipartite network connects two distinct

S. Kharroubi (✉) · Y. Dahmani
Faculty of Mathematics and Computer Science, Ibn Khaldoun University, Tiaret, Algeria
e-mail: sahraoui.kharroubi@univ-tiaret.dz

O. Nouali
Software Laboratory, C.E.R.I.S.T., Algiers, Algeria
e-mail: onouali@cerist.dz

S. Sedkaoui et al. (eds.), *International Conference on Managing Business Through Web Analytics*, https://doi.org/10.1007/978-3-031-06971-0_30

sets of users and items to model the recommendation process. A link prediction task in a bipartite network is the recommendation task in the recommendation system. In recommender systems, we've come up with an item share propagation method (ISpLR). Using this method, the user can predict the weighting of a link in relation to an item's ranking value. Forward-backward projection is used to link items to users in order to build up a database of collected data.

The process can be summarized as follows:

1. Item share propagation for link ranking in a weighted bipartite network (ISpLR).
2. Accumulation of shares linearly to preserve information propagated with a double projection (forward-backward).
3. Implementation of the method without adjustment parameters and with different dense datasets.

This paper is organized as follows: the background is given in Sect. 2, a detailed description of the proposal is presented in Sect. 3, a description of the experimental phase, and discussion of the results is given in Sect. 4 and 5 respectively. In the conclusion, we make some suggestions for future work.

2 Background

When it comes to creating personalized recommendations, collaborative filtering (CF) is the most popular and effective method (Chun-Yang et al. 2019). Through the sharing of similar experiences and opinions, CF is able to create user and/or product communities (neighborhoods). For items that have not yet been judged, the community is used to predict their relevance. Algorithms (Nesreen et al. 2018; Kim and Segev 2018; Shams and Haratizadeh 2016) have been proposed over the last two decades and can be divided into three levels.

2.1 *Traditional Approaches*

They are based on the user's "rating matrix" and use statistical methods to predict what the user will do next. Assumedly, users who have similar ratings will continue to have similar preferences in the future as well. There are two types of collaborative filtering: memory-collaborative (memory CF) and model-collaborative (model CF) (Kovkov and Lemtyuzhnikova 2018). The rating matrix scores is used to predict the current user's activity level. A portion of the "rating matrix" is used to estimate or learn a model that generates the predictions in model CF, in order to reduce the computational complexity.

2.2 Context-Aware and Semantic Approaches

This extra information, which can be categorized into dimensions, helps us to better understand the user or the item in question. Personal data (such as gender, occupation, and age) and professional and social data (such as activities, interests, interactions with other users, preferences, and so on) are included, as are environmental factors (such as weather, noise, and so on), geographic coordinates (such as latitude and longitude), and time (Kharroubi et al. 2018). The goal is to make use of this data and find similar ones in order to make an accurate prediction. Tuzhilin and Adomavicius (CARS context-aware recommender systems) (Adomavicius et al. 2011), using the mobile phone and the Global Positioning System (GPS) can recommend a restaurant or hotel to a tourist based on previous choices of its similar neighbors by context (Kashevnik and Ponomarev 2017).

2.3 Graph-Based Approaches

A network has a specific topology because each node is connected to every other node. Like social networks like Facebook and Twitter, the unipartite topology is represented by a set of nodes and a set of edges. PageRank (Page et al. 1999) and *HITS* (Giannoulakis and Tsapatsoulis 2019) algorithms use a random walk or an iterative process to determine the importance score of each node in a network of web pages. The EdgeRank algorithm takes care of everything (Birkbak and Carlsen 2016): a user's *Facebook* profile through the parameters of affinity, type, and freshness. The objective is to take advantage of the network's structure in order to gather future links. Several studies (Chun-Yang et al. 2019; Kharroubi et al. 2018) are based on a bipartite structure to improve recommendation algorithms. Nesreen K (Nesreen et al. 2018). implemented the *SimAdapt* algorithm to measure the similarity between two nodes according to the number of common neighbors in a bipartite graph. Jung-Hun Kim et al. (Kim and Segev 2018) used a bipartite network of textual titles and keywords in a physics research article to conduct their investigation.

In this paper, a weighted bipartite network recommendation method is presented. Predicting new links and recommending relevant items to users are done based on the item-rating history. In order to do this, both items and users are counted as having a cumulative share of the total.

3 Approach

3.1 Problem and Motivation

In many complex systems, a network model is used that consists of two parts: nodes and links. On a single-party network, these systems tend to work best (social networks, links between web pages, atoms and molecules, etc.). A single set of nodes connected by links makes up a single-party network. Using network model algorithms is a simple process. Bipartite networks are a type of network that is commonly used in recommendation systems.

3.2 Modeling with a Weighted Bipartite Network

A network is bipartite if the set of nodes is divided into two distinct sets I and U. Each link has one end in the first set and the other end in the second set (no links in the same set), and k determines the weight of the link. Formally, let $B\ (N, L\ )$ be a bipartite network where N is the set of nodes and L is the set of links. B has the following properties:

$$N = I \cup U$$

$$L = \{(i, u, k)\ , i \in I, u \in U, k \in N\}$$

$$I \cap U = \emptyset$$

3.3 Item Link Prediction

Consider a recommender system which is composed of a set of items $\{I_1, I_2, \ldots I_N\ \}$ and a set of users $\{U_1, U_2, \ldots U_M\ \}$ linked by weighted links $\{(i, u, k), i \in I, u \in U, k \in N\}$. From the previous interactions of users with the items (the rating history), let's try to deduce a nonexistent link weight. In another way, can we recommend a relevant planned item to a user? This decision is equivalent to valuing the link weight between the item and the user, like (I_3, U_1) (Fig. 1).

To make such a prediction, we need to know how many items each user owns and how many of those items are shared. It is necessary to calculate the shares of items among users in two steps, one forward and one backward. Since there are N items and M users, as well as $R(i, j)$, which represents the weight of the link (i, j), these calculations can be broken down into two steps (rating value).

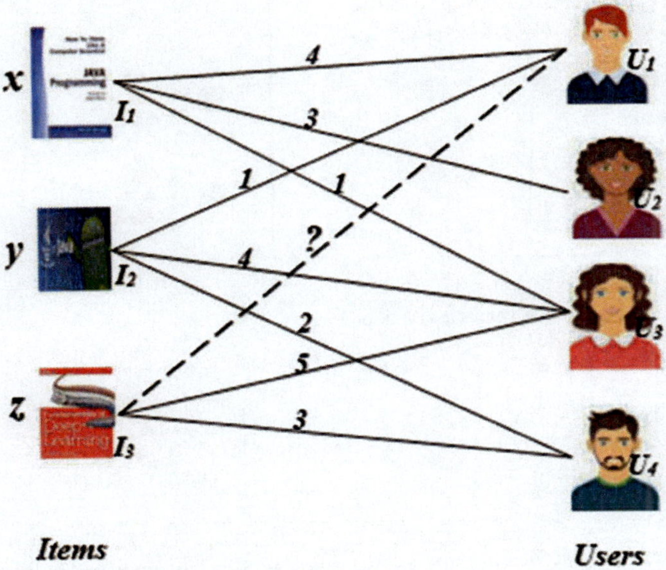

Fig. 1 Weighted link prediction

First, a total of all of the items is tallied up.

$$IC_i = \sum_{j=1}^{M} R(i, j) \tag{1}$$

The share of user j:

$$q_u(j) = \sum\nolimits_{i=1}^{N} \frac{R(i, j)}{IC_i} \tag{2}$$

The second step involves a user's total rating, which is calculated as follows:

$$UC_j = \sum_{i=1}^{N} R(i, j) \tag{3}$$

The share of item i:

$$q_i(i) = \sum\nolimits_{j=1}^{M} \frac{R(i, j)}{UC_j} \tag{4}$$

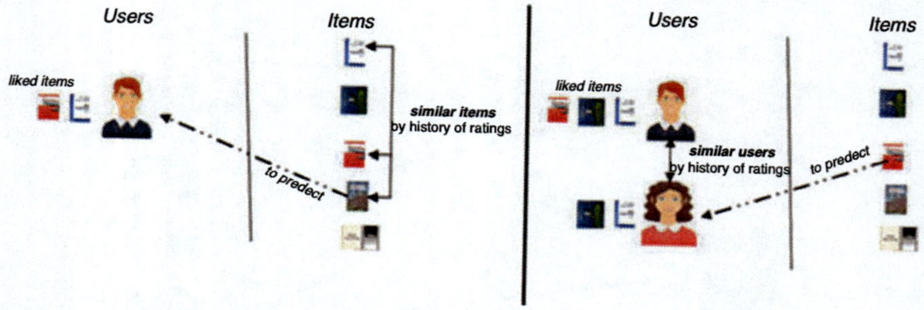

Fig. 2 Item link prediction versus user link prediction

Table 1 Item link prediction

	U_1	U_1	U_1	U_1
I_1	4	3	1	**0.5304**
I_2	1	**0.45**	4	2
I_3	**0.7071**	**0.1875**	5	3

As a result of this formula, the value of link prediction for item t to user s is:

$$P(t, s) = \sum_{j=1}^{M} \frac{R(t, j)}{\sum_{i=1}^{N} R(i, j)} * \left(\sum_{i=1}^{N} R(i, j) * \frac{R(i, s)}{\sum_{j=1}^{M} R(i, j)} \right) \quad (5)$$

We can get the following result by swapping out item cumulate (1) and user cumulate (3):

$$P(t, s) = \sum_{j=1}^{M} \frac{R(t, j)}{\sum_{i=1}^{N} UC_j} * \left(\sum_{i=1}^{N} R(i, j) * \frac{R(i, s)}{\sum_{j=1}^{M} IC_i} \right) \quad (6)$$

Using a bipartite topology, the information shared between the nodes can be preserved in a bidirectional manner. The degree to which users are satisfied can be determined by taking into account the link weights. Rather than using a user-based reasoning approach, this method employs an item-based approach (Fig. 2).

Item link prediction is shown in the following example, with the prediction performance between them being described in detail. Table 1's link values are based on the data shown in this figure.

Either three items I_1, I_2, I_3, with initial inputs x, y, z, respectively.

First step (forward): determining user's share:

$$\begin{cases} U_1 : \frac{4}{8}x + \frac{1}{7}y \\ U_2 : \frac{3}{8}x \\ U_3 : \frac{1}{8}x + \frac{4}{7}y + \frac{5}{8}z \\ U_4 : \frac{2}{7}y + \frac{3}{8}z \end{cases}$$

Table 2 MAE of item-based predictions

	U_1	U_1	U_1	U_1	MAE I_i
I_1	3.3214	2.3625	1.7857	0.5304	0.7006
I_2	0.9714	0.4500	3.6357	1.9429	0.1500
I_3	0.7071	0.1875	4.5786	2.5268	0.4473
	MAE				0.4326

Second step (backward): determining item's share:

$$\begin{cases} I_1 : \frac{4}{5}U_1 + \frac{3}{3}U_2 + \frac{1}{10}U_3 + \frac{0}{5}U_4 \\ I_2 : \frac{1}{5}U_1 + \frac{0}{3}U_2 + \frac{4}{10}U_3 + \frac{2}{5}U_4 \\ I_3 : \frac{0}{5}U_1 + \frac{0}{3}U_2 + \frac{5}{10}U_3 + \frac{3}{5}U_4 \end{cases}$$

which gives us:

$$\begin{cases} I_1 : \frac{63}{80}x + \frac{6}{35}y + \frac{1}{16}z \\ I_2 : \frac{3}{20}x + \frac{13}{35}y + \frac{2}{5}z \\ I_3 : \frac{1}{16}x + \frac{16}{35}y + \frac{43}{80}z \end{cases}$$

So, the prediction value of the $link\ (I_3, U_1) = {}^1\!/_{16}\,(4) + {}^{16}\!/_{35}\,(1) + {}^{43}\!/_{80}\,(0) =$ 0.7071.

Further, the prediction value of the $link\ (I_1, U_4) = {}^{63}\!/_{80}\,(0) + {}^6\!/_{35}\,(2) + {}^1\!/_{16}\,(3) =$ 0.5304.

It is necessary to extrapolate the results from previous votes (Table 2) in order to derive the mean absolute error (MAE) (see Sect. 4.3).

It is possible to distinguish between the rating value provided by users and the prediction value generated by formula using the MAE measure (6).

For item I_2 and I_3:

$$\text{MAE}\,(I_2) = 1/3\,(|1 - 0.9714| + |4 - 3.6357| + |2 - 1.9429|) = 0.1500$$

$$\text{MAE}\,(I_3) = 1/2\,(|5 - 4.5786| + |3 - 2.5268|) = 0.4473$$

3.4 Recommendation Task

A recommender system analyzes a large number of items and chooses the most relevant to recommend to users. Based on the system's predicted value, a particular item may be recommended. As an input parameter, the algorithm compares this value to a threshold. The formula for determining a user's *R*(*t*, *s*) for an item t is as follows:

$$R(t, s) = \begin{cases} t \text{ recommended to s if } p(t, s) \geq \delta \\ t \text{ not recomended to s otherwise} \end{cases}$$

$p(t, s)$: prediction value of item t to user s.

The threshold δ is bound to the rating scale as a degree of user satisfaction, for example, for MovieLens and Amazon, the rating scale is limited to 5. Below Algorithm summarizes the main sequences of a recommendation task by item link prediction.

Algorithm *weight prediction link's*

---

input *Bipartite Network* $B(N, L)$

$N = I \cup U$ *set of vertices*

$L = \{(t, s), t \in I, s \in U\}$ *set of links*

$W: L \rightarrow \mathcal{R}^+$ *weighted function*

$(t, s) \rightarrow w$

Output *weight link*

$w(t,s) = 0$

For j=1 to M

$UCs = 0;$

For i=1 to N

$UCs = UCs + R(i, j)$ *% $s^{ème}$ user cumulate*

$q_s = 0$

For i=1 to N

$IC_t = 0$

For k=1 to M

$IC_t = IC_t + R(i, k)$

$q_s = q_s + \frac{R(i,j)*R(i,s)}{IC_t}$ *% $t^{ème}$ item cumulate*

$w(t,s) = w(t,s) + \frac{R(t,j)*q_s}{UCs}$

Return $w(t, s)$

if (w(t,s)>=δ) recommend t to s

This item-based method ensures that shared information is propagated between the two types of nodes (items and users). Weighted bipartite network topology is used to model the recommendation task.

4 Experimentation

4.1 Datasets

To evaluate the effectiveness of the proposed approach, tests were performed on four real-world datasets:

Amazon Books: a free download dataset http://jmcauley.ucsd.edu/data/amazon/for research purposes.

MovieLens: a movie evaluation dataset https://grouplens.org/datasets/movielens/ provided by the GroupLens research group. The datasets are used for research purposes, in particular, to test the performance of collaborative filtering algorithms.

Yelp2018 dataset: a set of data provided by the Yelp recommendation platform https://www.yelp.com/dataset. It includes statistics and user ratings for different items such as restaurants, hotels, shopping centers, etc.

Yahoo! Song: is a database of user-rated music tracks and albums. The Yahoo Research Alliance's Webscope program has made this dataset available. http://webscope.sandbox.yahoo.com/.

4.2 Comparison Methods

The ISpLR proposal has been tested using three different methods, including the following:

- **Joint ranking of neighbors in a signed bipartite network analysis (SibRank)** (Shams and Haratizadeh 2016): The method exploits the relationships of a bipartite graph of positive or negative signs to infer a degree of similarity. The signs reflect trust or mistrust between users.
- **Neural Graph Collaborative Filtering (NGCF)** (Wang et al. 2019): It uses a hierarchical layered neural graph model. This work measures an aggregation of connectivity based on user/item interactions.
- **Spectral Collaborative Filtering (SpectralCF)** (Zheng et al. 2018): Exploits the proximities of the graph and extracts connectivity information hidden in a bipartite graph. The contribution is made to reduce the cold start problem.

4.3 Metrics

Several metrics, such as MAE (mean absolute error), NMAE (normalized MAE), MSE (mean square error), and RMSE, are employed in the literature on recommendation (root MSE). The MAE and the recall are the two most common:

$$\mathrm{MAE} = \frac{1}{n}\sum_{i=1}^{n}\left|r_{ij} - p_{ij}\right| \tag{7}$$

r_{ij} : rating of item i by user j
p_{ij} : prediction of item i by user j
n : rating number.

Recall: the ratio of relevant items returned to all relevant items in a search:

$$R = {N_{pr}}\big/{N_P} \tag{8}$$

MAE and recall metrics have a significant impact on user satisfaction, which is measured by these metrics.

5 Results and Discussion

We split the datasets in half, using 85% of them for training and only 15% for testing. Predictions for the test set are generated using the algorithm, which is trained on the training data. The number of recommendations is used by MAE to calculate the discrepancy between the generated predictions and the actual votes from the training set. On a scale of (Adomavicius and Tuzhilinand 2005; Breese et al. 1998; Musto et al. 2017; Yin et al. 2019; Chun-Yang et al. 2019), the recommendation parameter () is set to 3. According to the number of recommendations, the error rate is depicted in Fig. 3.

When compared to SpectralCF, SibRank, and NGCF, the MAE result (Fig. 3) shows that the proposed method (ISpLR) performs better. For example, MovieLens and Yahoo! Song are better datasets for the MAE, while Yelp2018 and Amazon-Book are worse. The following are the reasons for this:

1. Using ISpLR, information shared between items and users can be preserved. Information is transmitted between nodes in the network in a lossless manner, resulting in a low error rate. An algorithm called SpectralCF combines the hidden connectivity of the proximity nodes, which reduces the error slightly.
2. According to the sparsity rate (Formula 9; Table 3) of each dataset, the MAE is affected by the different datasets used:

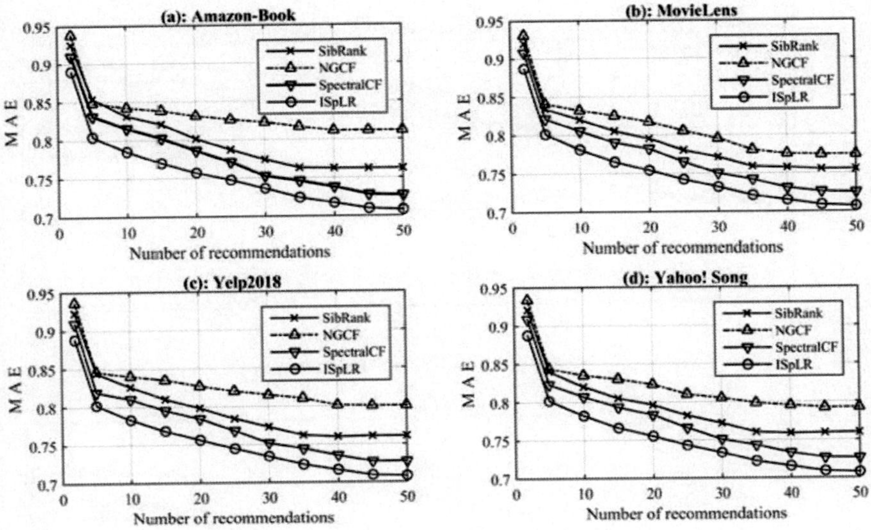

Fig. 3 Error rate depending on the length of the recommendation list.

Table 3 Sparsity rate

Dataset	MovieLens 1M	Yahoo! Song	Yelp2018	Amazon-Books
Sparsity rate	95.754%	97.976%	99.957%	99.988%

$$\text{sparsity rate} = 1 - \frac{|E|}{|U| * |I|} \tag{9}$$

$|E|$: number of edges

$|U|$: number of users

$|I|$: number of items

Error rates are directly proportional to sparsity rates (no ratings), and the quality of the recommendation system is strongly correlated to the number of judgments (ratings) of the users.

Figure 4 depicts the recall metric for Amazon Books (which has a higher sparsity) and MovieLens, two different datasets (lower sparsity). For the proposed method

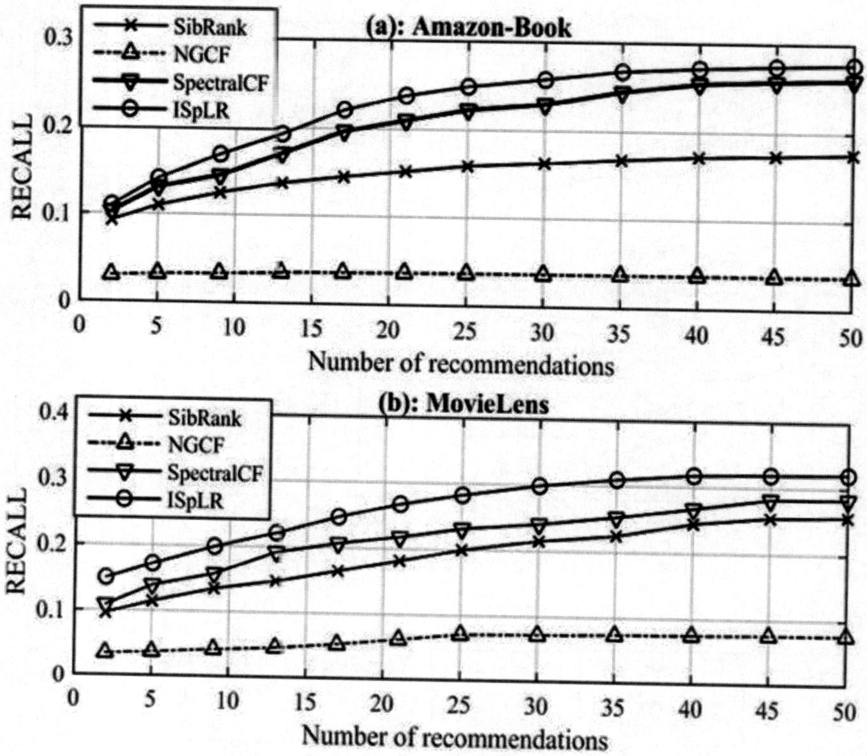

Fig. 4 Recovery rate depending on the length of the recommendation list

(ISpLR), the recall rate reached 30%, which represents the percentage of relevant items that were recommended from a few recommendations (20). (true positive rate).

NGCF has used several hierarchical layers to aggregate an order of connectivity that reduces the degree of similarity between users and the closest items, which negatively affects recall and accuracy.

It is reasonable to infer from these encouraging results that even with sparse datasets, ISpLR can produce high-quality predictions that are responsive to online suggestions. Using only a subset of the items on an e-commerce site, one can make predictions about the other items.

6 Conclusion and Future Work

For interpretation and implementation purposes, a recommendation system needs to be modeled in a network structure. The link prediction problem in a weighted bipartite network was addressed in this paper. For example, a user's perception of an item's predictive value can be calculated using the weightings of existing links. It's a linear and bidirectional approach to accumulating the shares of each user and each item. The term "share" refers to a portion of the item being shared with the user and the other way around. First, the information propagated between the items/users nodes is preserved when shares are accumulated; second, the link prediction calculation does not necessitate the use of additional adjustment parameters. It takes a lot of steps to get to the prediction phase in classic collaborative recommendation methods based on Pearson correlation or cosine similarity (Adomavicius and Tuzhilinand 2005; Breese et al. 1998). In the near future, we plan to investigate the method's scalability (combinatorial complexity) in comparison to currently used methods, as well as the accuracy of our predictions for online users.

References

Adomavicius, G., Tuzhilinand, A.: Toward the next generation of recommender systems: a survey of the state-of- the art and possible extensions. IEEE Trans. Know. Data Eng. **17**(6), 734–749 (2005). https://doi.org/10.1109/TKDE.2005.99

Adomavicius, G., Mobasher, B., Ricci, F., Tuzhilin, A.: Context aware recommender systems. Ai Magazine. **32**(3), 67–80 (2011). https://doi.org/10.1609/aimag.v32i3.2364

Birkbak, A., Carlsen, H.: The world of Edgerank: rhetorical justifications of Facebook's news feed algorithm. Comput. Cult. **2016**(5) (2016)

Breese, J., Heckerman, D., Kadie, C.: Empirical analysis of predictive algorithms for collaborative filtering. In: Proceedings of the 14th Conference on Uncertainty in Artificial Intelligence, pp. 43–52 (1998)

Chun-Yang, L., Yan-Zhen, Z., Ze-Qi, L., Bing, X.: Graph Embedding Based API Graph Search and Recommendation. J. Comput. Sci. Technol. **34**(5), 993–1006 (2019). https://doi.org/10.1007/s11390-019-1956-2

Giannoulakis, S., Tsapatsoulis, S. (2019). Filtering Instagram hashtags through crowd tagging and the HITS algorithm. IEEE Transactions on Computational Social Systems, 6(3): 592–603. https://doi.org/10.1109/TCSS.2019.2914080

Kashevnik, A.M., Ponomarev, A.V.: A multimodal context-aware tourism recommendation service: approach and architecture. J. Comput. Syst. Sci. Int. **56**(2), 245–258 (2017)

Kharroubi, S., Dahmani, Y., Nouali, O.: A semantic layer to improve collaborative filtering systems. Int. J. Comput. Sci. Eng. **17**(4), 365–376 (2018). https://doi.org/10.1504/IJCSE.2018.096024

Kim, J.H., Segev, A.: Hypothesis generation using link prediction in a bipartite graph. In: IEEE International Conference on Big Data, vol. 2018, pp. 2863–2867 (2018)

Kovkov, D.V., Lemtyuzhnikova, D.V.: Decomposition in multidimensional Boolean-optimization problems with sparse matrices. J. Comput. Syst. Sci. Int. **57**(1), 97–108 (2018)

Musto, C., Lops, P., Gemmis, M.D., Semeraro, G.: Semantics-aware recommender systems exploiting linked open data and graph-based features. Knowled. Based Syst. **136**(C), 1–14 (2017). https://doi.org/10.1016/j.knosys.2017.08.015

Nesreen, K.A., Nick, D., Liangzhen, X.: Estimating node similarity by sampling streaming bipartite graphs. In: Proceedings of Twenty-Seventh International Joint Conference on Artificial Intelligence, vol. 2018, pp. 3286–3292, Stockholm (2018)

Page, L., Brin, S., Motwani, R., Winograd, T. (1999). The PageRank Citation Ranking: Bringing Order to the Web. Stanford, CA, 94305–99025, Technical Report, Stanford InfoLab- Digital Libraries 1999

Shams, B., Haratizadeh, S.: SibRank: signed bipartite network analysis for neighbor-based collaborative ranking. Physica. A. **2016**(458), 364–377 (2016). https://doi.org/10.1016/j.physa.2016.04.025

Wang, X., He, X., Wang, M., Feng, F., Chua, T.S.: Neural graph collaborative filtering. In: ACM SIGIR International Conference on Research and Development in Information Retrieval, pp. 165–174 (2019)

Yin, R., Li, K., Zhang, G., Lu, J.A.: Deeper graph neural network for recommender systems. Knowl.-Based Syst. **2019**, 185(1) (2019). https://doi.org/10.1016/j.knosys.2019.105020

Zheng, L., Lu, C.T., Jiang, F., Zhang, J., Yu, P.S.: Spectral collaborative filtering. In: Twelfth ACM Conference on Recommender Systems, vol. 2018, pp. 311–319 (2018)

Index

S. Sedkaoui et al. (eds.), *International Conference on Managing Business Through Web Analytics*, https://doi.org/10.1007/978-3-031-06971-0

B

C

D

G

H

I